Global Change and Integrated Coastal Management

Coastal Systems and Continental Margins

VOLUME 10

Series Editor

Bilal U. Haq

Editorial Advisory Board

M. Collins, *Dept. of Oceanography, University of Southampton, U.K.*
D. Eisma, *Emeritus Professor, Utrecht University and Netherlands Institute for Sea Research, Texel, The Netherlands*
K.E. Louden, *Dept. of Oceanography, Dalhousie University, Halifax, NS, Canada*
J.D. Milliman, *School of Marine Science, The College of William & Mary, Gloucester Point, VA, U.S.A.*
H.W. Posamentier, *Anadarko Canada Corporation, Calgary, AB, Canada*
A. Watts, *Dept. of Earth Sciences, University of Oxford, U.K.*

The titles published in this series are listed at the end of this volume.

Global Change and Integrated Coastal Management

The Asia-Pacific Region

Edited by

Nick Harvey
University of Adelaide, Adelaide, Australia

Springer

A C.I.P Catalogue record for this book is available from the Library of Congress.

ISBN-10 1-4020-3627-2 (HB)
ISBN-13 978-1-4020-3627-9 (HB)
ISBN-10 1-4020-3628-0 (e-book)
ISBN-13 978-1-4020-3628-6 (e-book)

Published by Springer,
P.O. Box 17, 3300 AA Dordrecht, The Netherlands.

www.springer.com

Cover Illustrations
Main photo on left – Aerial photograph over the exposed ocean side of the southeastern rim of Tarawa Atoll, Kiribati (photograph Nick Harvey).

Smaller photo on top right – Small scale land reclamation on the protected lagoon side of the southern rim of Tarawa Atoll, Kiribati. Coral blocks have been used to construct a retaining wall which is being filled with rubble and sand (photograph Nick Harvey).

Smaller photograph at bottom right – Large scale reclamation and urban development on the coast of Kobe, Japan. The Osaka-Kobe megacity coastline has been significantly modified, with port development, reclamation, artificial islands and transport networks (photograph Nick Harvey).

Printed on acid-free paper

All Rights Reserved
© 2006 Springer
No part of this work may be reproduced, stored in a retrieval system, or transmitted
in any form or by any means, electronic, mechanical, photocopying, microfilming, recording
or otherwise, without written permission from the Publisher, with the exception
of any material supplied specifically for the purpose of being entered
and executed on a computer system, for exclusive use by the purchaser of the work.

CONTENTS

Preface — vii

Message from the APN Secretariat — ix

List of Contributors — xi

Part 1 – The Asia-Pacific Coastal Zone

1. Importance of Global Change for Coastal Management in the Asia-Pacific Region — 1
   Nick Harvey and Nobuo Mimura

2. State of the Environment in the Asia and Pacific Coastal Zones and Effects of Global Change — 17
   Nobuo Mimura

Part 2 – New Directions in Research

3. Coastal Management in the Asia-Pacific Region — 39
   Nick Harvey and Mike Hilton

4. Catchment–Coast Interactions in the Asia-Pacific Region — 67
   Shu Gao

5. Coastal Evolution in the Asia-Pacific Region — 93
   Patrick D. Nunn and Roselyn Kumar

6. Human Responses to Coastal Change in the Asia-Pacific Region — 117
   Patrick D. Nunn, Charles T. Keally, Caroline King, Jaya Wijaya and Renato Cruz

7. Hot Spots of Population Growth and Urbanisation in the Asia-Pacific Coastal Region — 163
   Poh Poh Wong, Lee Boon-Thong and Maggi W.H. Leung

8. Pressures on Rural Coasts in the Asia-Pacific Region — 197
   Liana Talaue-McManus

9. Impacts of Pollutants in the Asia-Pacific Region 231
*Kanayathu Koshy, Zafar Adeel, Murari Lal
and Melchior Mataki*

10. Landscape Variability and the Response of Asian Megadeltas
    to Environmental Change 277
    *Colin D. Woodroffe, Robert J. Nicholls, Yoshiki Saito,
    Zhongyuan Chen and Steven L. Goodbred*

11. New Directions for Global Change Research Related
    to Integrated Coastal Management in the Asia-Pacific Region 315
    Nick Harvey and Nobuo Mimura

Index 335

PREFACE

Most of the world's population lives close to the coast and is highly dependent on coastal resources, which are being exploited at unsustainable rates. These resources are being subject to further pressures associated with population increase and the globalization of coastal resource demand. This is particularly so for the Asia-Pacific region which contains almost two thirds of the world's population and most of the world's coastal megacities. The region has globally important atmospheric and oceanic phenomena, which affect world climate such as the Asian Monsoon and the El-Niño Southern Oscillation phenomena. The Asia-Pacific region also has highly significant marine diversity but over the last few decades, coastal resources such as mangroves, coral reefs and fisheries have experienced large-scale depletion. The need to find appropriate management solutions to these and other coastal issues is made more complex by the need to take account of international scientific predictions for global climate change and sea-level rise which will further impact on these coasts.

The idea for this book arose from a meeting of coastal scientists in Kobe, Japan in May 2003. The meeting was organized by the *Asia-Pacific Network for Global Change Research* (APN), an inter-governmental network, comprising 21 member countries, for the promotion of global change research and links between science and policy making in the region. The main purpose of the meeting was to organize a synthesis of scientific findings from previous APN-funded coastal research projects in the region and to assess the outputs of these projects in terms of the APN's goals and future directions. It soon became clear from this meeting that it would be difficult to write a synthesis to suit both scientific and non-scientific audiences. For this reason it was decided to have two separate outputs. First, a short APN synthesis report, suitable for a more general readership, was published in 2005 and also made available electronically through the APN website. Second a more comprehensive scientific manuscript dealing with global change impacts on coastal management in the region is published here as this book.

The project was facilitated by the support of the APN, particularly through Martin Rice who coordinated the project for two years and was instrumental in the production of the synthesis report in 2005. Martin along with Dr Linda Stevenson and Jody Chambers of the APN Secretariat also assisted with editing the synthesis report. Another person who deserves special mention is Professor Roger Mclean who contributed to the synthesis report and provided valuable input to the preparation of the scientific book at all of the key meetings. The project, which was fully funded by the APN, brought most of the lead authors together at meetings in Kobe, Japan (May 2003); Bangkok, Thailand (November 2003); and again in Kobe, Japan (November 2004).

The final stages of manuscript preparation were completed in 2005 while the editor was on study leave at the James Cook University of North Queensland. During this time a number of reviewers provided useful comments, which have considerably improved the final product. In particular I would like to thank Professor James Syvitski (University of Colorado) and Dr Scott Smithers (James Cook University) for agreeing to review sections of the book and to James Syvitski for providing additional input to Chapter Five. Special acknowledgement is also due for the editorial assistance provided by Heidi Schuttenberg (James Cook University), Kris James and Noelle Gall (University of Adelaide) and for the cartographic assistance provided by Chris Crothers (University of Adelaide).

The book is separated into two sections. The first section provides the background and rationale for the book; definitions of the region and the coast; and the state of the coastal environment focusing on current issues such as coastal erosion, sea-level rise, vulnerability of small islands, and pollution. The second and main section of the book, consists of nine chapters which focus on current research findings, address different global change coastal issues, and identify and discuss new directions for global change research related to coastal management. The authors for these chapters are an international mix of scientists selected for their coastal expertise in the Asia-Pacific region. These authors come from a wide range of countries including Australia, Canada, China, Fiji, Indonesia, Japan, Malaysia, New Zealand, Philippines, Singapore, the United Kingdom, and the United States of America.

The book is unique because it provides the first comprehensive treatment of global change as an important issue for coastal management in the Asia-Pacific region. There is a wealth of academic literature on coastal management in the region in addition to recent conference proceedings on the topic. However, none of these provides a focus on the importance of the global change issue for the Asia-Pacific region's coasts and there is very little discussion of any implications for future coastal management practices.

The importance of an integrated approach to coastal management has gained acceptance following the United Nations Conference on Environment and Development, known as the 'Earth Summit', held in Rio de Janeiro in 1992. Ten years later, the World Summit for Sustainable Development held in Johannesburg 2002, re-iterated the need for 'integrated management and sustainable development of coastal areas'. A key issue is how to develop better and relevant management frameworks and measures to ensure the sustainable development of coasts in the Asia-Pacific region. This book focuses on a systematic analysis of the future direction of global change coastal research in the region in order to better inform coastal management and policy makers.

Nick Harvey
Adelaide, South Australia, 2006

MESSAGE FROM THE APN SECRETARIAT

It is my great pleasure to announce the publication of APN's first book since its establishment in 1996. The endeavor to write a book in conjunction with APN's second synthesis of over 20 APN-funded activities on coastal zones and inland waters was challenging yet rewarding. The book not only highlights the synthesis of APN's projects but also outlines research gaps and identifies future research directions for coastal environments. Researchers, decision-makers and educators alike will find this book a useful resource. I would like to extend my sincere gratitude to everyone who contributed to the publication particularly Professor Nick Harvey, the Editor, and Professor Nobuo Mimura, APN's Scientific Planning Group Co-Chair, who were both lead authors of the book.

Hiroki Hashizume, Director, APN Secretariat
Kobe, Japan, 2006

LIST OF CONTRIBUTORS

Zafar Adeel
International Network on Water,
 Environment and Health
UNU-INWEH
McMaster University
1280 Main Street West
Hamilton, Ontario L8S 4L8
Canada
adeelz@inweh.unu.edu

Lee Boon-Thong
Department of Geography
University of Malaya
50603 Kuala Lumpur
Malaysia
leebt@um.edu.my

Zhongyuan Chen
Department of Geography
East China Normal University
Shanghai 200062
China
Z.Chen@gislab.ecnu.edu.cn

Renato Cruz
Department of Environment
 and Natural Resources
Manila
Philippines
renatodalmaciocruz@yahoo.com.ph

Shu Gao
Department of Geo-Ocean
 Sciences
Nanjing University
22 Hankou Road
Nanjing 210093, China
shugao@nju.edu.cn

Steven L. Goodbred
Department of Earth and Environmental
 Sciences
Vanderbilt University
2301 Vanderbilt Place
Nashville, TN 37235-1805
U.S.A
steven.goodbred@vanderbilt.edu

Nick Harvey
Geographical and Environmental
 Studies
The University of Adelaide
South Australia, 5005
nick.harvey@adelaide.edu.au

Mike Hilton
Department of Geography
University of Otago
85 Albany St
Dunedin
New Zealand
mjh@geography.otago.ac.nz

Charles T. Keally
Department of Comparative Culture
Sophia University
7-1 Kioi-Cho, Chiyoda-Ku
Tokyo 102-8554
Japan
c-keally@t-net.ne.jp

Caroline King
International Network on Water,
 Environment and Health
UNU-INWEH
McMaster University
1280 Main Street West
Hamilton, Ontario L8S 4L8
Canada
cking@inweh.unu.edu

Kanayathu Koshy
Pacific Centre for Environment
 and Sustainable Development
The University of the
 South Pacific
Suva
Fiji Islands
koshy_k@usp.ac.fj

Roselyn Kumar
Institute of Applied Sciences
The University of the South Pacific
Suva
Fiji Islands
roselyn802@yahoo.com

Murari Lal
Pacific Centre for Environment
 and Sustainable Development
The University of the South Pacific
Suva
Fiji Islands
lal_m@usp.ac.fj

Maggi W.H. Leung
Department of Geography
 and Resource Management
The Chinese University of
 Hong Kong
2nd Floor, Sino Building,
Shatin, N.T.
Hong Kong
maggileung@cuhk.edu.hk

Melchior Mataki
Pacific Centre for Environment and
 Sustainable Development
The University of the
 South Pacific
Suva
Fiji Islands
mataki_m@usp.ac.fj

Nobuo Mimura
Center for Water Environment Studies
Ibaraki University
4-12-1 Nakanarusawa
Hitachi
Ibaraki 316-8511
Japan
mimura@mx.ibaraki.ac.jp

Robert J. Nicholls
School of Civil Engineering
 and the Environment
University of Southampton
Highfield
Southampton SO17 1BJ
United Kingdom
R.J.Nicholls@soton.ac.uk

Patrick D. Nunn
Department of Geography
Faculty of Islands and Oceans
The University of the
 South Pacific
Suva
Fiji Islands
nunn_p@usp.ac.fj

Yoshiki Saito
Geological Survey of
 Japan, AIST
Central 7, Higashi 1-1-1
Tsukuba, Ibaraki, 305-8567
Japan
yoshiki.saito@aist.go.jp

Liana Talaue-McManus
Rosenstiel School for Marine and
 Atmospheric Science
4600 Rickenbacker Causeway
Miami, FL 33149-1098
USA
lmcmanus@rsmas.miami.edu

Jaya Wijaya
Department of Marine Affairs and
 Fisheries
Jl. Merdeka Timur No. 16 10th Floor
Jakarta
Indonesia

Poh Poh Wong
Department of Geography
National University of Singapore
21 Lower Kent Ridge Road
Singapore 119077
geowpp@nus.edu.sg

Colin D. Woodroffe
School of Earth and
 Environmental Sciences
University of Wollongong
Wollongong, NSW 2522
Australia
colin_woodroffe@uow.edu.au

CHAPTER 1

IMPORTANCE OF GLOBAL CHANGE FOR COASTAL MANAGEMENT IN THE ASIA-PACIFIC REGION

NICK HARVEY[1] AND NOBUO MIMURA[2]
[1]*Geographical and Environmental Studies, Adelaide University, Australia*
[2]*Center for Water Environment Studies, Ibaraki University, Japan*

1.1 INTRODUCTION

Over the last two decades there has been growing concern about the potential impact of global change on the world's coasts both from the biophysical effects of a projected climate change and from the impact of coastal development and the globalisation of economy and trade. In 1992 the United Nations Conference on Environment and Development (UNCED), known as the "Earth Summit", concluded that the world was living beyond its ecological means and that rapid action was necessary to avert future disaster (UNCED 1992). In the global context, the coast is particularly important because most of the world's population lives around the coast and is highly dependent on coastal resources. These resources are being exploited at unsustainable rates and are subject to increasing pressure associated with population increase and coastal resource dependency.

The United Nations' 21st century blueprint for sustainable development, *Agenda 21*, recognised the international imperative for effective coastal and marine management, in its directive that

Coastal states commit themselves to integrated management and sustainable development of coastal areas and the marine environment under their national jurisdiction (*Agenda 21, 1992, Chapter 17.5*)

In 2002, ten years after the UNCED, the World Summit for Sustainable Development (WSSD) was held in Johannesburg, South Africa, to review the progress made over the past ten years towards promoting actions for sustainable development of the global society. In its 'Plan of Implementation' (Section 29), the WSSD emphasized that

Ensuring the sustainable development of the oceans requires effective coordination and cooperation, including at the global and regional levels, between relevant bodies, and actions at all levels to:...
 promote the implementation of chapter 17 of Agenda 21 which provides the programme of action for achieving the sustainable development of oceans, coastal areas and seas through its programme areas of integrated management and sustainable development of coastal areas, including exclusive economic zones; marine environmental protection; sustainable use and conservation of marine living resources; addressing critical uncertainties for the management of the marine environment and climate change; strengthening international, including regional, cooperation and coordination; and sustainable development of small islands (*WSSD 2002*)

Agenda 21 (1992, Chapter 17.3) also noted that at that time (1992) more than half of the world's population lived within 60 km of the coast and it was projected that by the year 2020 that proportion could rise to two thirds. Hinrichsen (1998) quoted a figure of some 3.2 billion people living within 200 km of the coast, on about 10 percent of the earth's land area, and suggested that two thirds of the world's population was living within 400 km of the coast. Shi and Singh (2003) have shown that the coastal population has been increasing disproportionately to the global population increase since 1990 and have projected that this will become more marked in the future. There is also a marked decrease in population density away from the coast so that 40% of the 'near coastal' population occupies 4% of the land area with the greatest densities in Europe, Southeast, and East Asia (Nicholls and Small 2002).

Humans have a high dependency on coastal resources. Although the coastal ocean only accounts for 8% of the global ocean surface and less than 0.5% of its volume, it accounts for about 14% of its production, up to 50% of its denitrification, and up to 80% of global organic matter burial (Pernetta and Milliman 1995 p 16). It also accounts for 90% of global sedimentary mineralisation, 75–90% of the global sink of suspended river load and associated elements/pollutants, in excess of 50% of present day global carbonate deposition, and, it supplies approximately 90% of the world's fish catch (Pernetta and Milliman 1995 p 16).

Given the importance of the coastal zone for humans it is interesting to consider what value it has, in broad terms. Costanza et al. (1997) estimated the contribution of various ecosystems to the total economic value of the planet. They showed that the coastal zone contributed a global flow value of ecosystem goods and services amounting to 43% or $US12.6 trillion out of the total global flow value (Costanza et al. 1997). This is discussed further by Wilson et al. (2003) who provide examples of coastal values from the Asia-Pacific coastal region, such as fish and shrimp having the second highest export commodity value for Bangladesh; representing 7% of total export earnings and contributing 4.7% to GDP. They also note that often the market price of seafood is used as a proxy to determine the value of ecosystem goods such as the annual market value of seafood supported by mangroves, ranging from US$750–16,750 (1999 dollars). Wilson et al. (2003) quote an example of a study of mangrove valuation from Thailand where the value was assumed to be equivalent to the net income from the forests in terms of wood and non-wood products giving a mean annual value per household of around US$924 (1999 dollars). Yet another study from Sarawak estimated the value of timber products from mangroves in the region to be worth US$123,217 (Wilson et al. 2003). Adger et al. (2001) discuss the costs and benefits of mangrove conversion and restoration in Vietnam with a key benefit being the reduction in sea dyke maintenance and point out that in 'future decades this risk may change as a result of global warming' (Adger et al. 2001 p 260).

Wilson et al. (2003) note that the original global estimates of Costanza et al. (1997) were controversial and there has been subsequent debate over methodology. However, the global importance of coastal ecosystems is clear. What is less clear, is how values should be attributed and the relevance of these values for the countries of the Asia-Pacific region. In a review of the literature on the value of coastal ecosystem goods and service Wilson et al. (2003) conclude that there is a major discrepancy between those goods and services 'documented in the published valuation literature and those that could potentially contribute significantly to human welfare both directly and indirectly' (Wilson et al. 2003 p 17). Given these limitations and the fact that there is not a good global coverage of valuation studies for coastal ecosystem goods and services, it is not possible to produce meaningful results for the Asia-Pacific region. However, it does indicate the need for further research in this area.

Special attention to coastal management is needed because the coast is naturally dynamic at a variety of time scales and changes occur both naturally and by human use. This is very relevant when considering potential global change because it is sometimes difficult to separate rapid natural coastal change from human induced change on the coast.

The idea for this book arose from a recognition that global change, in both a bio-physical sense and also in terms of the globalization of resource use, had major implications for coastal management in the Asia-Pacific region. There is already a recognition that this region is important in terms of approaches to coastal management as demonstrated by a book on the topic *Coastal Management in the Asia-Pacific Region: Issues and Approaches* (Hotta and Dutton 1995). Chapter Six of Hotta and Dutton's (1995) book touches on the coastal implications of climate change and comments on problems of linking climate change science to coastal management policy (Holmes 1995). More recently, there has been a renewed interest in Asia-Pacific coasts indicated by the two international conferences titled *Coastal Zone Asia-Pacific* (CZAP), the first held in Bangkok (CZAP 2002) and the second in Brisbane (CZAP 2004. These conferences in part dealt with coastal management but neither specifically addressed the important issue of coastal management within the context of global change. In a brief review based on the two CZAP conferences, Smith and Thomsen (2005) analysed the trends in coastal management for the Asia-Pacific region since the CZAP 2002 conference. However, none of the themes, emerging issues or priority actions identified by Smith and Thomsen (2005) made specific reference to global change as an important issue for coastal management in the region (see Chapter Three for further discussion).

A first step toward addressing the global change coastal issue was taken by the Asia-Pacific Network for Global Change Research (APN) in its release of a *Global Change Coastal Zone Management Synthesis Report* (Harvey et al. 2005). The APN is a 21 member country (Table 1.1) intergovernmental network with a primary aim of fostering global change research in the Asia-Pacific region, increasing developing country participation in that research, and strengthening links between the science community and policy makers. As a major target area for research, APN has promoted and supported projects dealing with global change issues impacting on the coast. APN evaluated its past performance and needs for a future research direction relevant to the Asia-Pacific region in its Synthesis Report (Harvey et al. 2005).

Table 1.1. Member countries of APN

Australia	Japan	Philippines
Bangladesh	Laos*	Republic of Korea
Cambodia	Malaysia	Russian Federation
China	Mongolia*	Sri Lanka
Fiji	Nepal*	Thailand
India	New Zealand	United States
Indonesia	Pakistan	Viet Nam

landlocked countries.

The Synthesis Report has revealed that the Asia-Pacific coastal region faces a range of challenges, such as the interaction between catchment and coast, increasing pressure of population and urbanization, issues of rural coasts, and management of coastal resources, which closely interact with global changes. This also leads to the recognition that a key issue is how to develop better and relevant management frameworks and measures to ensure the sustainable development of coasts in the Asia-Pacific region. This book is one outcome of the APN Synthesis Report but unlike that report, it is a detailed scientific text which focuses on a systematic analysis of the future direction of global change coastal research in the region in order to better inform coastal management and policy makers.

1.2   IMPORTANCE OF THE ASIA-PACIFIC COASTAL REGION

The Asia-Pacific is an important region for understanding global environmental problems for a number of reasons. First, important atmospheric and oceanic phenomena occur here, such as the Asian Monsoon and the El Nino phenomena, which affect world climate. Second, the region has a diversity of ecosystems such as tropical forests, deserts, mountains, and globally significant marine ecosystems of corals, seagrasses, and mangroves. The southeast Asia marine region has been referred to as the epicenter of global marine diversity (Burke et al. 2002). This is illustrated in Figure 1.1 showing the diversity and distribution of corals, mangroves and seagrasses

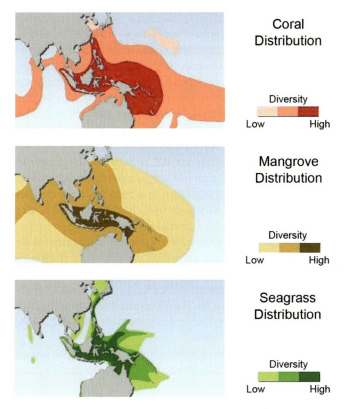

Figure 1.1. Distribution and diversity of coral, mangrove and seagrass in the Asia-Pacific region (Source: adapted from UNEP-WCMC (2001))

in the Asia-Pacific region. For example, it has been estimated that the southeast Asia region alone contains over 600 out of almost 800 reef-building coral reef species in the world (Burke et al. 2002, Veron and Stafford-Smith 2000). The southeast Asian region also holds almost 75% of the world's mangrove species (Burke et al. 2002, Spalding et al. 1997) and over 45% of the world's seagrass species (Burke et al. 2002).

Another reason why the Asia-Pacific region is important is because of the size of its population which is over 3.7 billion or roughly 60% of the world's population (see Chapter 7). In addition, its economic growth rate is the highest of any region in the world. Given the rapid population and economic growth, the region contributes to global climate change in a significant way. Degradation of the environment, such as deforestation and desertification, is becoming a matter of great concern, as are natural disasters, such as floods and droughts, which occur as a result of this degradation. Thus observation, monitoring and research on global change in the Asia-Pacific region are very important to an understanding of environmental changes taking place on a global scale. In addition, stronger links are needed between the science community and policy makers.

Global population concentrations in coastal areas have been illustrated graphically through the population density maps of the Global Demography Project (Tobler et al. 1995), satellite images of the earth by night and recent computer generated images of population distribution (NASA http://www.earthobservatory.nasa.gov) clearly depicting regional differences with major concentrations around the coasts of the Indian and Asian subcontinents (Figure 1.2). The majority of the world's largest cities are located on the coast and in some areas have expanded into coastal megacities which in 2000 comprised 17 of the world's total 24 megacities. Most of these megacities are located in southeast Asia (see Chapter Seven). One effect of population concentration on the coast is increased pressure of human activities on the coastal and marine environments. Rough proxy measures of this pressure can be obtained using an insularity index which illustrates the relative importance of the coast compared to a country's total area (see Chapter Five), or by dividing a country's total population by its length of coastline, or by attempting more detailed population distribution estimates for the area close to a country's coastline. The World Resources Institute (http://www.wri.org) has calculated the population density per kilometre of coastline and while there is a large error margin for these figures it appears that around one third of the total Asian population live within 100 km of the coast. Even with the high insularity indices for the Pacific island countries (Chapter Five) and assuming that most of the Pacific population lives within 100 km of the coast, the total population figure for Pacific countries is less than 10% of the Asian population living within 100 km of the coast.

Coastal societies in the Asia-Pacific region have been highly dependent on coastal resources in many ways. However, it is repeatedly pointed out that in the past 30 years,

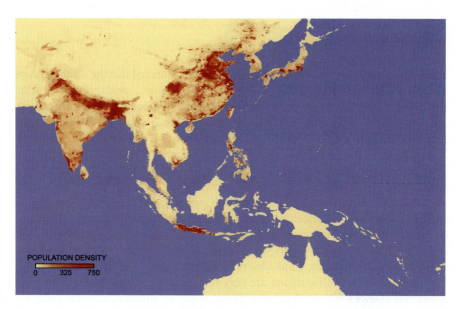

Figure 1.2. Population concentration in the Asia-Pacific region (Source: adapted from NASA (2006))

coastal resources such as mangroves, coral reefs and fisheries have experienced large-scale depletion and this has become a critical issue for the region (UNEP 1999). For example, more than 60% of Asia's mangroves have been converted to aquaculture farms (ESCAP and ADB 2000), while coral reefs are threatened by the combined effects of higher temperatures and seawater pollution. It has been estimated that 80–90% of coral reefs are threatened in the Philippines, Vietnam, Singapore, Cambodia, and Taiwan (Burke et al. 2002). Development of river basins (see Chapter Four) and marine resources is impacting on the coastal zone so that in many parts of the region the coastal resource base which supports coastal community life and future economic development is being eroded, along with the rich biodiversity of the coastal zone.

1.3   APN AND ITS COASTAL ZONE MANAGEMENT SYNTHESIS

The APN was established in 1995 as a regional organization parallel to similar organizations elsewhere in the world such as IAI (Inter-American Institute for Global Change research). The establishment of these networks originated from the United States, White House Conference on Science and Economics Research related to Global Change, 1990. At this conference, President Bush (Snr) invited other countries to join the US in developing three regional research institutes which could link the interests and capabilities of developing countries and their scientific communities on broadening global change research in developing countries, provide support for multi-disciplinary research and education, and to encourage a sound scientific framework to underpin national and international policy making needs. Against such a background, the APN established six main goals as documented in the First Strategic Plan (1999–2004). These were revised and updated in its Second Strategic Plan (2005–2010). The five goals identified in its Second Strategic Plan are as follows:
- Supporting regional cooperation in global change research on issues particularly relevant to the region;
- Strengthening appropriate interactions among scientists and policy-makers, and providing scientific input to policy decision-making and scientific knowledge to the public;
- Improving the scientific and technical capabilities of nations in the region;
- Cooperating with other global change networks and organisations; and
- Facilitating the development of research infrastructure and the transfer of know-how and technology.

To achieve these goals, the APN supports research on global change in complex climate, ocean, and terrestrial systems, and on physical, chemical, biological, and socio-economic processes. The research areas focused on in the First Strategic Plan were:
- Changes in atmospheric composition;
- Changes in coastal zones and inland waters;
- Changes in terrestrial ecosystems and biodiversity;
- Climate change and variability; and
- Human dimensions of global change.

The APN also has strong linkages with a number of international global change research organizations particularly through START (SysTem for Analysis, Research, and Training), which provides regionally based capacity building and global change research support. In turn, START is supported by three key international scientific global change research organizations: World Climate Research Programme (WCRP), International Geosphere-Biosphere Programme (IGBP), and International Human Dimensions Global Environmental Change Programme (IHDP). Through APN it is possible for governments of the region to collectively or individually provide support for scientific research and enhance international collaboration with the other inter-governmental networks and international scientific organizations for global change research.

The 7th APN Inter-Governmental Meeting in 2002 endorsed the production of synthesis reports to assess how the results of APN-funded projects have contributed to the Asia-Pacific region in terms of the main APN goals. A plan was then established to carry out Global Change Coastal Zone Management Synthesis studies following the first attempt of the LUCC (Land use and climate change) synthesis report (APN 2002).

The Global Change Coastal Zone Management Synthesis commenced in 2003, aiming to identify urgent research needs and a future research direction for coastal environments relevant to the region through evaluation of achievements of APN coastal zone projects and a review of present status and major problems of the coastal environment. The synthesis results have been published as a short monograph *Global Change Coastal Zone Management Synthesis Report* (Harvey et al. 2005) designed to be accessible to governments and policy makers. In addition, it was also intended to publish the results as a scientific book to promote future research activities in a wider research community.

1.4   SCOPE OF THIS BOOK

Although the initiative for this book came from APN, its scope is not restricted to the 21 member countries of APN, of which three do not have a coast (see Table 1.1). The APN Synthesis Report (Harvey et al. 2005) has provided an overview of APN-sponsored coastal research but it is recognized that this only represents part of a much broader body of coastal research related to global change in the region. In order to produce a detailed scientific assessment of issues and research needs, it is first necessary to define both the region and also what is meant by the coastal zone.

1.4.1   Definition of the Asia-Pacific Coastal Region

In their book on coastal management in the Asia-Pacific region, Hotta and Dutton (1995) use a broad definition for the region to include Asia from India in the west to Japan in the north, all Pacific nations including New Zealand plus the continent of Australia. Whilst recognizing the arbitrary nature of these boundaries they note that their definition is 'broader than those identified in most regional action and multilateral coastal management programmes (which normally separate Asia from the Pacific)' (Hotta and Dutton 1995 p 12).

The scope of the Asia-Pacific coastal region used for the current book is broadly similar but has some key differences. The 15 APN member countries on the Asian coast were used as a base for defining the limits of the region on the eastern and southern margins of the Asian continent. The five non-APN member countries of this part of Asia (Brunei, East Timor, Burma, North Korea, and Singapore) were also included so that the region extends from the Pakistan/Iranian coastal border in the west to the Russian coast just north of Japan. This restricted the study area to the more populated parts of the Asian coast where coastal management was likely to be a higher priority. It also provided a biophysical focus for the book on tropical to temperate coastal ecosystems. The Pacific part of the region was taken to include all of the Pacific countries including the small island states of the central Pacific such as Tuvalu, Kiribati, and the Marshall Islands and extending east to French Polynesia.

Three of the APN-member countries, Australia, New Zealand and the US were excluded, for the purposes of this book, to provide a greater focus on developing nations. The definition of the region is problematic, given that other regional organizations such as the Asia-Pacific Economic Cooperation (APEC) includes the Americas. The global change research community, however, has its own separate organization, the Inter-American Institute for Global Change Research (IAI), which is the APN's sister network. For that reason the United States has been excluded.

Thus the region is defined here to include the coast of 20 Asian countries from Pakistan in the west to part of the Russian coast in the northeast, extending into the Pacific to include 20 Pacific Island countries and dependencies (note that Hawaii, as part of the US, has been excluded). This group of 40 countries comprise mostly developing or low-income countries with the exception of Brunei, Japan, Republic of Korea, and Singapore. The countries included as part of the region covered by this book are shown in Figure 1.3.

1.4.2   Definition of the Coastal Zone

There is no clear scientific definition of the 'coast' because definitions become broader as they become more inclusive of catchment-related or marine-related physical processes which impact on the boundary zone between the sea and the land. Similarly, definitions from government or bureaucratic organizations within the Asia-Pacific region will vary considerably because each has its own purpose for defining the coast.

For example, the International Geosphere-Biosphere Programme (IGBP) includes a coastal scientific core project investigating Land-Ocean Interaction in the Coastal Zone (LOICZ), which defines the coastal zone as 'extending from the coastal plains to the outer edge of the continental shelves, approximately matching the region that has been alternately flooded and exposed during the sea level fluctuations of the late Quaternary period' (Holligan and de Boois 1993 p 9). In the LOICZ Implementation

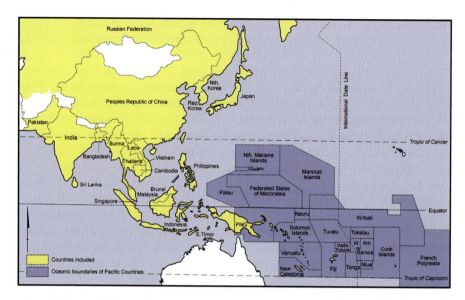

Figure 1.3. Scope of Asia-Pacific coastal region

Plan (Pernetta and Milliman 1995 p 16) this coastal domain is referred to as being from 200 metres above to 200 metres below sea level. Another coastal process-related definition of the coast is given by Sorensen (1997) who defines the coast as 'that part of the land affected by its proximity to the sea and that part of the ocean affected by its proximity to the land ... an area in which processes depending on the interaction between land and sea are most intense'. These broad scientific definitions attempt to be inclusive of the various processes influencing the coast with the LOICZ definition clearly recognising the importance of past processes, particularly the last 2 million years, in shaping our modern coastline.

However, scientific definitions of the coastal zone based on coastal processes or coastal form are not usually practical for management purposes so that another suite of coastal definitions is designed for policy purposes. For example, the Environment Directorate of the international Organisation for Economic Cooperation and Development (OECD) suggests that the definition of the coastal zone should vary according to the nature of the problem being examined and the objectives of the management. This approach is reflected in the Australian Commonwealth Government's definition of the coastal zone which states that: 'The boundaries of the coastal zone extend as far inland and as far seaward as necessary to achieve the policy objectives, with a primary focus on the land/sea interface' (Commonwealth of Australia 1992 p 2). This definition is quite different to various Australian State government definitions (Harvey and Caton 2003). Similarly, there is variation in selected official definitions of the coastal zone in Asia-Pacific countries as shown in Table 1.2. This makes it virtually impossible to make comparisons between countries of the region which all have different definitions of the coastal zone for policy purposes.

Importance of Global Change for Coastal Management

Table 1.2. Selected definitions of the coastal zone in the Asia-Pacific region

Country	Definition of coastal zone
China	an area 10 km inland and seaward out to the 15m isobath from the mean high water tide line (*Source*: http://www.globaloceans.org/country/China.html)

Malaysia	a) an area 5 km inland from the shoreline dependent on the following:
	  1. If the coastal zone is lined with mangroves/nipah swamps, then it is 5 km from the inner boundary (landward side) of the swamp; and
	  2. If the coastal area is covered by peat swamps, it starts from the shoreline to the inner limit (landward side) of the swamp that could be more than 5 km in width.
	b) The seaward limit is the Exclusive Economic Zone (EEZ). i.e., 200 nautical miles from the shoreline (*Source*: Basiron 1995)

Sri Lanka	a 2 km wide band of ocean and an adjoining strip of land extending 300 m inland. In the event a water body connected to the sea occurs the zone extends two kilometres inland from the mouth of the water body (*Source*: http://www.rrcap.unep.org/reports/soe/srilanka_coastal.pdf)

Republic of Korea	the area of coastal waters and coastal lands defined as follows
	a) 'coastal waters' are either
	  1. transitional and intertidal areas from the high tide line to the first line of a cadastral map; or
	  2. waters within the territorial seas outer limit from the high tide line.
	b) coastal lands falls under any of the following definitions
	  1. islands with no residents
	  2. areas set by the integrated coastal management plan under Article 5 within 500 meters from the landward boundary of coastal waters
	  3. designated port, fishing harbor of I or III Category, or industrial complex.
	(*Source*: http://www.globaloceans.org/laws/koreanczma.html)

1.5 STRUCTURE OF THIS BOOK

This book is separated into two sections. The first section, including this introductory first chapter by Harvey and Mimura, provides the background and rationale for the book along with definitions of the region and the coastal zone. The second chapter by Mimura provides an assessment of the state of the coastal environment in the region focusing on current issues such as coastal erosion, sea-level rise, vulnerability of small islands, and pollution. He gives an overview of the significance and vulnerability of key coastal ecosystems such as mangroves, coral reefs, and seagrasses and outlines the importance of fisheries and aquaculture in the region. Mimura then discusses population growth and economic development as human drivers of many coastal problems and places this in the context of projected changes arising from the biophysical drivers of climate change and sea-level rise. He then concludes with comments on the need for appropriate coastal management strategies to combat the current and projected coastal problems. Thus the first section of the book outlines the imperatives for considering global change issues within coastal management in the region and sets the scene for some of the current and future coastal issues in the region.

The second and main section of the book, consists of eight chapters which focus on current research findings, address different global change coastal issues, and identify and discuss new directions for global change research related to coastal management. The authors for these chapters are scientists selected for their coastal expertise in the Asia-Pacific region. These authors come from a wide range of countries including Australia, Canada, China, Fiji, Indonesia, Japan, Malaysia, New Zealand, Philippines, Singapore, the United Kingdom, and the United States of America. The final chapter (Eleven) provides a synthesis of global change issues in the Asia-Pacific region and discusses research priorities of relevance to coastal management.

In Chapter Three, Harvey and Hilton note the diversity of coastal management practices and point out that global change issues have so far received very little attention in international forums dealing with coastal management in the Asia-Pacific region. They outline the global adoption of integrated coastal management (ICM) and discuss its appropriateness for different countries of the region. Harvey and Hilton examine the various drivers for coastal management in the region such as international conventions and agreements, globalization, NGOs, climate change science, coastal resource degradation from human pressure, natural coastal erosion and hazards, and the roles of different communities. They then give an overview of the various approaches to ICM in the region and attempt to assess its effectiveness using different criteria. In conclusion, Harvey and Hilton identify a number of future research directions such as developing better criteria for assessing ICM outcomes, methods for better linkage of science to coastal policy, and new approaches to education and awareness raising of coastal issues.

Gao, in Chapter Four, deals with the important issue of the catchment-coast interaction. He discusses the variable discharges of water, sediment, nutrient, and pollutants to regions' coasts and attributes much of the problem to intense human activity in the catchment basins, ranging from construction of numerous dams to the large-scale utilization of chemical fertilizers. Gao notes that changes of material fluxes will result in morphological, environmental and ecosystem changes in catchment areas, estuaries, and adjacent coastal waters. He proposes that future research should include *in situ* monitoring and measurements; new methods and techniques for the prediction of future changes, and the application of findings to future catchment-coast management practices.

Chapter Five by Nunn and Kumar outlines the geologically recent evolution of the region's coasts and discusses the geomorphic processes behind these changes, since the postglacial rise of global sea level over the last 18,000 years. They explain the effects of sea-level change and put this in context of other processes affecting coastal evolution such as sediment supply from the major rivers and the effect of geotectonics, which is highly variable throughout the region. Nunn and Kumar discuss the relative importance of coasts to individual countries and provide a particular focus on island coasts. They conclude the chapter with a section on directions for future research and note the regional imbalances in our understanding of coastal evolutionary processes.

The following chapter (Six) by Nunn et al. complements Chapter Five by examining more recent coastal changes with a focus on the human response to these changes. The authors provide a context of early (up to 50,000 years ago) human interactions with the coast through to more recent human response to changing sea levels, particularly in the Pacific, and to current problems of coastal vulnerability including natural hazards. This chapter uses seven case studies from different countries (Japan, Indonesia, Pacific Islands, Papua New Guinea, The Philippines, and Bangladesh) to illustrate different human response to various aspects of coastal change. Nunn et al. conclude by examining institutional responses to coastal change, the need for new research into coastal vulnerability, and ways to strengthen existing coastal management systems.

Chapter Seven by Wong et al. provides an insight into the coastal pressures arising from population growth and urbanization, noting that the Asia-Pacific region contains almost two-thirds of the global population. It begins with a discussion of the pattern of population growth including demographic changes and migration, both internal and intra-regional. Wong et al. then focus on the urbanization process, particularly the growth of megacities and mega-urban areas. They conclude with an overview of current research into urbanization and megacities and stress the need for future research directed toward a more sustainable approach-involving comprehensive, integrated, and forward-looking strategic planning and management.

Talaue-McManus, in Chapter Eight, provides a nice balance to the previous urbanization chapter by examining the coastal resource-dependent rural economies of the region and resultant issues of population pressure, poverty, and health. She discusses the major livelihoods of farming, fishing, and tourism. The author presents data that demonstrates the importance of rural coasts of the region and puts the data in a global context with comparisons from elsewhere. Talaue-McManus uses current global data to show that the majority of the world's fishers are in the Asia-Pacific region which is dominant in terms of global fish catch and aquaculture production. She also makes some important comparisons between the developing and developed world in terms of the differences in the quality of fish consumption. Talaue-McManus provides a similar overview, supported by detailed comparative figures, for the farming industry along the region's rural coasts. This chapter highlights aspects of the rural sector such as income generation, environmental costs, and the all-prevailing conditions of poverty.

Chapter Nine, by Koshy et al. deals with the important issue of chemical pollution of the region's coasts. They note that part of the problem is the diversity of pollutant sources including sewage discharges, industrial effluents, rubbish dumps, surface runoff, and intentional dumping. Another problem is that many of the pollutants which end up on the coast may be sourced from hundreds of kilometers away. Koshy et al. echo the comments from Chapter Eight with their observations on the social aspects of pollution and direct and indirect linkages to poverty in the region. They discuss organic and inorganic pollution separately from nuclear waste and industrial effluents. Their chapter concludes by examining institutional responses to coastal pollution and recommend new research aimed at methods for achieving integrated pollution management in the region.

Chapter Ten by Woodroffe et al. examines the important issue of the variability of the megadeltas of Asia and their response to anticipated climate change and sea-level rise. They examine the deltas of The Indus, Ganges-Brahmaputra-Meghna, Irrawaddy, Chao Phraya, Mekong, Red, Pearl, Changjiang, and Huanghe Rivers. Woodroffe et al. note that the Asian megadeltas have a history of human exploitation, and support large populations with strong urbanisation trends. The megadeltas have been affected by water control from flood mitigation and groundwater extraction, to irrigation and channelisation, exacerbating compaction, subsidence, and acid sulphate soil development. Woodroffe et al. note that upstream dams, water abstractions, and diversions have impacted on sediment supply to the deltas causing increased coastal erosion. They stress the need for a better understanding of natural processes and global change impacts in order to design long-term management strategies for the megadeltas.

The final chapter in this book, Chapter Eleven, by Harvey and Mimura attempts to provide a synthesis of the discussion in the previous chapters by reflecting on the lessons from ICM practice in the Asia-Pacific region, lessons from past global changes and the recent history of human impacts on the region's coasts. They then examine future directions for global change research, which are applicable to improving future coastal management in the region. Harvey and Mimura conclude with a set of priority research directions.

REFERENCES

Adger WN, Kelly PM, Tri NH (2001) Costs and benefits of mangrove conversion and restoration. In: Turner RK, Bateman IJ, Adger WN (eds) Economics of Coastal and Water Resources: Valuing Environmental Functions. Kluwer Academic Publishers, Dordrecht, The Netherlands

Agenda 21 (1992) Agenda 21 and the UNCED Proceedings. Oceania Publications, New York

APN (2002) APN Land Use and Cover Change Initial Synthesis Report. APN

Basiron, MN (1995) http://www.mima.gov.my/mima/htmls/papers/pdf/MNB/

Burke L, Selig E, Spalding M (2002) Reefs at risk in Southern Asia. World Resources Institute, 72p

Commonwealth of Australia (1992) Draft Coastal Policy. AGPS, Canberra

Costanza R, d'Arge R, de Groot R, Farber S, Grasso M, Hannon B, Limburg K, Naeem S, O'Neill RV, Paruelo J, Raskin RG, Sutton P, van den Belt M (1997) The value of the world's ecosystems and natural capital. Nature 387:253–260

CZAP (2002) Proceedings of the Coastal Zone Asia-Pacific Conference '02, Bangkok, Thailand

CZAP (2004) Proceedings of the Coastal Zone Asia-Pacific Conference '04: Improving the Quality of Life in Coastal Areas, 5–9 September, 2004, Brisbane, Australia

ESCAP, ADB (2000) State of the Environment in Asia and Pacific 2000. Economic and Social Commission for Asia and the Pacific and Asian Development Bank, United Nations, New York

Harvey N, Caton B (2003) Coastal Management in Australia. Oxford University Press, Melbourne, Australia

Harvey N, Rice M, Stevenson L (Eds) (2005) Global Change Coastal Zone Management Synthesis Report. Asia-Pacific Network for Global Change Research, Kobe, Japan

Hinrichsen D (1998) Coastal Waters of the World: Trends, Threats and Strategies. Island Press, Washington

Holligan PM, de Boois H (1993) Land-Ocean Interactions in the Coastal Zone (LOICZ) Science Plan. International Geosphere-Biosphere Programme. IGBP Report No. 25, Stockholm

Holmes N (1995) Coastal dynamics and global change: Implications for coastal management. In: Hotta K, Dutton IM (eds) Coastal Management in the Asia-Pacific Region: Issues and Approaches. Japan International Marine Science and Technology Federation, Tokyo

Hotta K, Dutton IM (1995) Coastal Management in the Asia-Pacific Region: Issues and Approaches. Japan International Marine Science and Technology Federation, Tokyo
NASA (2006) Earth Observatory. Online at http://www.earthobservatory.nasa.gov
Nicholls RJ, Small C (2002) Improved estimates of coastal population and exposure to hazards released. EOS Transactions of the American Geophysical Union 83:301–305
Pernetta JC, Milliman JD (1995) Land-Ocean Interactions in the Coastal Zone: Implementation Plan. IGBP Report No 33, Stockholm
Shi H, Singh A (2003) Status and interconnections of selected environmental issues in the global coastal zones. Ambio 32:145–152
Smith T, Thomsen D (2005) Coastal management trends in the Asia-Pacific region. Water 32:38–39
Sorensen J (1997) National and international efforts at integrated coastal zone management: Definitions, achievements, and lessons. Coastal Management 25:3–41
Spalding, MD, Blasco, F Fields CD (eds) (1997) World Mangrove Atlas, The International Society for Mangrove Ecosystems, Okinawa
Tobler W, Deichmann U, Gottsegen J, Maloy K (1995) The Global Demography Project. National Center for Geographic Information and Analysis, Technical Report 95-6, Santa Barbara, California
UNCED (United Nations Conference on Environment and Development) (1992) Agenda 21 and the UNCED Proceedings. Oceania Publications, New York
UNEP (1999) GEO-2000. United Nations Environment Programme, Earthscan
UNEP-WCMC (2001) United Nations Environment Programme and World Conservation Monitoring Centre online at http://www.unep.org/vitalwater/33.htm
Veron JEN, Stafford-Smith M (2000) Corals of the World, Australian Institute of Marine Science
Wilson MA, Costanza R, Boumans R, Liu S (2003) Integrating assessment and valuation of ecosystem goods and services provided by coastal systems. In: Wilson JG (ed) The Intertidal Ecosystem. Royal Irish Academy, Dublin, pp 1–28
WSSD (2002) Plan of Implementation of the World Summit on Sustainable Development, http://www.un.org/esa/sustdev/documents/WSSD_POI_PD/English/WSSD_PlanImpl.pdf

CHAPTER 2

STATE OF THE ENVIRONMENT IN THE ASIA AND PACIFIC COASTAL ZONES AND EFFECTS OF GLOBAL CHANGE

NOBUO MIMURA
Centre for Water Environment Studies, Ibaraki University, Japan
Email: mimura@mx.ibaraki.ac.jp

2.1 INTRODUCTION

Coastal zones in Asia and the Pacific are extremely diverse in their characteristics. They extend from temperate to tropical regions and are influenced by climatic conditions ranging from short-term weather events, such as monsoons and cyclones, to inter-annual variability resulting from the El Niño-Southern Oscillation (ENSO). Various geomorphological features also exist, including deltas formed by large rivers, islands, and rocky coasts. Numerous large rivers in Asia discharge huge amounts of fresh water and sediments to the sea, accounting for about 70% of the world's sediment transport flux. Coastal ecosystems, including mangroves, coral reefs, and seagrass beds, are home to rich and diverse biological resources, which support important commercial and subsistence fisheries.

Another distinguishing characteristic of the region's coastal environment is the long-standing and significant influence of human activity. The Asia and Pacific region accommodates over half of the world's population, about 60% of which exists on or near the coasts. During the Asian countries' dramatic economic development over the past 30 years, anthropogenic pressures have increased tremendously. Economic development and changes in land cover and land use in the river basins have also affected the coastal zones through discharge of wastes, including organic matter, nutrients, and hazardous chemicals, into rivers and outfalls.

In addition, global environmental changes, particularly climate change induced by global warming, will widely increase threats on the coasts. Higher air and seawater temperatures, sea-level rise, and changes in precipitation, tropical cyclones, and marine conditions will all exert various impacts on the health of the biophysical environment. These impacts weaken the ability of coastal systems to adapt to the existing human-induced pressures and vice versa. The result is that coastal zones in Asia and the Pacific are under multiple and interrelated stresses.

Past and ongoing changes in coastal environments are the result of natural and socioeconomic influences and should be understood by analyzing changes in those drivers. This knowledge provides a basis for estimating future trends in the coastal environment. It is also needed to supply information necessary for ensuring sustainable development of the coastal zone in the region. However, gaining a critical understanding of past and ongoing changes is challenging and involves both identifying critical driving factors and mutually correlating them.

To date, several attempts have been made in this direction, including ESCAP and ADB (2000), UNEP (2002), and the Millennium Ecosystem Assessment (2003). Following on these efforts, the International Center for the Environmental Management of Enclosed Coastal Seas (EMECS) is working to grasp the state of the coastal environment in the region, and the results will be published as a book in parallel with this volume. This work is an overview of the current understanding of the states of the environment and drivers in the Asia and Pacific region that will be presented in the EMECS book.

2.2 OVERVIEW OF THE STATE OF THE ENVIRONMENT

2.2.1 Coastal Changes

Erosion of Asian deltas

Asian coasts are characterized by large deltas, defined as a convex coastal topography formed by seaward shoreline migration. After decades of coastal erosion, however, many Asian deltas do not exhibit natural deltaic features (Table 2.1). This erosion is caused by a decrease of sediment supplied by rivers as a result of river catchment development, such as dam construction.

For example, after the Hoa Binh Dam was constructed in 1989 on the upper reaches of the Song Hong in Vietnam, sediment delivery was decreased by more than 30% from levels before dam construction. Sediment supply to the mouth of the main distributary of the Song Hong changed from about 26 million $t\ a^{-1}$ in 1949 to 11 million $t\ a^{-1}$ in 2000, engendering severe coastal erosion (Tanabe et al. 2003b). The Mekong River also has several dams in its drainage basin and more than ten additional dams planned or under construction. After the Manwan Dam began operation in 1993 in the upper reaches of the Mekong in China, the sediment load in Laos declined approximately 35 million $t\ a^{-1}$ (MRC 2003). The Huanghe River in China, once the second largest river in the world in terms of sediment discharge, delivers less than 10% of its past sediment load because of dam construction and irrigation. This reduction created conditions that have allowed serious coastal erosion around the river mouth in the Bohai Sea.

Table 2.1 Shoreline migration of major Asian deltas during the last 2000 years. (Data source: Saito et al., 2001, Hori et al., 2002, Ta et al., 2002, Tanabe et al., 2003a, b)

	shoreline migration during the last 2000 years	average rate (m/y)
Huanghe (Yellow River)	ca. 80 km	ca. 40 m
Changjiang (Yangtze River)	100–150 km	50–75 m
Song Hong (Red River)	20–30 km	10–15 m
Mekong River	30–40 km	15–20 m
Chao Phraya River	10–25 km	5–13 m

Change in relative sea level

Many places face relative sea-level change, a process that combines global and local changes in mean sea level relative to the land, including land subsidence and crustal motion. From a global change viewpoint, acceleration of beach erosion attributable to the rising sea level is a salient concern. The IPCC WGI (2001) suggested that the mean sea level has risen 10–20 cm during the 20th century, possibly exacerbating coastal erosion globally. By 2100, sea level is estimated to rise 9–88 cm globally, which is up to a four-fold increase in the past trend. Consistent with global trends, Asian coasts will generally face further serious erosion in this century. This threat will be particularly severe on the coasts, where the effects of rising sea level will be compounded by those from ground subsidence and decreased sediment supply from rivers.

The northern coast of the Gulf of Thailand is an example of the impact of a relative sea-level rise on a muddy coast. Pumping up groundwater has caused land subsidence around the Bangkok area and the Chao Phraya river mouth, which have undergone more than 60 cm of subsidence between the 1960s and 1980s. This has resulted in severe coastal erosion, and the shoreline retreat amounted to 700 m until the early 1990s (Vongvisessomjai et al. 1996). That subsidence has been stabilized since 1993 by regulation of groundwater pumping. Although the shoreline has been stabilized in the river mouth areas, coastal erosion continues to propagate westward and eastward from the river mouth.

Vulnerability of small island coasts

As with the Asian mainland, erosion poses a serious threat to small islands in the Pacific. The vulnerability of these islands is determined by the character of their coastline, the influence of large-scale environmental changes, and the condition of natural features that afford coastal protection. The coasts of most small islands in the Pacific are either cliffed or low-lying. Low-lying coasts, which are composed of sediments, are far more vulnerable to most environmental changes than are the cliffed-island coasts. Large-scale influences that can significantly increase erosion include sea-level rise and changes in wave heights and incident directions caused by El Niño.

Increased coastal erosion results from the combined effects of large-scale environmental changes, such as El Niño and sea-level rise, with loss of natural protection. Coral reefs, mangrove forests, and beach rocks protect many island coasts from destructive wave forces. Most Pacific island coasts are surrounded by coral reefs known as fringing and barrier reefs, which provide physical protection from large waves. Atolls are rings of coral reefs that enclose a shallow lagoon where a volcanic island has subsided. Many atoll reefs host islands of largely unconsolidated calcareous sand and gravel. Particularly for atolls, beach sediments are biological products, and the sediment supply is extremely limited. Where broad mangrove forests exist, they serve as a powerful protective buffer for the shoreline. However, many mangrove areas have been cleared in the past 150 years. Today, the shorelines of most low-lying coasts in the small islands are generally only lightly vegetated to allow access to the shore (Nunn and Mimura, 2005).

2.2.2 Water and Sediment Pollution

Organic pollution and eutrophication

Rapid growth of population and industrial activities has degraded environmental quality. Water pollution is the most serious problem of many countries in the region, and sewage is the major source of organic pollution in populous coastal areas. Whereas varying degrees of treatment are employed in some localities, untreated sewage is commonly disposed of either directly or indirectly. For instance in Thailand, pollutants discharged from the Chao Phraya River are the major contributor to coastal water pollution in the Upper Gulf of Thailand. The estimated biological oxygen demand (BOD) reaching the Gulf of Thailand via this river alone is 114.7 t d^{-1}. The estimate for all rivers emptying into the Gulf is 305.2 t d^{-1} (Thailand 1984).

Figure 2.1 shows BOD levels of East Asian coastal countries. It was estimated that 6 million tons of BOD are generated by coastal populations of countries neighboring the South China Sea, including Cambodia, China, Indonesia, Malaysia, Philippines, Thailand and Vietnam. Sewage treatment removes only about 11% of the generated BOD. A need to raise the volume removed by sewage treatment clearly exists, especially for large urban areas.

Eutrophication in coastal waters receiving high organic inputs from domestic and industrial effluents is common in the region. Eutrophication is associated with blooms of phytoplankton known as "red tides". In particular, red tides have been reported frequently in Malaysia, the Philippines, Thailand, Hong Kong, and Seto Inland Sea in Japan. In Hong Kong, an eight-fold increase was reported in the number of red tides per year in Hong Kong Harbor during 1976–1986, which was attributed to the 6-fold increase in population and a concurrent 2.5-fold increase in nutrient loading (Lam and Ho 1989).

Japan's Seto Inland Sea also suffered from severe water pollution and negative effects of eutrophication about 30–40 years ago. During the period of rapid national economic growth in the mid-1960s to mid-1970s, increased industrial activities and expansion of landfills in waterfronts caused rapid increases in water pollution, reduction of shallow water areas, and destruction of marine habitats. Concurrently, there was an increase in the frequency of red tide events (Figure 2.2)

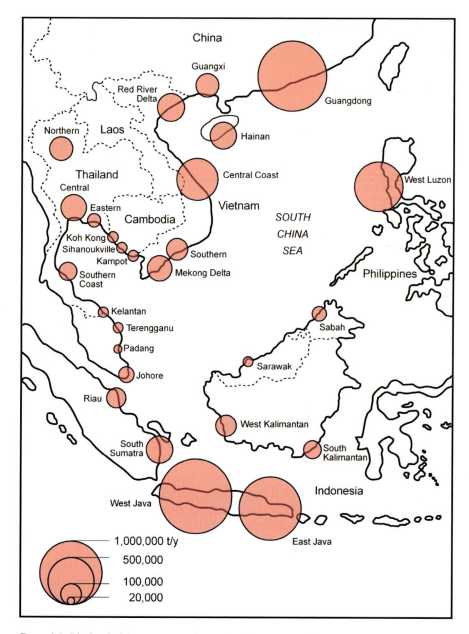

Figure 2.1. Biochemical Oxygen Demand (BOD) loading from domestic sources in East Asian Seas (*Source*: UNEP, 2000).

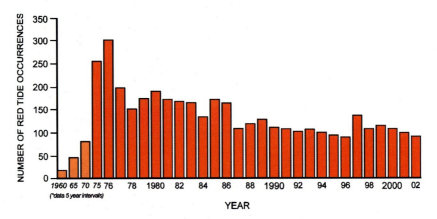

Figure 2.2. Change in the number of occurrences of red tide in the Seto Inland Sea, Japan

with notable consequences for aquaculture activities in the affected areas. For example, around 300 occurrences were recorded in 1976 causing mass mortality of caged fish cultures. In 1972, 14 million cultured yellowtails were killed by red tide, resulting in economic losses of 7.1 billion yen. In recent years, as a consequence of various environmental conservation measures, red tide bloom incidents have been reduced to around 100 events annually.

During that period of heavy pollution, the Seto Inland Sea was called "the dying sea", but it has slowly recovered by virtue of the efforts of various groups and the support of a strong legal framework. The Law on Temporary Measures for the Environmental Conservation of the Seto Inland Sea was enacted in 1973. This law became permanent in 1978 and has played an important role in environmental conservation of the area. Through total Chemical Oxygen Demand (COD) load control enforced by the law, COD discharged in the coastal zone of the Seto Inland Sea was remarkably reduced from 1,700 t d^{-1} in 1972 to 718 t d^{-1} in 1996 (Figure 2.3). Unfortunately, the achievements recorded in relation to the Seto Inland Sea are rare cases, and the region needs greater efforts to reduce the discharge of organic pollutants and nutrients in order to control the occurrence of red tide.

Hazardous chemicals

Discharge of chemicals into the sea by industrial effluents and enhanced use of agrochemicals on land is a common difficulty. An estimated 1,800 tons of pesticides were transported into the Bay of Bengal (Holmgren 1994). Aquaculture has also contributed to the discharge of pesticides, antibiotics and hormones, as well as nutrients (ESCAP and ADB 2000).

In 2001, the Stockholm Convention on Persistent Organic Pollutants (POPs) was adopted to safeguard human health and the environment. The acronym, POPs, is a collective term used to describe organic chemicals that remain in the environment for long periods, are widely distributed, accumulate in fatty tissues of living organisms,

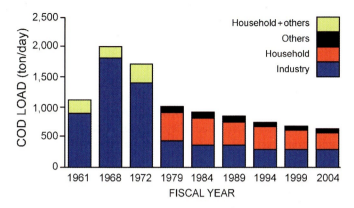

Figure 2.3. Trend in the change of total amount of Chemical Oxygen Demand (COD) load in the Seto Inland Sea, Japan

and are toxic to animals and humans. Among them, organochlorine compounds (OCs) have been recognized as endocrine disrupters, which pose serious effects on marine animals through bioaccumulation. The levels of dichlorodiphenyltrichloroethane (DDT) and polychlorinated biphenyls (PCBs) in dolphins are considerably higher than other OCs because they are highly persistent, relatively lipophilic and less biodegradable (Tanabe and Tatsukawa 1991). The maximum level of DDT, ranging from 80–96 μg/g wet weight, was found in northern right whale dolphins from temperate waters. Since DDT was discharged from tropical countries, where it remains in use to control malaria, it is of interest that similar concentrations of DDT were observed in animals of tropical waters.

Butyltin compounds (BTs) are another typical hazardous chemical. Since the 1960s, BTs have been used worldwide for various purposes such as anti-fouling agents in paints for boats and aquaculture nets. The International Maritime Organization (IMO) prohibited the use of harmful organotins in anti-fouling paints used on ships in 2001. BTs concentrations in the marine mammals are higher for the coastal waters than for pelagic waters (Tanabe et al. 1998), and animals in waters of developed countries showed higher concentrations than those from waters surrounding developing countries, reflecting the retention of BTs used in the past in these countries. Regarding BT concentrations in fish, Asia and Oceania showed lower values than Japan, Canada, or the USA.

Marine pollution of heavy metals, such as mercury and lead, is also a concern for the region. Mercury contamination has been found in bottom sediments of many semi-enclosed bays surrounded by industrial bases and coastal waters, such as the Sea of Japan. The people around Minamata city, Japan, are known to have suffered from severe mercury pollution called "Minamata Disease" from the 1950s to the 1970s. After 22 years of controversial discussions and a judicial ruling in 2004, they still face grim human health problems. Recently, mercury pollution from gold mining processes was also found in the city of Manado and the Ratatok area of North Sulawesi, Indonesia. Local people in these areas face the risk of severe and irreversible health problems from eating fish and marine animals (Harada et al. 2001; Limbong et al. 2003 and 2004).

Oil Spills

Accidental oil spills have occurred frequently along oil transport routes from the Persian Gulf and at points of discharge and loading. Oil spills cause severe pollution in ports in Bangladesh, Pakistan, and countries in Southeast Asia, and Sri Lanka's south-eastern beaches have frequently been affected by tar-balls (UNEP, 2002). The frequency and wide distribution of oil spills has led to the development of strict regulations in many countries of the Southeast Asian region. Countries in East Asia have also faced oil-spill problems. For example, accidents are quite frequent, particularly if small-scale spills are included in the accounting. In Japanese waters alone, spills of over 2 kl occur 400 times or more a year, according to annual reports by the Japan Coast Guard in the past ten years. Table 2.2 shows the list of large-scale accidents that have occurred in Japan, South Korea, China, and Taiwan over a recent 10-year period.

A large-scale accident is often devastating to coastal environments and can require a resource-intensive recovery effort. Here, we present the case of the *Sea Prince* in Korea. In July, 1995, the oil tanker *Sea Prince* loaded with 260,000 tons of crude oil from Saudi Arabia grounded on submerged rocks near Sori Island because of a typhoon. After the collision, a fire started in the engine room (Figure 2.4) and 5,035 tons of oil was spilled. Oil from the ship spread more than 200 km along the shoreline, which was characterized by rocky coasts and sandy beaches. The most seriously affected area was along Sori Island and the neighboring coasts of Busan and Ulsan. Nineteen days were required to recover 1,390 kl of the discharged oil.

Table 2.2 List of oil spills occurring around the Northwest Pacific region (Sawano, 2003)

Date (YMD)	Country	Place	Vessel	Amount	Type of oil
1992.5.1	Japan	Kushiro, Hokkaido	Shell Oil base	246 kl	Unknown
1993.9.27	S. Korea	Jeonnam Yeocheonsi, east coast of Myo Island	Gumdong No. 5	1,228 kl	Heavy B-C
1994.10.17	China	Qinhuangdao, Hebei	Fwa Hai No. 5	Unknown	Unknown
1995.7.23	S. Korea	Jeonnam Yeocheonsi Sori Island	Sea Prince	5,035 kl	Crude/ bunker
1995.9.20	S. Korea	Busan South Hoyongie Island	No. 1 Yuilu	Unknown	Unknown
1995.11.17	S. Korea	Jeonnamyeosu Honam Oil Refinery berth	Honam Sapphire	1,402 kl	Crude
1996.9.19	S. Korea	Nine miles from Jeonnam Yoso Island	Ocean Joedo	207 kl	Heavy B-C
1997.1.2	Japan	Near Oki Island, Shimane Pref.	Nakhodka	8,660 kl [*1]	Heavy C
2001.1.17	Taiwan	Kenting National Park	Amorgos	1,150 kl	Bunker
2001.3.30	China	Mouth of Yangtze River	Deiyong	700 kl	Styrene

[*1] – *This volume is based on Sao (1998), official value announced by Japan Coast Guard was 6,240 kl.*

State of the Environment in Asia and Pacific Coastal Zones 25

Figure 2.4. Firing Sea Prince near Yosu, Sori Island This photograph is quoted from Korean Ministry of Maritime Affairs and Fishery (2002)

Police and local residents cleaned contaminated areas, mostly manually. Cleanup operations along the shorelines continued for five months (Lee 2001).

2.2.3 Coastal Ecosystems

Mangroves

Mangrove forests are widely distributed throughout Asia and the Pacific islands with approximately 40% of the world's mangroves found in Asia. Large areas of mangrove forests exist in India, Bangladesh, Burma, Thailand, Vietnam, Malaysia, the Philippines, and Indonesia. Among them, Indonesia has the largest area in the region because of its long coastline, whereas the Sundarbans, located in India and Bangladesh, have the largest continuous area of mangrove forests in the world. Mangrove forests are extremely biodiverse and act as a nursery and spawning ground for numerous species of fish, crustaceans, molluscs, and reptiles. They are also important for the daily lives and economies of local societies, providing a myriad of goods and services.

Asian countries are estimated to have lost about 26% of their mangrove area during the 20 years period from 1980 to 2000 (FAO 2003a; Kashio 2004). The annual rates of decrease are 1.6% from 1980 to 1990, and 1.4% from 1990 to 2000. Causes of mangrove degradation include conversion to shrimp ponds, felling of timber for charcoal, firewood, wood chips and pulp production, development of human settlements, ports, agriculture, industries, roads, and other infrastructure, as well as excessive siltation. Shrimp culture has been the most serious cause of mangrove conversion, particularly in Southeast Asian countries such as Indonesia, Vietnam, the Philippines, Malaysia, and Thailand. It is estimated that over 60% of mangrove forests have already been converted to aquaculture ponds (ESCAP and ADB 2000).

Clearing and degradation creates serious impacts on mangrove ecosystems and neighbouring coastal areas. Destruction of mangroves directly degrades biodiversity of both flora and fauna, lowers productivity, and causes the eventual loss of fish and shrimp. Clearing mangroves also alters soil characteristics in many ways, such as through increased sediment transport and soil erosion. Importantly, mangroves serve as a filter between land and ocean. Therefore, the disappearance of mangrove forests increases the discharge of land-based pollutants, such as organic matter and toxic chemicals, into coastal areas, which adversely affects water quality. These impacts eventually result in the loss of productivity of inshore and near-shore fisheries and threaten coastal communities that depend on the fisheries for both commerce and subsistence.

Coral reefs

Coral reefs are another precious ecosystem of coastal zones. They are sites of rich biodiversity and provide resources for human use. The Asia-Pacific region has been recognized as the global centre of tropical marine biodiversity. About four-fifths of the world's coral reefs are in the region, with about half in the Pacific, one-third in the Indian Ocean, and the remainder in South Asia (ESCAP and ADB 2000). Fifty of seventy coral genera found in the world occur in the Indian and western Pacific Oceans. Coral reefs in Southeast Asia are biologically diverse, holding an estimated 34% of the earth's coral reefs (Burke et al. 2002).

In the Asia-Pacific region, coral reefs are threatened by a range of human activities, including coastal development, exploitation and destructive fishing practices, as well as land-based and marine-based pollution. Coastal development, including tourism facilities, often causes not only direct destruction, but also indirect impacts, such as discharge of sediments and nutrients that cause high turbidity in the sea. Destructive fishing practices, such as the use of explosives and cyanide, still constitute a major problem. Growth of the live fish trade in the region has fuelled the wide use of cyanide fishing, which has severe effects on juvenile fish and coral organisms. In addition, coral reefs are often damaged physically by fishing equipment, diver interactions, and anchors. Dredging and filling for ports and fishery harbours are other causes.

In Southeast Asia, 88% of the coral reefs face a medium to very high threat from human impacts. In Indonesia and the Philippines only 30% of the coral reefs are in good or excellent condition. The reefs of the South Pacific region are under less immediate threat than those of Southeast Asia. About 40% of the Pacific reefs are classified as threatened, and 10% face a high risk (WRI 1999).

Global environmental changes, particularly global warming, will increasingly threaten coral reefs as demonstrated by the spread of coral bleaching in 1997–1998. Bleaching occurred globally during that period, including in the Pacific, Indian Ocean, and Caribbean regions (Figure 2.5). Although the immediate trigger for this event was increased sea temperatures related to the 1997 El Niño, long-term rises in seawater temperature potentially contributed to it (Walther et al. 2002). Damage by coral bleaching continued after the 1998 event. Most reefs damaged in 1998 have been recovering in Japan. However, bleaching occurred again in 2001 when some reefs experienced about 50% mortality. In the South Pacific, severe coral bleaching occurred in 2000 and 2002, especially in Fiji, Tuvalu, and Vanuatu. It is also suggested that rising sea level would inundate coral reefs if the rate of sea level rise were faster than the coral reefs' upward growth rate. Healthy coral reefs are better able to keep up with rising sea level and recover from coral bleaching damage.

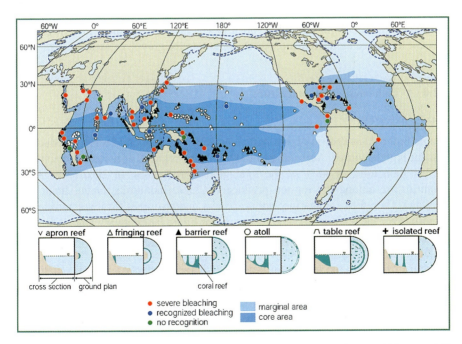

Figure 2.5. Bleaching of coral reefs reported from 1997 to 1998 (Source: Mimura and Harasawa (ed.) (2000). Revised from Wilkinson (1998) and Hori (1980))

Therefore, the combined stress of local human activities and global warming should be given attention by management.

Sea grass beds

Seagrass beds grow best in estuaries and lagoons where they are often associated with mangrove forests and coral reefs. Because seagrass meadows support rich biodiversity and provide primary refugia for marine organisms, most major commercial fisheries in the region are situated adjacent to seagrass beds. Globally, seagrass beds occupy an area of about 600,000 km^2, contributing 12% of the total carbon storage in the ocean (Duarte and Cebrian 1996). Seagrass diversity is highest in East Asia, reaching up to southern Japan, followed by the Red Sea and East Africa. So far, 16 species of seagrasses have been identified in Philippine waters, and 14 species have been reported from Indonesia. Australia's 30 seagrass species represents the highest diversity in the region (Fortes 1989, Kuo and Larkum 1989). Seagrass beds can also exist in other areas, although they are often less dense.

In Southeast Asia, seagrasses are under threat from the loss of mangroves, which act as a buffer that filters discharged sediments. They are also threatened by agricultural runoff, industrial wastes, and sewage discharges. In Indonesia, about 30–40% of the seagrass beds have been lost in the last 50 years, including a loss of 60% for the seagrass beds around Java. In Singapore, the patchy seagrass habitats have suffered severe damage largely through burial under landfill operations. Losses of the beds respectively amount to 20–30% and 30–50% in Thailand and the Philippines. Coastal eutrophication is a major long-term threat to seagrass ecosystems in Southeast Asia because it reduces light exposure, thereby retarding the growth of many plants.

2.2.4 Fisheries and Aquaculture

Marine fisheries

People in the Asia-Pacific region have long relied on fish and fishery products as a major protein source. According to FAO data, the region's fish catch amounted to 44.7 million tons, or 48% of the total global production, in 2002. Five Asian countries are among the top-ten producers: China, Indonesia, Japan, India, and Thailand. Therefore, the marine capture fishery and resources for it are particularly important for the region. Potential resources for future landing vary among regions. Of the world's fisheries, the Indian Ocean fishery might offer the greatest potential for future development (FAO 1997).

Figure 2.6 shows the long-term trend of fisheries production in Asia, excluding China. During the 1950s to the 1970s, fisheries in the region showed rapid development through structural changes in production and technologies. Through modernization of fishing technology and the use of engines for boats, many traditional and subsistence fisheries evolved into a productive and market-oriented industry.

In the 1990s, the total fisheries production in the Asia-Pacific region fell into stagnation after having reached a peak of 29 millions tons in 1989. The largest

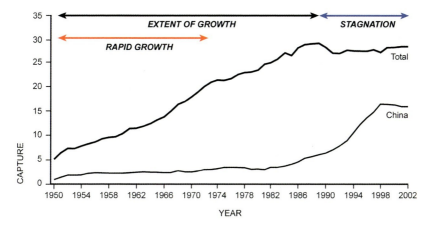

Figure 2.6. Production in capture fisheries in China and Asia excluding China between 1950 and 2002

component, the marine fisheries, had started full-scale development in the early 1960s, and reached a turning point in the late 1980s. The marine fisheries production peaked at 24.7 million tons in 1989 and fluctuated after that. The fractions of pelagic and demersal fish catches in the marine fishery have also changed. The trend in pelagic landings shows a moderate increase from about 2 million to 5.5 million tons between 1950 and 1973, followed by a period of rapid growth that reached a peak of nearly 11.7 million tons in 1988. Thereafter, the fish catch declined to 9.4 million tons by 2002 (after FAO 2003b). Simultaneously, the catch of demersal fish increased gradually from 1.5 million tons in 1950 to 5.2 million tons in 1974, then decreased to nearly 4 million tons in 1983. The rapid increase of marine catches after the 1960s in Southeast Asia was attributed to the development of trawl fisheries. Fishery activities in Southeast Asia have increased dramatically during the past two decades. However, the region faces heavy overfishing, which has raised the question of how to maintain fishery resources for future sustainable use.

Aquaculture development

During its growth throughout the Asia-Pacific region, aquaculture production has increased at a rate four times faster than production from capture fisheries. In addition, aquaculture's share of all fish landings increased from 20.7% in 1984 to 38% in 1995 (FAO 1997). Shrimp culture is the most widely expanded practice. Today, Asia accounts for 87% of global aquaculture production by weight (IFPIC and WFC 2004). Freshwater culture, rather than marine culture, used to dominate the industry. Since the 1980s, however, technological innovation has advanced rapidly in marine and brackish water aquaculture, thereby contributing to the rapid increase in production over the Asia-Pacific region.

Such rapid development in aquacultural production has adversely affected the coastal environment, including mangrove clearing for shrimp culture and water pollution. An additional environmental concern associated with aquaculture is the potential hazard of the accidental release of exotic species and the spread of diseases from aquacultural facilities to the surrounding natural environment (ESCAP and ADB 2000).

2.3 POPULATION GROWTH AND ECONOMIC DEVELOPMENT AS DRIVERS OF ENVIRONMENTAL PROBLEMS

Coastal zones in Asia and the Pacific region continue to experience tremendous and rapid changes, as shown in the previous section. They have been driven by various factors, including natural forces of climatic and oceanographic phenomena, human activities, including discharge of land-based and sea-based pollutants, and processes controlling the flows of material and energy within the coastal ecosystems. Among the drivers of those changes pressures from human activities that are conducted in the coastal zones appear predominant.

Coastal areas are among the most crowded and developed in the world. The overwhelming bulk of humanity is concentrated along or near coasts, which comprise just 10% of the earth's land surface. Recent data (Population Reference Bureau 2004) show that the Asia-Pacific region is home to over half of the world's population with expectations that this ratio will remain in a similar range throughout the 21st century. More than half of all Asian people reside in the coastal zone. For example, of China's 1.3 billion people, close to 60% of the population live in coastal provinces. Furthermore, the region's coastal populations are growing faster than those in inland areas. Migration is a key factor affecting population growth in coastal zones. Indonesia and Vietnam are two typical examples of Asia's population shift. In Indonesia, 65% of people live on the main island of Java, on just 7% of country's land area. Similarly, in Vietnam coastal populations are growing 20% faster than the remainder of the country (Hinrichsen 1998).

Population growth has exacerbated the trend of an increasing number of megacities (those with 10 million inhabitants or more), as shown in Table 2.3. In 1950, New York was the world's sole megacity. Megacities multiplied from five in 1975 to 17 in 2001. Among them, nine urban agglomerations were located in Asia's coastal zones. With 26.5 million inhabitants, Tokyo was the most populous city in the world in 2001. It is projected that the number of coastal megacities in Asia will increase to at least 10 of the world's 21 by the year 2015. Tokyo, Dhaka, and Mumbai (Bombay) are each expected to hold more than 20 million inhabitants (United Nations Population Division 2001).

The Asia-Pacific region has emerged as a leading growth region. Concurrent with increased economic activity has been an increased mutual dependence between the countries of the region in terms of resources and trade. Over the past 30 years, the region has gradually moved from a subsistence lifestyle towards a consumer society, with rapid rates of urbanization and westernization that have occurring concomitant with the population increase. The ASEAN countries and coastal areas of China are undergoing rapid economic growth and will continue to be a center of the world's economic growth.

Table 2.3 Population of cities with 10 million inhabitants or more, 1950, 1975, 2001 and 2015 (millions)

City	Population	City	Population	City	Population
Year of 1950		Year of 2001		Year of 2015	
1 New York	12.3	1 Tokyo	26.5	1 Tokyo	27.2
		2 Sao Paulo	18.3	2 Dhaka	22.8
Year of 1975		3 Mexico City	18.3	3 Mumbai	
		4 New York	16.8	(Bombay)	22.6
1 Tokyo	19.8	5 Mumbai		4 Sao Paulo	21.2
2 New York	15.9	(Bombay)	16.5	5 Delhi	20.9
3 Shanghai	11.4	6 Los Angeles	13.3	6 Mexico City	20.4
4 Mexico City	10.7	7 Calcutta	13.3	7 New York	17.9
5 Sao Paulo	10.3	8 Dhaka	13.2	8 Jakarta	17.3
		9 Delhi	13.0	9 Calcutta	16.7
		10 Shanghai	12.8	10 Karachi	16.2
		11 Buenos Aires	12.1	11 Lagos	16.0
		12 Jakarta	11.4	12 Los Angeles	14.5
		13 Osaka	11.0	13 Shanghai	13.6
		14 Beijing	10.8	14 Buenos Aires	13.2
		15 Rio de Janeiro	10.8	15 Metro Manila	12.6
		16 Karachi	10.4	16 Beijing	11.7
		17 Metro Manila	10.1	17 Rio de Janeiro	11.5
				18 Cairo	11.5
				19 Istanbul	11.4
				20 Osaka	11.0
				21 Tianjin	10.3

Note: Shaded mega-cities are located in coastal zones of Asia
(Sources: United Nations Population Division, 2001) World Urbanization Prospects: The 2001 Revision)

Given the relentless and cumulative processes of demographic changes, urbanization and industrial development, trade and transport demands, and lifestyle changes, the coastal zones of Asia are under increasing anthropogenic pressures (Turner et al. 1996). The high concentration of people and the rapid economic growth in coastal regions has produced many economic benefits, including improved transportation links, industrial and urban development, revenue from tourism, and food production. However, the combined effects of booming population growth and economic and technological development are threatening the ecosystems that provide a basis for them. All coastal areas are facing an increasing range of stresses and shocks. These trends are expected to continue and intensify in the future.

2.4 EFFECTS OF CLIMATE CHANGE AND SEA-LEVEL RISE

Another major factor driving future coastal change will be global environmental changes, in particular climate change induced by global warming. Some effects of global warming have already been felt, although, it is difficult to isolate these effects in observations. As warming proceeds, however, its effects will become more profound in this century.

The IPCC WGI (2001) indicated that the global mean surface temperature has already risen by 0.6°C ± 0.2°C during the past century. The degree of global warming varies with the emission of greenhouse gases (GHGs) such as CO_2. Therefore, the future extent of warming will ultimately depend on how human society and its economic activities develop in the future. The IPCC has developed future socioeconomic scenarios called SRES scenarios, which give GHG emission conditions for climate models to predict climate change in this century (Nakicenovic et al. 2000). Results of those estimates summarized in the IPCC Third Assessment Report (IPCC WG1 2001) show that the global mean surface temperature will rise 1.4°C to 5.8°C by the year 2100. It is noteworthy that the range of that estimate is wide because of the different socioeconomic scenarios and climate models used.

Figure 2.7 shows spatial distributions of increased mean temperatures for northern winter (DJF; December–February) and northern summer (JJA; June–August) during 2081–2100, which are obtained by averaging predictions of 14 different climate models using an IPCC SRES emission scenario. This scenario shows that higher latitudes have larger warming than the lower latitudes, and that the land surface is warmed more than over the oceans. In northern winters, increased warming is projected over Eastern Asia, whereas in northern summer, stronger warming appears over the semi-arid Middle East and western China. Warming in the coastal zones is light compared with that of the inland regions because of the influence of oceans.

Projections of precipitation derived by the climate models show large variability, particularly within the region. General trends are increased precipitation in the tropics and in mid-latitudes and high-latitudes, and decreased precipitation in the subtropics. In East Asia, precipitation might increase in the warmer season (April–September), whereas it might decrease in the colder season

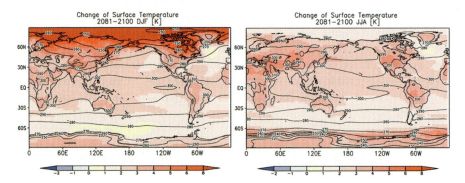

Figure 2.7. Global distribution of mean surface air temperature for northern winter (DJF; December–February) and northern summer (JJA; June–August). The present-day model climate (1981–2000) is shown by contours and its changes for 2081–2100 are by shading

(November–February) (Min et al. 2004). Global warming would also increase the amplitude and frequency of strong rainfalls.

Another concern is changes in tropical cyclones, specifically their frequency, intensity, seasonality, and geographical range. Although convincing evidence of change is not available in the observed records of tropical cyclone behaviour, the frequency of maximum-intensity tropical cyclones is expected to increase by 5–10% by around 2050. Similarly, the frequency of mean-intensity tropical cyclones is also expected to increase. Consequently, peak precipitation rates are estimated to increase by 25%. No significant change is expected as far as the geographic regions in which cyclones form; however, the formation rate may change in some regions. This formation is also strongly influenced by ENSO (Walsh 2004). These estimates indicate an increase in the frequency of floods and landslides resulting in damage to socio-economic sectors and potential impacts on the coastal zones in the region through various mechanisms.

The sea level is anticipated to rise between 9 cm and 88 cm by 2100 (IPCC WGI 2001). Because the past rate of mean sea-level rise is about 2 cm per decade in the Pacific, the estimated maximum increase is some four times greater. In addition, the changes in the sea level in the coastal zone are attributable to vertical land movement as well. This relative change to the land elevation is a salient factor affecting the coastal environment. If the relative sea-level rise becomes large, serious effects will appear such as inundation of coastal low-lying coastal plains and wetlands, exacerbation of flooding attributable to storm surges and river floods, accelerated coastal erosion, and saltwater intrusion into rivers and aquifers. We should note that these adverse effects appear throughout the region, and exacerbate existing environmental problems in each coast.

A risk assessment of coastal areas serves to illustrate the potential effects of climate change. There have been a number of country studies on the vulnerability of climate change and sea-level rise in the Asia-Pacific region. However, difficulties in gaining a comprehensive picture of future threats based on country studies prompted Mimura (2000) to perform a region-wide assessment. He investigated the impacts of sea-level rise and storm surges using global datasets on climatic, environmental, and societal information targeting the whole Asia-Pacific region. This study area extends from 30°E to 165°S and from 90°N to 60°S and includes a land area of 6.5 million km^2. The population within this area is estimated using Bos et al. (1994) to increase from 3.8 billion people in 1994 to 7.6 billion in 2100.

By combining sea-level rise, astronomical tide, and storm surge, future scenarios of sea level were set at two points in time: 2000 and 2100. An analysis of tropical cyclones data observed from 1949 to 1989 was used to determine the highest storm surge level at each coastal segment in the region. The study assumes that global warming will not change tropical cyclones, thus storm surges, from those of the past 40 years. The study was further constrained by the accuracy of land elevation in the base maps used and the need to treat storm surge uniformly in determining the flooded area, although, storm surge flooding seldom penetrates far inland. In spite of such constraints, these results indicate the degree and scale of possible impacts from future sea-level rise and climate change for the Asia-Pacific region.

This study has revealed that, even today, the land area of the region below high tide level and storm surge level, (i.e. inundated and flooded areas) is 311,000 km^2, or 0.48% of the total land area, and 611,000 km^2, or 0.94%, respectively. The land area increase to 618,000 km^2 and 858,000 km^2 (0.98% and 1.32%) with a 1 m sea-level rise, resulting in an 247,000 km^2 increase in flooded area.

The study also showed that people in the Asia-Pacific region are already threatened by storm surges. Today, about 47 million people, or 1.21% of the total population, live in areas below high tide level, while 270 million people or 5.33% live below storm surge level. If the mean sea level rise of 1 m and the population growth estimates for 2100 are taken into account, the affected population becomes approximately 200 and 450 million people. The areas that would be severely affected are distributed in the deltas of the Mekong River, Ganges and Brahmaputra Rivers, and Yangtze River, and the southern part of Papua New Guinea. The countries and areas where more than 10% of the national population may be affected include Vietnam, Taiwan, Cambodia, Brunei, Bangladesh, and small islands in the Pacific.

2.5 CONCLUSIONS: RESPONSE BY INTEGRATED COASTAL MANAGEMENT

We have seen that coastal zones in the Asia-Pacific region continue to experience unprecedented environmental changes that are driven both by pressures of local human societies and global environmental changes. Given the plausible estimates for future growth in the region's population and economies, in combination with progressing global warming and climate change, it appears that increasingly ominous consequences loom ahead. As the home of the world's richest biodiversity and a vast assortment of biogeophysical processes, the coastal environment in Asia and the Pacific is a global heritage. The environment also provides a common foundation for economic, cultural, and aesthetic resources for local communities throughout the region. Therefore, relevant management of the coastal zones is of essential importance for the region's sustainable development.

What is an appropriate response to the increasing pressures and threats facing coastal societies and the environment? We have not yet resolved this conundrum. The processes to respond to this question might consist of several components. Primarily, we must understand the past and ongoing phenomena through observations and scientific studies on coastal processes. With this understanding, we can predict future changes and their impacts in order to establish response strategies and options for management policies. Fostering practical policies and actions requires both collecting and interpreting scientific information and building the capacity to support these activities. Subsequently, we must implement appropriate responses and evaluate their effects. Responsibility for such efforts is not limited to government, and responses should be pursued at the regional, national, and even local levels of society.

In the face of the upwelling of local and global environmental problems, Integrated Coastal Management (ICM) has been proposed as a relevant framework for management of coastal zones. ICM has drawn attention globally as a policy tool that engenders a comprehensive framework to address multiple management issues. Following Agenda 21, which was adopted at the Earth Summit (UNCED) held in 1992, major international organizations, such as the World Bank, The World Conservation Union (IUCN), UNEP, and the OECD (1995) successively published ICM guidelines. The common understanding behind them is that the coastal zones are a unique resource system that demands special management measures.

Along with the global movement, Asian countries have moved toward introducing and establishing ICM at both national and regional levels. For example, in China the government has developed a framework for ICM to respond to increasing pressures of population and economic development on the coasts. Chinese Ocean Agenda 21 was developed in 1996, and the National Sea Area Use Management Law, the basis of today's coastal zone management, was enacted in 2001. Korea also has developed a framework for ICM that establishes an integrated coastal and ocean governance system strengthened by legislative actions, such as the Coastal Management Act, the Marine Pollution Prevention Act, and the Wetland Conservation Act (1999) (PEMSEA, 2003). As an initiative for collective international efforts in the East Asia region, Partnerships in Environmental Management for the Seas of East Asia (PEMSEA) was established in 1994 with the support of the governments in the region and international organizations such as the Global Environment Facility (GEF), UNDP, and IMO. These activities signal that the concept of ICM has been widely accepted by governments and regional organizations.

In addition to national and regional efforts, countries of this region are increasingly relying on community-based management of coastal zones, which involves local government, local enterprises, self-employed individuals, and inhabitants. Community-based management recognizes that local organizations and inhabitants often relate to coastal environments as both resource users and ecosystem stewards because the community's well-being is closely linked to coastal resource condition. In some countries, this reliance is reflected in a range of traditional customs or sea tenure systems that still governs access to coastal resources at a local level. These traditional approaches have shown remarkable resilience over time. They are often flexible and responsive to local circumstances. The goal of these approaches is to sustain resources, such as fisheries, by modifying rates and patterns of harvest depending on local resource availability. This approach is consistent with modern concepts of sustainability (Harvey et al. 2005).

Different types of coastal management with different concepts and background are being practiced in the region. Through these practices each country and each local community will find the most relevant scheme to manage the valuable coastal resources of the Asia-Pacific region.

ACKNOWLEDGEMENTS

This overview is the main result of contributions to the book *Asian-Pacific Coasts and Their Management: States of Environment*, prepared by the International EMECS Center. I sincerely appreciate the authors of the book, who prepared relevant drafts to the sections of this paper indicated in parentheses: Tetsuo Yanagi (2.1), Yoshiki Saito (2.2.1), Piamsak Menasveta (2.2.2), Osamu Matsuda (2.2.2), Nobuyuki Miyazaki (2.2.2), Nobuhiro Sawano (2.2.2), Toyohiko Miyagi (2.2.3), Thamasak Yeemin (2.2.3), Miguel Fortes (2.2.3), Masahiro Yamao (2.2.4), Jingshan Yu (2.3.1), Haruyuki Kojima (2.3.1), Akio Kito (2.3.2), and Kwangwoo Cho (2.3.2). I also express my appreciation to the members of the book's Editorial Committee and the Secretariat of the International EMECS Center.

REFERENCES

Bos, E., Vu, M.T., Massiah, E. and Bulatao, R. 1994: World populations 1994–95: Estimates and projections with related demographic statistics, Johns Hopkins University Press for the World Bank, Baltimore.

Burke, L., Selig, E. and Spalding, M. 2002: Reefs at Risk in Southeast Asia. World Resources Institute, Washington D.C., 72pp.

Duarte, C.M. and Cebrian, J. 1996: The fate of marine autotrophic production. *Limnology and Oceanography* 41, 1758–66.

ESCAP and ADB, 2000: State of the Environment in Asia and Pacific 2000, Economic and Social Commission for Asia and the Pacific and Asian Development Bank.

FAO, 1997: Review of the state of world fishery resources: Marine fisheries, FAO *Fisheries Circular*, No. 920 FIRM/C920.

FAO, 2003a: FAO's Database on Mangrove Area Estimates. Forest Resources Assessment Working Paper No. 62.

FAO, 2003b: Year Book of Fishery Statistics 2003.

Fortes, M.D. 1989: Seagrasses: A resource unknown in the ASEAN region. ICLARM Education Series 5, 46 pp. International Center for Living Aquatic Resources Management, Manila, Philippines.

Harada, M., Nakanishi, J., Yasoda, E., Pinheiro, M.D.N., Oikawa, T., Guimaraes, G.D., Cardoso, B.D., Kizaki, T. and Ohno, H. 2001: Mercury pollution in the Tapajos River basin, Amazon, mercury level of head hair and health effects. *Environment International* 27(4), 285–290.

Harvey, N., Rice, M. and Stevenson, L. 2004: Global Change Coastal Management Synthesis Report, APN, 37 p.

Hinrichsen, D., 1998: Coastal Waters of the World: Trends, Threats, and Strategies, Island Press, Washington, D.C., 276p.

Hori, N. 1980: Coral reef in Japan, Science Journal KAGAKU, 50, 111–122 (in Japanese).

Hori, K., Saito, Y., Zhao, Q. and Wang, P. 2002: Evolution of the coastal depositional systems of the Changjiang (Yangtze) River in response to Late Pleistocene–Holocene sea-level changes. *Journal of Sedimentary Research* 72, 884–897.

International Food Policy Research Institute (IFPRI) and World Fish Center (WFC), 2004: Fish to 2020, World Fish Center, 1–152.

IPCC WGI, 2001. Climate Change 2001, Scientific Basis, Cambridge Univ. Press, Cambridge, 881p.

Kashio, M. 2004: Mangrove resources and management in Southeast Asia. In: Bhandari, B.B., Kashio, M. and Nakamura, R. Mangrove in Southeast Asia: Status, Issues and Challenges. Ramsar Center Japan/IGES, Tokyo, 266 p.

Korean Ministry of Maritime Affairs and Fisheries, 2002: White papers on marine pollution of *Sea Prince* oil spill accident (Korean official report of the accident).

Kuo, J. and Larkum, A.W.D. 1989: Seagrass taxonomy, structure and development, pp. 6–73. In Larkum, A.W.D., McComb, A.J. and Shepherd, S.A. (eds.), Biology of Seagrasses. Elsevier, Amsterdam, 841 p.

Lam, C.W.Y. and Ho, K.C. 1989: Red tides in Tolo Harbour, Hong Kong. In: Okaichi, T., Anderson, D.M. and Nemoto, T. (eds.), Red tides: Biology, Environmental Science and Toxicology, Elsevier Science Publishing, New York, pp. 49–52.

Lee, B. 2001: Changes of oil spill response system after Sea Prince accident. Reports of international Symposium of Changes on OSR System in Major Countries/Recent Movement of Compensation Scheme. March 1–2, 2001. Petroleum Association of Japan, Tokyo (in Japanese).

Limbong, D., Kumampung, J., Rimper, J., Arai, T. and Miyazaki, N. 2003: Emissions and environmental implications of mercury from artisanal gold mining in north Sulawesi, Indonesia. *Science of the Total Environment* 302, 227–236.

Limbong, D., Kumampung, J., Ayhuan, D., Arai, T. and Miyazaki, N. 2004: *Coastal Marine Science* 29(1), 69–74.

Millennium Ecosystem Assessment, 2003: Ecosystems and Human Well-being, Island Press, Washington, D.C., 212p.

Mimura, N. 2000: Distribution of vulnerability and adaptation in the Asia and Pacific Region, Proc. APN/SURVAS/LOICZ Joint Conference on the Coastal Impacts of Climate Change and Adaptation in the Asia-Pacific Region, APN and Ibaraki University, 21–25.

Mimura, N. and Harasawa, H. (eds) 2000: Data Book of Sea-Level Rise 2000, Center for Global Environmental Research, National Institute of Environmental Studies, 128p.

Min, S.-K., Park, E.-H. and Kwon, W.-T. 2004: Future projections of East Asian climate change from multi-AOGCM ensembles of IPCC SRES scenario simulations. Journal of the Metrological Society of Japan, 82, 1187–1211.

MRC, 2003: State of the Basin Report. Mekong River Commission, 300p.

Nunn, P. D. and Mimura, N. 2006: Promoting sustainability on vulnerable island coasts: a case study of the smaller Pacific Islands, in McFadden, L., Nicholls, R.J. and Penning-Rowsell, E. (eds.) *Managing Coastal Vulnerability: An Integrated Approach*. Elsevier (In press).

Saito, Y., Yang, Z. and Hori, K. 2001: The Huanghe (Yellow River) and Changjiang (Yangtze River) deltas: A review on their characteristics, evolution and sediment discharge during the Holocene. *Geomorphology* 41, 219–231.

Sao, K. 1998: Cutting down oil spill response system. Heavy oil pollution: For tomorrow can "Nakhodka" change Japan? Ocean Engineering Research Inc., 372–431 (in Japanese).

Sawano, N. 2003: Marine pollution of the Japan Sea. White paper of the Asian environment 2003/04. Toyo-Keizai-Junpou-sha, 158–173 (in Japanese).

Ta, T.K.O., Nguyen, V.L., Tateishi, M., Kobayashi, I., Tanabe, S. and Saito, Y. 2002: Holocene delta evolution and sediment discharge of the Mekong River, southern Vietnam. *Quaternary Science Reviews* 21, 1807–1819.

Tanabe, S. and Tatsukawa, R. 1991: Persistent organochlorines in marine mammals. In: Jones, K.C. (ed.), Organic Contaminants in the Environment: Environmental Pathways and Effects. Elsevier, Amsterdam, pp. 275–289.

Tanabe, S., Prudente, M., Mizuno, T., Hasegawa, J., Iwata, H. and Miyazaki, N., 1998: Butyltin contamination in marine mammals from North Pacific and Asian coastal waters. *Environ. Sci. Technol.* 32(2), 193–198.

Tanabe, S., Saito, Y., Sato, Y., Suzuki, Y., Sinsakul, S., Tiyapairach, N. and Chaimanee, N. 2003a: Stratigraphy and Holocene evolution of the mud-dominated Chao Phraya delta, Thailand. *Quaternary Science Reviews* 22, 789–807.

Tanabe, S., Hori, K., Saito, Y., Haruyama, S., Vu, V.P. and Kitamura, A. 2003b: Song Hong (Red River) delta evolution related to millennium-scale Holocene sea-level changes. *Quaternary Science Reviews* 22, 2345–2361.

Thailand, Department of Health, 1984: Survey of water quality in the estuary. In *Proceedings of the Third Seminar on Water Quality and the Quality of Living Resources in Thai Waters*. Bangkok, March 26–28, 1984. Bangkok, National Research Council, pp. 62–78.

Turner, R.K., Subak, S. and Adger, W.N. 1996: Pressures, trends and impacts in coastal zones: Interactions between socio-economic and natural systems. *Environmental Management* 20, 159–173.

UNEP, 2000: Overview on Land-Based Sources and Activities Affecting the Marine Environment in the East Asian Seas. UNEP Reg. Seas Rep and Stud. 173.

UNEP, 2003: GEO 3, United Nations Environment Programme, Earthscan.

United Nations Population Division, 2001: World Urbanization Prospects: The 2001 Revision.

Vongvisessomjai, S., Polsi, R., Manotham, C., Srisaengthong, D. and Charulukkana, S. 1996: Coastal erosion in the Gulf of Thailand. In: Milliman, J.D. and Haq, B.U. (eds.), Sea-Level Rise and Coastal Subsidence, Kluwer Academic Publ., Dordrecht, 131–150.

Walsh, K. 2004: Tropical cyclones and climate change: Unresolved issues, *Climate Research* 27, 77–83

Wilkinson, C. (ed.) 1998: Status of Coral Reefs of the World, Australian Institute of marine Science, 184p

WRI, 1999: East Asia, http://www.wri.org/wri/indictrs/reefasia.html.

CHAPTER 3

COASTAL MANAGEMENT IN THE ASIA-PACIFIC REGION

NICK HARVEY[1] AND MIKE HILTON[2]
[1] *Geographical and Environmental Studies, Adelaide University, Australia*
[2] *Department of Geography, University of Otago, New Zealand*

3.1 INTRODUCTION

As noted in Chapter One, the only consolidated book on coastal management for the Asia-Pacific region was written over a decade ago (Hotta and Dutton 1995) and although one chapter deals with coastal dynamics and global change, this refers mostly to time scales, past changes and coastal evolution. A small section of the chapter examines coastal implications of predicted climate changes but makes no specific reference to the relevance of these for the Asia-Pacific region. The scope of the book is described to include south Asia (Hotta and Dutton 1995 p 10) but none of the 18 chapters examining national approaches to coastal management include south Asian countries such as Bangladesh, Burma, India, Pakistan or Sri Lanka. In addition, the ethno-cultural grouping of most Pacific countries into two chapters on Micronesia and Melanesia provides an incomplete coverage.

The more recent international conferences *Coastal Zone Asia Pacific* (CZAP 2002, 2004) include coastal management within their conference themes but do not provide a consolidated update of what is happening with coastal management practices in the region. Chuenpagdee and Paully (2004) draw on many of the papers from the first of these conferences (CZAP 2002) to discuss how various approaches to research, education, information sharing, and coastal policies were presented with the aim of improving the state of the coast. Selected papers from the second conference (CZAP 2004) have been subject to peer review for a special issue of the *Journal of Ocean and Coastal Management* (forthcoming) but otherwise the quality of papers in both proceedings is variable. There has also been an attempt to use the proceedings of these two conferences as a basis for identifying coastal management trends in the Asia-Pacific region (Smith and Thomsen 2005). Smith and Thomsen identify 6 major issues raised at CZAP 2002 and 10 themes from the CZAP 2004 conference:

Table 3.1. Coastal management Issues; CZAP conferences, 2002 and 2004

CZAP 2002 *"6 issues of major concern"*	CZAP 2004 *"10 coastal management themes"*
1) uses of coastal areas and resources	1) livelihoods
2) roles of the community in integrated coastal management	2) communities
	3) modeling and assessment methods
3) education programs for coastal communities	4) sustainable industries
4) human resource development	5) protecting cultural heritage
5) national & regional frameworks for coastal management	6) international agreements and collaboration
	7) integrated policy and planning
6) data collection and information sharing	8) monitoring and evaluation
	9) adaptive management and innovation
	10) stakeholder partnerships and collaboration

Source: Smith and Thomsen (2005).

What is interesting from this analysis is the fact that global change does not feature in either the issues of major concern (CZAP 2002) or the coastal management themes (CZAP 2004). A closer examination of Smith and Thomsen's (2005) breakdown of the 10 themes from CZAP 2004 into emerging issues and priority actions reveals no reference to global change. In fact the only mention of global change issues is contained in one of 10 examples (almost half of which come from Australia) relating to these themes, emerging issues and priority actions. The example from Bangladesh refers to income and food production as a method for reducing vulnerability to climate change.

Apart from these examples much of the literature on coastal management in the Asia-Pacific region is scattered through scientific journals in addition to a wealth of grey literature from government and NGO sources. It is against this background that, this chapter has been written. One of the problems as noted by Hotta and Dutton (1995) is that most regional and multi-lateral coastal management programmes usually treat Asia separately from the Pacific Island nations.

Another problem is the fact that coastal management mechanisms and practice in the Asia-Pacific Region are as diverse as the states are numerous. These states include some of the smallest and most populous on the planet, some of the poorest and wealthiest, states with long histories of colonisation or political isolation, and with a range of contemporary political institutions. Rates of economic growth of nations and regions, over the last two decades, range from negative to the highest recorded. Yet most are coastal states with populations concentrated on coastal plains, and all depend, to a greater or lesser degree, on coastal resources. Most also share a number of problems centred on or near the coast, including rapidly growing coastal settlements, deteriorating environmental quality, loss of critical habitats (particularly coral reefs, estuaries and mangroves), depletion of fisheries, reduced biodiversity, increasing vulnerability to natural hazards and susceptibility to the impact of global warming (Hinrichsen 1998, Westmacott 2002).

3.2   INTEGRATED COASTAL MANAGEMENT

As noted in Chapter One there are international imperatives for integrated coastal management (ICM) as stated in the *Agenda 21* (Chapter 17) document produced by the United Nations Conference on Environment and Development (UNCED 1992) and re-iterated in the World Summit for Sustainable Development (WSSD 2002), Plan of Implementation 29. International agreements have driven the widespread adoption of the concepts of 'sustainable development' and 'integrated coastal management' (ICM) in the Asia-Pacific region. Coastal management, specifically ICM has been promoted as a tool to address problems of resource depletion in the Asia-Pacific region, and as a component of sustainable development. This has been done through international agreements and donor agencies, including NGOs.

Prior to the existence of ICM as an approach to resource management, traditional approaches existed for the regulation of access to coastal resources in the Asia-Pacific Region. For example, most pre-colonial Pacific and Asian societies employed traditional management systems, based on customs and taboos, to regulate access to fisheries (e.g. Ruddle 1989, Johannes 1978, Panayotu 1985). Such practices persist in some societies, but these have been widely displaced by open access regimes since World War II, particularly in Asia. The remainder of this section deals with the phenomenon of coastal management as it has developed over the last three decades driven, in large part, by problems of resource depletion.

Although the concept of coastal management has been around for at least three decades the concept of 'integrated coastal management' (ICM) is more recent. There has been debate over the use of related terms such as integrated coastal zone management (ICZM), integrated coastal area management (ICAM), and integrated coastal and marine area management (ICMAM) although Cicin-Sain and Knecht (1998) and Burbridge (1999) argue that there has been a shift in emphasis away from 'zone' or 'area' management to 'integrated coastal management' (ICM). In fact Sorensen (1997) distinguishes between ICM as a concept and ICZM as a program which has the task of defining the boundaries of the coastal 'zone'. More recently, Visser (2004) suggests an alternative term integrated coastal 'development' in order to focus on transdisciplinary ways of examining the interface between people and the sea. However, Visser's arguments fail to provide a balance across the land-ocean interface and she seems unaware of the literature discussing the use and appropriateness of different terms. As noted by Burbridge (1999) and Harvey (2004) ICM is a more accepted term today.

The term ICM is used in this chapter except where there is a direct quote using another term. Cicin-Cain and Knecht (1998) demonstrate that there has been a global acceptance of ICM as demonstrated by separate ICM guidelines produced by agencies such as the OECD (1991), IUCN (Pernetta and Elder, 1993), World Bank (1993), IPCC (1994) and UNEP (1995). ICM as defined in *The Encyclopedia of Geomorphology* is given as

Integrated Coastal Management is a continuous and dynamic process incorporating feedback loops which aims to manage human use of coastal resources in a sustainable manner by adopting a holistic and integrative approach between terrestrial and marine environments; levels and sectors of government; government and community; science and management; and sectors of the economy (Harvey 2004 p 568).

Most recent definitions of coastal management emphasize the need for holistic, integrated, management. For example, the OECD (1993) define 'integrated coastal management' as *management of the coastal zone as a whole in relation to local, regional, national and international goals. It implies a particular focus on the interactions between the various activities and resource demands that occur within the coastal zone and between coastal zone activities and activities in other regions*. Coastal management comprises several complementary elements, including the regulation of human use of resources (Holmes 1995), integration of decision making between sectors and levels of government (Cicin-Sain and Knecht 1998), the sustained quality of life of coastal peoples (Olsen 1993), recognition of catchment-coast and coast-oceans interactions (Chua 2003), and community participation in decision-making (Hildebrand 1997). The World Coast Conference definition includes another element *taking into account traditional, cultural and historical perspectives and conflicting interests and uses (IPCC 1994 p 40)*, which is particularly relevant for the Asia-Pacific region. The goal of coastal management is usually identified as 'sustainable development' for example, ... *it is a continuous and evolutionary process for achieving sustainable development uses (IPCC 1994 p 40)* and, *to control the types, scope and rates of human uses of resources so as to achieve sustainable rates of exploitation* (Holmes 1995).

Sustainable development is generally accepted as the central objective of ICM which implies that patterns and intensities of coastal resource use today, should not compromise the ability of future generations to meet their needs. ICM has been widely promoted as a means towards this end (e.g. Cicin-Sain and Knecht 1998). Olsen (2003) interprets this goal of sustainable coastal development to translate into two project and program goals; 1) improvements in the bio-physical environment, and 2) improvement in the quality of life of the human population in the area of concern. Chua et al. (2004) attempts to take this further by suggesting that *the ultimate goal of ICM is to improve the quality of life of coastal inhabitants* (Chua et al. 2004 p 109). This anthropocentric centred interpretation of ICM appears to place too much emphasis on development, as with Visser's (2004) argument, rather than the protection of the environment. It is more appropriate to emphasise the importance of ICM as a mechanism for minimizing human impact in our role as custodians of coastal resources for future generations. This distinction is subtle but important.

The key elements of integration in coastal management have been defined as follows:
- Intergovernmental integration (vertical integration) between different levels of government such as national, provincial or state and local governments;
- Intersectoral integration (horizontal integration) between different government sectors: such as industry, conservation, recreation, tourism, beach protection and integration of policies between different sectors of the economy;
- Community integration with government producing effective community participation and involvement in coastal management;
- Spatial integration between management of the land, ocean and coast;
- Integration between science and management particularly between different disciplines; scientists and managers; including economic, technical and legal approaches to coastal management; and
- International integration between nations on trans-boundary coastal management issues (Harvey 2004 p 569).

Another key aspect of the process is that ICM is iterative and requires some time in order to produce effective outcomes. This has been expressed as a cyclical approach (GESAMP 1996, Olsen 2003) which is not unique to ICM but shares similar features to adaptive management techniques (Figure 3.1).

Nichols (1999) presents an alternative perspective of ICM. In this view the United Nations environmental regulatory regime legitimises state-centredness and centralises the objective of reorganising society and space, for the primary purpose of stimulating national and international economic development. Pre-existing resource management

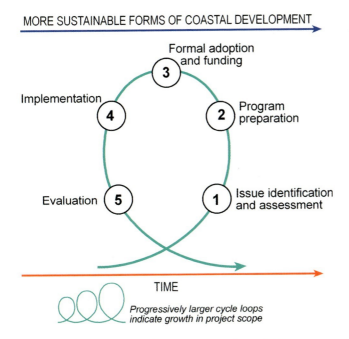

Figure 3.1. ICM cycles of development (Source: Olsen (2003))

regimes are displaced by ICM regimes, which facilitate the further subdivision of community resource management systems by encouraging national and global capital penetration. In short, Nichols argues that the ICM regulation creates space for the new global economy, by organising the coast into new arenas for investment and by politically and spatially marginalizing pre-existing resource users. This process has been reinforced by the United States Agency for International Development (USAID) and the World Bank. The World Wildlife Fund (WWF) and the World Conservation Union (IUCN) promote ICM as one of their five policy objectives (WWF and IUCN 1998).

It is useful to recognise the underlying assumptions that have guided the development of environmental management, in general, over the last few decades. Bryant and Wilson (1998) critique the traditional state-centric basis of the process of environmental management, where the state is assumed to be the key actor. Environmental management, they argue, has become synonymous with the development of large bureaucracies and an associated top-down approach to environmental problems. State-affiliated 'experts', trained in western positivist science, are relied upon to address selected environmental problems. Finally, environmental management is seen as a 'problem-solving' endeavour, in which scientific expertise and the latest technology are applied to a specific environmental problem. Fundamentally, it is assumed that environmental problems are amenable to technological resolution without any need to modify substantially broader political, economic or social forces. Environmental management is seen as a process designed gradually and selectively to alter the status quo so as to solve environmental problems, without upsetting prevailing political or economic interests (Bryant and Wilson 1998). Coastal management has evolved with these assumptions, to a greater or lesser degree, depending on the context.

It is important for the purposes of this chapter to be aware of these challenges to the efficacy of the ICM process. The success or failure of ICM to deliver outcomes for the Asia-Pacific coast will be discussed in Section 3.6 below but it should be emphasized that any ICM efforts will also operate in the context of a diversity of governance regimes throughout the region. This is relevant in terms of the levels of community involvement/displacement discussed by Nichols (1999). There will be a variety of different levels of community involvement which can be represented on a continuum as shown in Figure 3.2.

Figure 3.2. Bottom up-top down community involvement (Source: Adapted from Clarke (2003))

3.3 DRIVERS OF COASTAL MANAGEMENT

Recent changes to coastal management practices in the Asia-Pacific Region have been driven by both local and external (global) forces (Figure 3.3). These forces may be seen to have largely socio-economic and biophysical dimensions, though the reality is of course more complex. They also have positive and negative dimensions, depending on the environmental context. For example, the transfer of technology and growth in international commerce has fueled the growth of economies and provided new opportunities to increase living standards. However, global forces also contribute to a great many coastal problems – including global climate change and eustatic sea-level rise. The consequences of catastrophic events, such as the Sumatran tsunami of 2004, coupled with concern over future sea-level rise, has naturally raised awareness of coastal management in the Region, particularly among the atoll states of the Indian and Pacific Oceans.

From a socio-economic perspective, globalisation and the rapid growth of regional economies have driven changes in patterns of resource access and consumption,

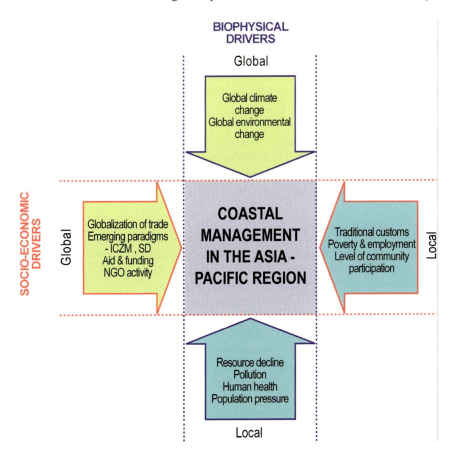

Figure 3.3. Drivers of coastal management

resulting in severe depletion of resources or degradation of critical habitats. The political institutions inherited from the colonial period might also be considered external influences on the development of coastal management in the Region. Local drivers may be no less important, including local concern for resource depletion and environmental quality, persistence of traditional management practices, the political involvement of local communities in decision making and employment opportunities. The relative importance of these drivers will vary from state to state within the Region and within states, vary between generations and between individuals based on their economic circumstances and education. The synthesis of the following drivers will be unique for each location and community.

3.3.1   International Conventions and Agreements

A number of international agreements have been adopted and/or ratified by countries in the Region, although their affect has not been substantial. The 1982 United Nations Law of the Sea came into force on 16 November 1993 has, arguably, been the most influential, since it directly concerns access to resources. The Law of the Sea is expected to give impetus to coastal zone management, because each state with a coastal zone can now claim jurisdiction over an Exclusive Economic Zone (EEZ) that extends 200 nautical miles from the coastline. However, the Convention does not compel coastal states to better manage marine resources within their EEZ (Hinrichsen 1998). Agenda 21 and the Biodiversity Convention, products of the United Nations Conference on Environment and Development 1992, have raised awareness of environmental problems, but this awareness has not been followed by significant action in many nations. Significant progress has been made in several areas, however, including the establishment of marine protected areas and increased opportunities for the participation of local communities and indigenous peoples.

3.3.2   Globalisation

Changes in consumption and use patterns brought about by economic globalisation are exerting chronic stress on coastal resources (Chua 2003). Demand for coastal space is increasing, leading to the privatisation of coastal lands and the decline of common property resource management (Adger et al. 2001). This process is exemplified by the rapid and ongoing growth of modern shrimp mariculture in Thailand, Vietnam and elsewhere in the Region. Farming of shrimp, crab and fish is a traditional practice in Asia, however, the new techniques, developed in Taiwan in the 1970s, are considerably more intensive. Ponds are stocked intensively with larvae of a single species, and fed a high-nutrient meal made primarily from fish bycatch, in specially constructed ponds (Fahn 2003). The industry has been supported by substantial government subsidies in many countries and finance from the World Bank and the Asian Development Bank. Asia now supplies about three quarters of the world's farmed shrimp, most exported to Japan and the United

States. The global retail value of cultured shrimp in 1996 was over US$18 billion (Flaherty et al. 1999).

The social and environmental impact of modern shrimp farming has been severe, particularly in the Gulf of Thailand. "Shrimp fever" drove farmers to convert publicly owned mangroves or private land into shrimp ponds during the 1980s (Fahn 2003). By 1994 Thailand became the worlds largest shrimp exporter and accounted for one third of the world's production (Charles et al. 1997). Shrimp became the countries second largest export after rice in 1995. Shrimp production subsequently declined from 259,000 metric tonnes in 1995 to 215,000 metric tonnes in 1997 as a result of problems with water quality and disease. Salinisation has contaminated coastal lands and prevented farmers reverting to lower return, albeit lower risk, crops such as rice.

"Shrimp fever" has fuelled the rapid decline of mangrove forests in Southeast Asia, although a range of pressures have contributed at most sites. This process commenced following World War II in Singapore and Peninsula Malaysia (Hilton and Manning 1997), however, it has accelerated and spread throughout Southeast Asia since the early 1980s, and particularly since the collapse of the Taiwanese shrimp industry in 1988. The area of mangroves in Thailand declined from 370,000 ha to 170,000 ha between 1961 and 1996 (Sukpanich 1999). Moreover, the modern shrimp farm can usually only be operated for a period of 4–10 years before the accumulation of organic wastes and sedimentation renders them useless. Mangrove forests have been cleared and associated natural resources lost for a short-term gain. The farm was likely owned by urban entrepreneurs. The net result in areas suitable for shrimp farming has been loss of local control over vital coastal resources (Bailey 2000).

3.3.3 Non-Governmental Organisations

The involvement of NGOs in coastal management in the Region was given impetus by UNCED, although many NGOs were active before 1992. In some countries, these NGOs have provided the catalyst for, or significantly supported coastal management, by initiating local programs, sharing information or by direct funding of programs. In some cases these are large international groups based in the west, formed in response to perceived threats to biodiversity and coastal resources. For example, the Mangrove Action Plan (MAP), founded in 1992, has been instrumental in publicising problems of mangrove conversion in Southeast Asia. MAP (i) coordinates a network of NGOs concerned with the conservation of mangroves, (ii) promotes public awareness of mangrove forest issues, (iii) develops technical and financial support for NGO projects, and (iv) helps publicise within developing nations the basic needs and struggles of Third World coastal fishing and farming communities.

The involvement of western NGOs has led, in many instances, to the formation of locally based NGOs. For example, The Nature Conservancy (TNC), a US-based NGO, established its first office in Paulau in 1990. TNC collaborated with the Division of Marine Resources to complete surveys of crocodile and dugong populations and a Rapid Ecological Assessment of the Palau Islands. TNC subsequently supported the establishment of the Palau Conservation Society in 1995, which now employs 11 staff and works with local villages to promote coastal conservation.

Local NGOs have evolved in the Region, independent of international influence. Some function in societies with limited democratic traditions. The Singapore Nature Society, for example, formerly the Singapore Branch of the Malayan Nature Society (1954), was established in 1991. Since this time it has persistently advocated for increased conservation of threatened wildlife and critical coastal habitats in Singapore. The Society has identified and documented areas of high conservation value, conducted baseline inventories of avifauna and coastal resource, and publicised the serious state of conservation in Singapore and neighbouring countries. This work has occurred with scant Government support or approval and within a society not used to overt public opposition. The recent inclusion of Society representatives on a Government committee reviewing the Singapore Green Plan may signal greater recognition of the Society's work.

3.3.4 Climate Change Science, Coastal Hazards and Coastal Erosion

This is both a global and local driver – climate change science has given significant impetus to coastal management in the Region. Awareness of the magnitude of change is due in large part to publicity gained by the Intergovernmental Panel on Climate Change (IPCC). The exposure of the atoll states of the Region to eustatic sea-level rise is now widely known, albeit this awareness has not translated into rapid progress in reducing greenhouse gas emissions. There is also increasing recognition of coastal processes occurring at annual to decadal time scales. Recent research has established relationships between El Niño and La Niña and climate extremes in the Region, lasting from a season to a year in the tropics and subtropics (Kelly et al. 2000). For example, the link between El Niño and cyclone frequency in Vietnam (Kelly et al. 2001).

Within the Region there is growing awareness of the problem of persistent coastal erosion. Coastal erosion is widespread in the Gulf of Thailand, for example. Severe erosion, with rates of at least 5 m/y, occurs along a distance of 181 km or about 11% of the total coastline. Moderate erosion, with rates of 1–5m, comprises about 302 km or 18%. Depositional coastlines account for 127 km, or 8% of the coast. The remaining 63% appear stable with seasonal changes (Chaimanee et al. 2003). High rates of erosion have given impetus to coastal management throughout the Region. The National Coastal Zone Management Plan (1991) of Sri Lanka, for example, identified coastal erosion as one of four policy imperatives, after a long history of reactive management.

Finally, many states have a heightened concern for the security of coastal communities and infrastructure following the devastating 2004 Sumatran tsunami. Attention has been belatedly focused on the need to conserve or restore mangrove forests, sand dunes and fringing coral reefs. These habitats afforded some locations significant protection during the Sumatran event. Their conservation is also critical for the protection of coasts from storm surge and eustatic sea-level rise. Of course they are equally important as fisheries resources.

3.3.5 Scientific Collaboration within the Region

There has been an unprecedented growth in programmes and projects that aim to generate and disseminate knowledge of coastal systems and coastal management and capacity development. These include the Secretariat for the Pacific Regional Environment Programme (SPREP); the United Nations Environment Programme (UNEP) Regional Seas initiative; the Programme of Partnerships in Environmental Management for the Seas of East Asia (PEMSEA) supported by the Global Environment Fund (GEF), United Nations Development Programme (UNDP) and the International Meteorological Organisation; the SysTem for Analysis, Research and Training (START) in Global Change, supported by the World Meteorological Organisation (WMO), International Geosphere-Biosphere Programme (IGBP) and International Human Dimensions Programme (IHDP); and the Asia-Pacific Network (APN) for Global Change Research. In the main these collaborations originate outside the region, however, they provide valuable opportunities for scientists and practitioners within the Region to network and share knowledge. This process has helped develop management capacity, but it has also contributed to a second agenda – the need to collaborate across state boundaries to facilitate effective management of shared resources.

3.3.6 Coastal Environment Degradation & Resource Depletion

The coasts and shallow seas of the Region are subject to a range of pressures, as reviewed in Chapter Two of this volume. Degradation of water quality, over-harvest of fisheries and conversion of littoral habitats, arguably the key issues, are particularly associated with those Asian countries that experienced rapid post-war industrial growth. These pressures have led to resource degradation or total loss at all spatial scales, but the impact of these pressures has been particularly conspicuous at the local level – with a corresponding impact on local communities most dependent on associated resources. Such communities may have few options in developing nations. In the case of fisheries they may find themselves trapped in a cycle of increasing fishing effort, often involving destructive methods, and declining yield. The problem is compounded when local and traditional management systems are displaced by central authority, resulting in disempowerment, or where population pressure is critical.

3.3.7 Community Participation & Recognition of Indigenous Peoples Rights and Traditional Practices

Globalisation has greatly increased pressure on coastal resources throughout the Region. Globalisation has also resulted in management authority shifting from resource users to governments and their institutions (Renard 1991). Loss of community management responsibility and authority has often involved the transfer of marine tenure from the community to the government, effectively placing the resources under an open-access regime because of the ineffectiveness of enforcement measures (White et al. 1994). For example, numerous marine reserves were established during the 1980s in an attempt to arrest habitat and species decline of coral reefs. Many of these attempts were driven by western scientists with experience of national parks and marine protected areas in developed western nations. The need to involve all stakeholders, particularly local communities, was largely overlooked. The need to involve those stakeholders dependent on coral reef productivity is now widely appreciated, in part because of the prominence of principles of community involvement in *Agenda 21* and in part because of the failure of many marine protected areas in the Region. The need to shift the emphasis from a regulatory, legalistic, and sectoral approach, to education and voluntary compliance, has been recognised for some time (e.g. Wells and Hanna 1992), but seldom implemented.

White et al. (1994) identify processes and factors that contribute to the success of community involvement in coastal management:
(a) utilisation of popular ecological knowledge;
(b) effective integration of traditional and new management systems;
(c) ensure local ownership of resources;
(d) integration of management initiatives;
(e) respond to community needs;
(f) accept community solutions;
(g) efficiency and efficacy of management;
(h) community empowerment;
(i) diversity of solutions; and
(j) cultural diversity.

The value and wisdom of Community Based – Coastal Resource Management (CB-CRM) lies in its recognition that communities are potentially the best resource managers, since they have the biggest stake in the sustainability of natural resources (Rivera and Newkirk 1997).

3.4 COASTAL MANAGEMENT PRACTICES IN THE REGION

There is diversity of coastal management approaches throughout the 41 countries of the Asia-Pacific region. Since Hotta and Dutton (1995) compiled 18 chapters reporting on comparative coastal management practices in the region there have been a number of changes to coastal management and an increasing adoption of ICM as a preferred model. It is not possible here to report on all relevant coastal management legislation in each country but there is an attempt here to synthesise trends of current ICM practice using specific criteria. In the last ten years there have been papers in scientific journals relating specifically to coastal management practices within the Region. For example, an overview of papers from CZAP 2002 appeared in *Coastal Management* (Chuenpagdee and Pauly 2004) and a special issue of *Ocean and Coastal Management* (2005) contained a number of papers relating to sustainability and ICM practices in Indonesia and the Philippines. Papers from CZAP 2004 have yet to appear in a special issue of *Ocean and Coastal Management*.

The large number of countries and the diversity of coastal management practices in the Asia-Pacific region make it difficult to present a meaningful summary for the whole region. However, it is useful to examine two key groupings of coastal management practices that stand out for the region: a) The Pacific Islands, and b) South East Asia. These are discussed as separate case studies before an attempt is made to provide a broader overview of coastal management practices across the whole region.

3.4.1 Coastal Management in the Pacific

There are differences of opinion on the success of ICM and its applicability to small island states. For example, Kaluwin and Smith (1997) discuss ICM in the context of coastal vulnerability assessment and suggest that it is *the most effective and realistic strategy for sea-level rise, coupled with natural variability, for the small island states of the Pacific region* (p 100). However, they go on to discuss some of the shortcomings in the application of ICM to the region particularly the concept of zoning which they point out is *usually culturally inappropriate and administratively difficult if not impossible to enforce, particularly where customary ownership of land and marine areas takes precedent over the western concepts of eminent domain and common property ownership* (Kaluwin and Smith 1997 p 102). Similarly Huang (1997) notes that the concept of ICM is based largely on the experiences of developed nations, where the balance of power between national, state and local authorities is well established. He suggests that the model may not be suited to Pacific Island countries (PICs) where traditional cultures are much stronger and there is a need to merge modern assessment and management tools with traditional systems.

Aston (1999) suggests that ICM is particularly suited to the island member countries of the South Pacific Regional Environmental Program (SPREP) because of their size, interconnectedness of coast and terrestrial areas and the high degree of community involvement with the management of coastal systems. While several of the PICs have drafted legislation for dealing with environmental problems, most countries need help in ratifying the relevant treaties and conventions so that top-down initiatives are more appropriate at the regional level for the PICs whereas, at the national and community level, bottom up approaches tend to be more successful, especially where all the stakeholders can collaborate and contribute to finding solutions (Aston 1999).

Aston (1999) uses a case study from the Solomon Islands to demonstrate this. In 1975, the government of the Solomon Islands tried to establish a marine protected area in the Arnavon Islands to preserve the breeding habitat of the Hawksbill turtle and green turtles. The islands are legally owned by the government, but two tribal groups claimed traditional ownership forcing the government to use a different approach with the help of the Nature Conservancy. In 1993 several consultative workshops and household surveys involving the resource users and adjacent communities resulted in the support of the local village communities. Subsequently the Arnavon Marine

Conservation Area is one of the most successful examples of a marine conservation area in the Pacific. The AMCA is managed by a management committee comprising representatives from the three adjacent communities, as well as members of the Environment and Conservation Division, Ministry of Forests, Environment and Conservation and the Fisheries Division of the Ministry of Agriculture and Fisheries (Aston 1999).

At the local and national levels, scientific research and analysis in the PICs is generally non-strategic and uncoordinated (Aston 1999). This means that many of the decisions regarding coastal management are not based on the best available data. While there is a lot of traditional knowledge on coastal resources, this is rarely in a useable form. On the other hand, scientific reports produced by external consultants have little meaning for local communities. In order to break down these communication barriers and make use of both forms of information, scientists, policy makers and members of the local community need to collaborate and pool their knowledge. The institutional framework in PICs is generally poorly developed, with few consultative mechanisms for dealing with coastal problems in an integrated way (Aston 1999). Operating budgets and staffing levels are often inadequate to deal with complex programmes, and lack of coordination and cooperation between government agencies has led to several problems:

- conservation and development are still seen as conflicting ideals;
- overlapping mandates have led to confusion over who has authority to make the final decisions; and
- the division of responsibilities among the various levels of government is not clearly defined.

These institutional constraints are currently holding up the adoption of ICM in PICs. However, some PIC governments are attempting to address this by devolving some of their NRM responsibilities to local communities. Aston (1999) outlines the roles of various stakeholders in PICs (See Table 3.2).

Table 3.2. Key Stakeholder Groups in the Pacific

STAKEHOLDER	ROLE
Environmental NGOs	Local, regional and international NGOs question power base and decision-making processes of government. Tend to support participatory approaches to NRM. Knowledge base of these groups can be limited – often regional bodies and government officers provide supporting *scientific advice*.
Customary Leaders	Conservation generally undervalued by indigenous communities in PICs, – focus on immediate needs rather than future benefits. Often responsibility of chiefs to reconcile needs. NRM arrangements in many PICs fall under customary tenure systems, eg in American Samoa, the government implements policy through the leadership of the village elders and chiefs (*who still command a great deal of respect within the community*).
Regional Organisations	Regional agencies who provide services to PICs to help them manage their coastal sectors: SPREP, South Pacific Applied Geo-science Commission, Forum Fisheries Agency, USP, Secretariat of the Pacific Community, Tourism Council of the South Pacific, the Pacific Islands Development

	Programme. Organisations important for capacity building as they have specialist technical skills and process skills for developing government policy.
Governments	Governments of PICs often do not fully appreciate the environmental impacts of developments on the coastal zone. Incentive to raise revenue through foreign investment often stronger than incentive to protect the environment. Strategic planning, EIA and compliance monitoring is needed to make sure that coastal development is sustainable.

Source: summarized from Aston (1999 p 494).

Huang (1997) comments that without the support and involvement of the local community, coastal management initiatives are rarely successful. Under traditional systems, kinship networks provide for community-based resource sharing and management. Many communities use seasonal fishing bans, and close off areas of reef. While these practices have worked well in the past, there is some doubt as to their continued effectiveness. As coastal communities undergo development and modernisation, educated people, who accept scientific knowledge over superstition and custom, may not follow traditional leaders. Huang suggests that ICM should be based 'on the scale of the most effective management unit, be it the village, district or the whole island'. While national approaches may be appropriate in some cases, for those islands with strong traditional systems, traditional management practices should be used as the basis for ICM programmes, and individual villages should be encouraged to develop their own resource management plans. Most PICs already have mechanisms for managing coastal resources that can be adapted to fit the ICM framework. Huang (1997) also recognises the need for a regional approach. Many of the problems facing PICs are common across the whole region, and individual countries rarely have access to critical management information and tools such as climate data and coastal models. For these reasons, a regional ICM effort seems logical. SPREP was set up to assist PICs manage their *shared* environment, and to answer the combined concerned of most islands by providing technical assistance and coordination efforts.

3.4.2   Coastal Management in Southeast Asia

Over the last twenty years international funding agencies have supported the development of a number of ICM programs in Southeast Asia. Chua (1998) reports on eight such programs, six which began under the ASEAN/US (see glossary for acronyms) funded Coastal Resources Management (CRM) Project (1986–1992). Another two programs were funded under the GEF/UNDP/IMO Regional Programme for the Prevention and Management of Marine pollution in the East Asia Seas (1994) which aimed to support participating governments from ten countries in the region in combatting marine pollution at national and sub-national levels. The six CRM pilot sites were located in the Lingayen Gulf,

Philippines: Upper South, Thailand: Segara Anakan-Cilicap, Indonesia; South Johore, Malaysia; the main and southern islands of Singapore and Brunei Daurussalem. The two GEF/UNDP/IMO demonstration sites were located at Xiamen, China and Batangas Bay, Philippines as described by Chua (1998).

Subsequently the GEF/UNDP/IMO program moved into a new phase with eleven countries (Brunei Darussalam, Cambodia, North Korea, South Korea, Indonesia, Malaysia, China, Philippines, Singapore, Thailand and Vietnam) working together in Partnerships in Environmental Management for the Seas of East Asia (PEMSEA). PEMSEA was launched in October 1999 and has developed more demonstration sites (Figure 3.4) in selected locations where PEMSEA co-fund and implement Integrated Coastal Management strategies. PEMSEA has also targeted Hot spot program sites (Figure 3.4) which have been identified as sites having major pollution problems. PEMSEA co-funds and implements these in order to reduce pollution and re-habilitate these sites. In addition PEMSEA has developed Parallel Sites (Figure 3.4) as second demonstration sites in which PEMSEA gives free technical expertise to any willing country, but the funding through the program has to be supported by the country itself.

Balgos (2005) in the *globaloceans.org* website draws together some of the lessons learned in implementing ICM in the Southeast Asia region based on two major programs (CRMP and PEMSEA) for the region including the need for a national policy on natural resource management with prescribed guidelines, the early involvement of local authorities, support from policymakers and community, and multisectoral and interagency participation. She also stresses the need for a common understanding of ICM and appropriate program development and implementation cycle with performance monitoring.

A key point Balgos makes is that the participatory nature of ICM is strong within the Region and has been important in the widespread adoption of ICM in the Philippines where the community-based approach is well recognised. Balgos (2005) notes that the lessons learned from regional, national and local ICM programs have served as best practice examples from which to replicate ICM projects in other countries such as the expansion of the PEMSEA sites; development of a National Training Program on ICM in the Philippines, and the expansion of community-based programs

3.4.3 Comparative Assessment of ICM Practice Throughout the Asia-Pacific Region

The diversity of nations in the Asia-Pacific Region limits the value of any generalizations concerning integrated coastal management practice – with the possible exception that few formal ICM initiatives are sustained beyond the implementation phase. Several authors have suggested evaluation frameworks for ICM programs at the nation-state level. Cicin-Sain and Knecht (1998), for example, compare 22 selected nations (including several in the Asia-Pacific Region) in relation to a number of criteria (see Section 3.5 below). They recognized a

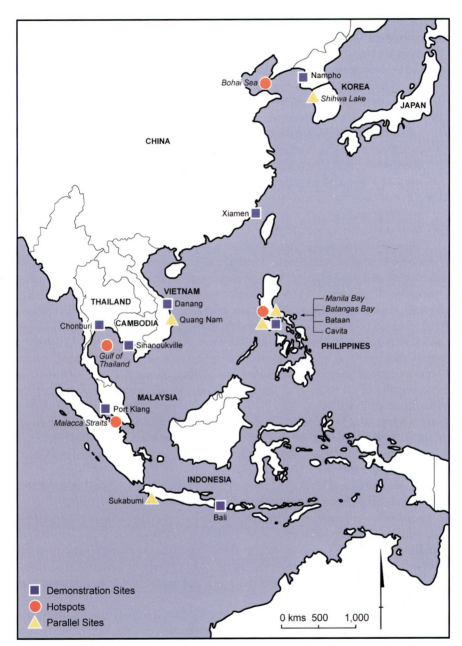

Figure 3.4. PEMSEA ICM sites in Southeast Asia (Source: Adapted from PEMSEA (2005))

number of similarities among the developed, semi-developed and developing nations studied, namely;
  (i) most nations shared common problems of environmental degradation;
  (ii) the development of environmentally-oriented institutions (e.g. departments of environment) in the 1970s and 1980s, partly in response to international conventions such as UNCED;
  (iii) multiple levels of government are concerned with ICM in most nations, with the national government often having prime responsibility;
  (iv) insufficient data is available to assess the implementation of ICM in most countries (versus ICM activity);
  (v) most nations adopt a similar range of regulatory and planning processes;
  (vi) external assistance and funding has played an important role in many countries; and
  (vii) there is some evidence of a movement towards greater integration (intersectoral, intergovernmental and interdisciplinary).

A number of authors discuss levels of integration and the need for criteria to assess the achievement of ICM in practice. Some of these are referenced in Section 3.5 below discussing the effectiveness of ICM. In order to provide an overview of ICM practices across the region it was decided to use a set of criteria presented in Hotta and Dutton's (1995) original book on coastal management in the region. In that book, Hildreth and Gale (1995) offer a hierarchical perspective of ICM practice in the Asia-Pacific Region, comprising six approaches with the highest being the most integrated:

1. traditional practices, based on custom and cultural norms;
2. community based approaches with modest institutional support provided only on an 'as needed' basis;
3. formal special area management plans for particular regions which sometimes are not well integrated with the existing social, legal and institutional structure;
4. special area management plans which integrate legal and institutional considerations on a sector by sector basis;
5. inter-ministerial and interagency coordinated approaches to coastal management on a national or sub-national basis; and
6. special coastal agencies with meaningful planning and regulatory authority at the national or sub-national levels.

This hierarchical approach was used to assess the extent to which coastal management in each country was more or less integrated. A review of available ICM literature was undertaken, and a judgment made as to the degree to which each nation has implemented each approach (Table 3.3).

No nation has a single agency responsible for ICM and only three (Singapore, Japan and Indonesia) have interagency or inter-ministerial coordination. It is not surprising, therefore, that progress implementing ICM has not been rapid. In comparison a great deal of effort has gone into developing special area programs or management plans, usually with external assistance. The ASEAN nations have adopted this process, consistent with the general approach of the ASEAN-US CRMP and subsequent PEMSEA program discussed above.

Table 3.3. Comparative Assessment of ICM practices in the Asia-Pacific region using criteria from Hildreth and Gale (1995)

	Traditional practices based on customs & cultural norms	Community based approaches with modest institutional support	Formal special area plans for regions	Special area management plans on a sector by sector basis	Inter-ministerial and inter-agency co-ordination on a national or sub-national basis	Special coastal agencies at the national or sub-national level
American Samoa						
Bangladesh						
Brunei						
Burma						
Cambodia						
China						
Cook Islands						
East Timor						
Fiji						
French Polynesia						
FS Micronesia						
India						
Indonesia						
Japan						
Kiribati						
Malaysia						
Maldives						
Mariana Islands						
Marshall Islands						
Nauru						
New Caledonia						
Niue						
North Korea						
Northern Mariana						
Pakistan						
Palau						
Papua New Guinea						
Philippines						
Republic of Korea						
Russian Federation						

Singapore							
Solomon Islands							
Sri Lanka							
Taiwan							
Thailand							
Tonga							
Tuvalu							
Vanuatu							
Viet Nam							
Wallis and Fortuna							
(Western) Samoa							

KEY	
	No governmental recognition or provision made for this approach
	Some governmental recognition or provision made for the approach (and/or data lacking)
	Explicit government support and implementation of the approach

Source: based on expert judgement by the authors from available literature on ICM.

Recognition of traditional customs and approaches to coastal resource management was not always evident. This is not to say that local communities are not actively involved in management, but that this involvement is not formally recognized and provided for by governments. Relatively few countries provide for such involvement, while in many cases traditional approaches appear to have been lost. This appears to be the case with Tonga, for example. Colonisation has imposed systems of regulation that in some cases have subsumed traditional approaches, to the detriment of local communities. The importance of local community-based management is a recurring theme in the literature. Hildreth and Gale (1995 p 30) conclude that . . . *the key player in successful coastal management in both developed and developing nations is the local community*. Johannes (2002) sees a renaissance of community-based marine resource management in Oceania, stimulated by a growing perception of scarcity, the re-strengthening of traditional village-based authority and better education. Evidence of increasing recognition of the importance of community involvement in coastal management is provided by the Joint Ministerial Statement of The 2nd APEC Ocean-related Ministerial Meeting (AOMM2) in Bali, Indonesia (the 'Bali Plan of Action') in September 2005. The declaration reinforces the role and responsibilities of local communities and the importance of traditional practices in ICM.

Some nations have failed to provide for significant community involvement or are unable to do so because of recent conflict (East Timor, Cambodia) or extreme poverty (North Korea). In other cases low or negligible levels of community involvement are the product of the prevailing political system, where political influence is highly centralized and democratic institutions are weak (Burma, China, Tonga) or incompletely developed (Singapore).

3.5 THE EFFECTIVENESS OF COASTAL MANAGEMENT IN THE REGION

There have been a number of attempts to examine the effectiveness of coastal management and specifically ICM but a major problem has been the lack of criteria on which to base any objective assessment. Sorensen (1997) warns of the danger of becoming over optimistic about achieving immediate beneficial outcomes from ICM as an alternative management process. Burbridge (1999) expresses concern at progress with ICM and notes that notwithstanding *the 30 year or so history of ICM there are few good examples of fully integrated coastal management strategies, plans or management practices beyond a local level or problem – specific scale* (Burbridge 1997 p 181). Similar comments are made by Sorensen (1997 p 13): *There is scant information derived from rigorous evaluations of ICM program outcomes*, and also Olsen et al. (1997 p 155): *There is very little information that demonstrates the success of ICM efforts and how the process of ICM has influenced outcomes*.

There have been attempts to conduct surveys of the global adoption of ICM (Cicin-Sain and Knecht 1998, Meltzer 1998, and Sorensen 2000) together with attempts to develop and apply assessment criteria (Olsen 2003, Chua 1998). In 1996 Cicin-Sain and Knecht undertook a comparative study of the ICM process in 22 developing and developed countries based on the following assessment criteria
- Level of government responsible for ICM;
- Top-down or bottom up approach;
- Reliance on external assistance;
- Degree of integration; and
- Effectiveness of integration.

They concluded that there was very little evidence on the effectiveness and extent of implementation of ICM (Cicin-Sain and Knecht 1998). Similarly Meltzer (1998) undertook a desk top study of ICM programs and concluded that it was difficult to find a successful model and given that ICM was still evolving it was too early to properly assess progress.

Perhaps the largest survey of ICM practice has been conducted by Sorensen who estimated that ICM activity had grown from 180 'efforts' in 1993 to around 385 in 2000, in 87 countries and semi-sovereign states. Clarke (2003) comments on the lack of critical analysis in Sorensen's data base and points out that *of the 51 initiatives listed as national efforts, in fact there were 15 programs, 24 policies and 12 feasibility studies* (Clarke 2003 p 10). She notes that ICM policies have not necessarily been implemented and feasibility studies tend to be site specific so that neither of these categories by themselves constitutes a successful ICM activity. Clarke (2003) points out that a 'program' is the highest level category because it includes supportive legislation or policy, a strategy or plan and a formal staffing or a governing body. For this reason it is important to apply criteria for ICM success to any such database.

Experience has demonstrated that there are significant barriers to implementing environmental policy towards ICM, even in the developed nations of the region. In Korea, these obstacles include insufficient investment, lack of information and skilled human resources, weak analytical tools, inadequate political will and fragmented institutional frameworks (Hong and Lee 1995). For many poor and marginalised groups living in the developing world, the notion of sustainable development as a realisable working concept remains a far-off dream, primarily because developed countries have failed to alleviate poverty in the World's poorest countries. At the national level, major disincentives to achieving sustainable patterns of resource use include:
- the incentive structure in both business and government does not reward or encourage those who pursue sustainable levels of development;
- the time horizons of elected officials that influence development policy are short;
- there are no accepted techniques for calculating net long-term benefits;
- the people who make decisions affecting how national resources are utilized in central government are seldom directly impacted by their decisions or the projects they launch;
- the development planning function in most nations is weak; and
- central government control over finances and decision-making means that local governments have little authority (Olsen et al. 1998).

Tropical ICM has passed through several phases – pre-colonial, centralised, community-based and co-management (Christie and White 1997). There are few examples of the successful implementation of ICM in the Region. A survey by Westmacott (2002) found that only a few locations (12%) are fully implementing ICM in the tropics, a further 12 percent have no management, while 38 percent are only starting to recognise there are management problems in the coastal zone. In this survey only 24 percent of the respondents identified the local community as an ICM stakeholder. The interests of coastal communities are often overlooked (Olsen and Hale 1994). It would seem that many countries in the Region have not progressed very far beyond the 'centralised' phase of coastal management development. Perhaps it is too early to assess the effectiveness of ICM. Olsen (2000) concluded;

If ICZM is to fulfill its promise as a means of progressing towards sustainable forms of development, at a time when the trends demonstrate we are usually moving in the opposite direction, we must become realistic on how long it takes to make progress. Project designs must recognise that measurable outcomes at significant geographic scales will take many years of sustained effort to achieve.

Olsen (2003) has since attempted to develop frameworks and indicators for assessing ICM initiatives. He stresses *that successful initiatives link the steps within a generation of management – particularly the need to bridge between steps devoted primarily to planning (Steps 1, 2 and 3) and a period of policy implementation (Step 4). Progress is further enhanced when completed cycles of management build upon each other and are expressions of purposeful learning* (Olsen 2003 p 357). These steps are illustrated in Figure 3.1. In the same article, Olsen discusses another framework for assessing effectiveness of ICM initiatives over multi-year time

frames. He proposes four orders of outcomes that group sequences of institutional, behavioral and social/environmental changes which lead to more sustainable coastal development. He lists three orders of intermediate coastal governance outcomes:
1. Enabling conditions;
2. Changes in behaviour; and
3. The harvest.

The first involves a commitment to a plan of action with a formalized commitment at the national level to an ICM program. This requires the funding and institutional capacity to undertake the program and the authority to make decisions on the coastal policies. The second order outcome is where there is evidence of change in the behaviour of institutions and stakeholder groups such as collaborative action and partnerships along with resultant investment in infrastructure. His third order outcome which he refers to as 'The Harvest' refers to the restoration or improvement of societal and environmental qualities relating to coastal resources and development. Olsen's end outcome is his fourth order outcome which he refers to as Sustainable Coastal Development where there is a *desirable and dynamic balance between social and environmental conditions is achieved* (Olsen 2003 p 349). Having described his four orders of coastal management outcomes, Olsen discusses in some detail the indicators for each of the different outcomes. He concludes that in many cases the limiting factors to progress in coastal management relate to the *capacity of the institutions most directly involved to instigate and sustain integrated and adaptive forms of management* (Olsen 2003 p 358).

At the regional level, Chua (1998) assessed the performance of 8 ICM projects (2 PEMSEA and 6 CRM funded) from Southeast Asia. His assessment examined management frameworks and planning and implementation process relative to outputs and impacts. Chua applied a suite of performance indicators, which broadly followed Olsen's work, to each of the 8 projects. He concluded that although ICM had been practiced for many years in the US, its application in Southeast Asia is relatively new and there are not many success stories. He showed that all of the 8 case studies had completed requirements for problem identification and project formulation but only 2 projects at Batangas Bay, Philippines and Xiamen, China had reached the implementation stage by 1998.

A more recent study by Christie et al. (2005) assessed the effectiveness of ICM projects in delivering sustainable outcomes in Indonesia and the Philippines. They examined factors such as linking management to biophysical change, stakeholder participation in ICM, legal policy framework, institutional continuity, project exit strategy and awareness raising. White et al. (2005) make an important observation related to donor funding of ICM projects which often fail in terms of sustainability because of the *dependence on external assistance creates both the potential for and the reality of non-sustainable ICM institutions and policies as projects are terminated and staff withdrawn* (White et al. 2005 p 272). They note that in the Philippines most community-based coastal resource management projects were not maintained after the funding and external assistance finished. They also note that in the Philippines between 1974–2000 US$230 million was invested in coastal resource management projects, of which 63% came from international donors. This raises the question about long-term sustainability of these projects.

3.6 FUTURE DIRECTIONS FOR COASTAL MANAGEMENT RESEARCH

A key research direction for coastal management is to use well-formulated criteria to assess the effectiveness of ICM programs in the Asia-Pacific region. It is interesting to note that Olsen's (2003) evaluation of ICM initiatives suggests that *availability of funding* is not a major factor in the lack of progress for donor sponsored initiatives, including Southeast Asia, but the problem lies more with the *capacity of the institutions* to deal effectively with *integrated and adaptive forms of management* (Olsen 2003 p 358). In contrast to this Christie et al. (2005) in their ICM sustainability studies in Indonesia and the Philippines suggest that the availability of funding 'is' a major factor in producing sustainable ICM outcomes. They imply that donor sponsored ICM initiatives would be more effective with longer-term funding. Given the importance of donor-funded ICM initiatives in the region it is important to conduct further research to assess its effectiveness in other countries and reasons for success or failure.

Another avenue of research is to examine the best methodology for incorporating global change issues into ICM practice. This may need to start with an assessment of why global change issues in the region have not yet gained prominence in recent major forums on coastal management in the Asia-Pacific region. In part this could be an awareness raising issue which may be helped with improved data sets on the state of critical coastal resources, particularly data on the current state of mangrove forests, seagrass beds and coral reefs (see Chapter Eleven). In this connection, there is ample scope to better utilize GIS systems and satellite imagery in an effort to obtain more and better quality data on the actual extent of these resources and their current condition at local, state and regional scales.

Given the diversity of governance structures, cultures and coastal resource usage throughout the region it is likely that research investigations into ICM effectiveness will produce a wide range of outcomes. For example, answers to the following research questions may vary considerably between countries reflecting the need for different approaches to achieve ICM effectiveness in different countries:
- How can effective coastal management programmes be facilitated by local communities?
- How can the top-down and bottom-up approaches be synthesized?
- How can traditional knowledge and practices be more effectively integrated with other forms of management?
- How can a better understanding of coastal processes and the interactions within coastal systems be achieved at local levels and among decision makers?
- How is it possible to increase political awareness and commitment to ICM?

- How can ICM capacity be increased?
- What institutions are needed in each country and do these differ between groups of countries?
- To what extent is the interaction between catchment and coast recognized?
- What mechanisms are successful in achieving effective interstate coastal management?

From the discussion in this chapter it is clear that within the Asia-Pacific region there are sub-regional differences in ICM approaches such as the Pacific islands and the Southeast Asia. The former has SPREP as an intergovernmental organization, which can provide the necessary collective capacity to support ICM programs, whereas many of the small island states lack this capacity. In Southeast Asia, a partnership of countries implementing ICM programs (PEMSEA) has produced some success such as the Xiamen case study quoted by Chua (2004). This raises the question of what has worked or failed and whether the implementation of ICM programs can be better supported through groupings of countries? Other research issues for coastal management in the region relate to issues such as the importance and/or enforcement of existing coastal laws and regulations; the status of coastal ownership and changing patterns of the issue of ownership; the relative importance of coastal resources and how the coast is valued.

However, an over-riding research question for this book is how to address science-policy linkages in order to facilitate the incorporation of global change issues in ICM within the Region. As noted from the APN Synthesis Report (Harvey et al. 2005) the development of science-policy linkages was one of the weakest outcomes from the global change coastal research projects, in part a result of poor research project design. The research challenge is now to investigate the most appropriate method(s) to incorporate global change issues into ICM programs within the Region. This may include research into awareness raising of the issue, appropriate education, science communication, and mechanisms for better science-policy linkages.

REFERENCES

Adger WN, Kelly PM, Ninh NH (eds) (2001) Living With Environmental Change: Social Vulnerability, Adaptation and Resilience in Vietnam. Routledge, London

Aston J (1999) Experiences of coastal management in the Pacific Islands. Ocean & Coastal Management 42:483–501

Bailey C (2000) Continuing concerns related to shrimp farming in the tropics. Intercoast Network 37:18–19

Balgos MC (2005) Coastal Zone Management in Southeast Asia. Available online from 'Global Web Service on Oceans, Coasts and Islands'. Last viewed19/05/2005. http://www.globaloceans.org/global-info/seasia/ICM_SEA.htm

Bryant RL, Wilson GA (1998) Progress in Human Geography 22, 3:321–343

Burbridge PR (1997) A generic framework for measuring success in integrated coastal management. Ocean and Coastal Management 37:175–189

Burbridge PR (1999) The challenge of demonstrating socio-economic benefits of integrated coastal management. In: Salomons W, Turner RK, de Lacerda LD, Ramachandran, S (eds) Perspectives on Integrated Coastal Zone Management. Springer, Berlin, Heidelberg, pp 35–53

Chaimanee N, Sinsakul S, Tiyapairach S (2003) Coastal Erosion Impact Along the Gulf of Thailand. Proceedings, 6th International Conference on The Environmental Management of Enclosed Coastal Seas, Bangkok, 18–21 November, 2003, p 83

Charles et al. (1997) The University of Rhode Island newsletter

Chua TE (1998) Lessons Learned from Practicing Integrated Coastal Management in Southeast Asia. Ambio 27(8)

Chua TE (2003) Challenges to Sustainable Coastal Development in East Asia. Proceedings, 6th International Conference on The Environmental Management of Enclosed Coastal Seas, Bangkok, 18–21 November, 2003, pp 3–6

Chua TE, Bonga D, Bermas N (2004) The Dynamics of Integrated Coastal Management and Improvement of the Quality of Life. Proceedings of the Coastal Zone Asia Pacific Conference, 5–9th September Brisbane, Australia

Chuenpagdee R, Pauly D (2004) Improving the State of Coastal Areas in the Asia-Pacific Region. Coastal Management 32:3–15

Christie P, Lowry K, White AT, Oracion, EG, Sievanen L, Pomeroy RS, Pollnac RB, Patlis JM, Eisma RV (2005) Key findings from a multidisciplinary examination of integrated coastal management process sustainability. Ocean and Coastal Management 48:468–483

Christie P, White AT (1997) Trends in development of coastal area management in tropical countries: From central to community orientation. Coastal Management 25:155–181

Cicin-Sain B, Knecht R (1998) Integrated Coastal and Ocean Management: Concepts and Practices. Island Press, United States

Clarke B (2003) Coastcare, Australia's Community-Based Coastal Management Program. An effective model of Integrated Coastal Management. PhD Thesis, Department of Geography and Environmental Studies, The University of Adelaide, SA

CZAP (2002) Proceedings of the Coastal Zone Asia-Pacific Conference '02: Bangkok, Thailand

CZAP (2004) Proceedings of the Coastal Zone Asia-Pacific Conference '04: Improving the Quality of Life in Coastal Areas, 5–9 September, 2004. Brisbane, Australia

Fahn JD (2003) A Land on fire: The Environmental Consequences of the Southeast Asian Boom. Westview Press, Colorado

Flaherty M, Vandergeest P (1999) Rice Paddy or Shrimp Pond: Tough Decisions in Rural Thailand. World Development 27:2045–2060

GESAMP (1996) The contributions of science to integrated coastal management. GESAMP Reports and Studies, 61

Harvey N (2004) Integrated coastal management. In: Goudie A (ed), Encyclopedia of Geomorphology, Routledge, London, New York

Harvey N, Rice M, Stevenson L (Eds) (2005) Global Change Coastal Zone Management Synthesis Report. Asia-Pacific Network for Global Change Research, Kobe, Japan

Hildebrand, LP (1997) Community-based coastal management: Developing experience from around the globe Pacific Coasts and Ports '97. Proceedings 1:33–37

Hildreth RG, Gale MK (1995) Institutional and legal arrangements for coastal management in the Asia-Pacific Region. In: Hotta K, Dutton IM (eds), Coastal Management in the Asia-Pacific Region: Issues and Approaches, Japan International Marine Science and Technology Federation, Tokyo, pp 21–38

Hilton MJ, Manning SS (1997) Conversion of coastal habitats in Singapore: Indications of Unsustainable development. Environmental Conservation 22(4):307–322

Hinrichsen D (1998) Coastal Waters of the World: Trends, Threats, and Strategies. Island Press, Washington, DC

Holmes N (1995) Coastal dynamics and global change: Implications for coastal management. In: Hotta K, Dutton IM (eds), Coastal Management in the Asia-Pacific Region: Issues and Approaches. Japan International Marine Science and Technology Federation, Tokyo

Hong SY, Lee J (1995) National level implementation of Chapter 17: The Korean example. Ocean & Coastal Management 29(1–3):231–245

Hotta K, Dutton JM (eds) (1995) Coastal Management in the Asia-Pacific Region: Issues and Approaches. Japan International Marine Science and Technology Federation, Tokyo

Huang JCK (1997) Climate change and integrated coastal management: a challenge for small island nations. Ocean & Coastal Management 37(1):95–107

IPCC (Intergovernmental Panel on Climate Change) (1994) Preparing to Meet the Coastal Challenges of the 21st Century. Report of the World Coast Conference, 1–5 November 1993, Ministry of Transport, Public Works and Water Management, The Hague, The Netherlands

Johannes RE (1978) Traditional marine conservation methods in Oceania and their demise. Annual Review of Ecological Systems 9:349–364

Johannes RE (2002) The renaissance of community-based marine resource management in Oceania. Annual Review of Ecological Systems 33:317–340

Kaluwin C, Smith, A (1997) Coastal vulnerability and integrated coastal zone management in the Pacific island region. Journal of Coastal Research 24:95

Kelly M, Granich S, Nguyen Huu Ninh (eds) (2000) Report of the Workshop 'The Impact of El Niño and La Niña on Southeast Asia', 12–23 February 2000, Center for Environmental Research Education and Development, Hanoi, Vietnam

Kelly PM, Tran Viet Lien, Hoang Minh Hien, Nguyen Huu Ninh, Adger WN (2001) Managing environmental change in Vietnam. In: Adger WN, Kelly PM, Nguyen Huu Ninh (eds), Living with Environmental Change: Social Vulnerability, Adaptation and Resilience in Vietnam, Routledge, London, pp 35–58

Meltzer E (1998) International review of integrated coastal zone management, Meltzer Research and Consulting. Oceans Conservation Report Series, Canada

Nichols K (1999) Coming to terms with "Integrated Coastal Management": Problems of meaning and method in a new arena of resource regulation. Professional Geographer 51(3):388–399

OECD (1991) Report on CZM: integrated policies and draft recommendations of the Council on Integrated Coastal Zone Management. Paris, Organization for Economic Cooperation and Development

OECD (1993) Coastal Zone Management–Integrated Policies. Paris: OECD, pp 19–124

Olsen S (1993) Will Integrated Coastal Management Programs be Sustainable: The Constituency Problem. Ocean & Coastal Management 21(1–3):201–225

Olsen SB, Tobey J, Hale L (1998) A learning-based approach to coastal management. Ambio 17(8)

Olsen SB (2000) The common methodology for learning: Ecuador's pioneering initiative in integrated coastal management. Coastal Resources Center Report 2227, University of Rhode Island, Narragansett, RI, USA

Olsen SB (2003) Frameworks and indicators for assessing progress in integrated coastal management initiatives. Ocean and Coastal Management 46:347–361

Olsen S, Hale L (1994) Coasts: The ethical dimension. People & the Planet 3(1):29–31

Olsen S, Tobey J, Kerr M (1997) A common framework for learning from ICM experience. Ocean and Coastal Management 37(2):155–174

Panayotu T (ed.) (1985) Small Scale Fisheries in Asia. Socioeconomic Analysis and Policy. IDRC, Ottawa

PEMSEA (Partnerships in Environmental Management for the Seas of East Asia) (2005) URL: http://www.pemsea.org

Pernetta J, Elder D (1993) Cross-sectoral, Integrated Coastal Area Planning (CICAP): Guidelines and Principles for Coastal Area Development. IUCN, Marine and Coastal Areas Programme

Renard Y (1991) Institutional challenges for community-based management in the Caribbean. Nature and Resources 27(4):4–9

Rivera RA, Newkirk GF (1997) Power from the people: A documentation of non-governmental organizations' experience in community-based coastal resource management in the Philippines. Ocean and Coastal Management 36:73–95

Ruddle K (1989) Solving the common-property dilemma: Village fisheries rights in Japanese coastal waters. In: Berkes F (ed.), Common Property Resources, Belhaven Press, London, UK, pp 168–184

Smith T, Thomsen D (2005) Coastal management trends in the Asia-Pacific region, Water 32:38–39

Sorensen J (1997) National and international efforts at integrated coastal management. Coastal Management 25:3–41

Sorensen JB (2000) Background Paper for Coastal Zone Canada 2000: Coastal Stewardship—Lessons Learned and the Paths Ahead. September 17–22, 2000, New Brunswick, Canada, http://www.sybertooth.ca/czczcc2000/

Sukpanich T (1999) The Unkindest Cut: Mangrove Forest Management. Bangkok Post, March 29,1999

UNCED (United Nations Conference on Environment and Development) (1992) Agenda 21 and the UNCED Proceedings. Oceania Publications, New York

UNEP (1995) Guidelines for Integrated Management of Coastal and Marine Areas with Special Reference to the Mediterranean Basin. UNEP Regional Seas Reports and Studies No 161

Visser L (ed.) (2004) Challenging Coasts: Transdisciplinary Excursions into Integrated Coastal Zone Development. University Press, Amsterdam

Wells S, Hanna N (1992) The Greenpeace Book of Coral Reefs. Sterling Publishing Co., New York, 160 pp

Westmacott S (2002) Where should the focus be in tropical integrated coastal mangement? Coastal Management 30:67–84

White et al. (1994) The need for community-based coral reef management. In: White AT, Hale LZ, Renard Y and Cortesi L (eds.), Part I in Collaborative and Community-Based Management of Coral Reefs: Lessons from Experience, Kumarian Press, Connecticut, pp 1–18

White AT, Christie P, D'Agnes H, Lowry K, Milne N (2005) Designing ICM projects for sustainability: Lessons from the Philippines and Indonesia. Ocean and Coastal Management 48:271–296

World Bank (1993) The Noordwikjk Guidelines for Integrated Coastal Zone Management, Washington DC: World Bank, Environment Department, Land Water and Natural Habitats Division

WSSD (2002) Plan of Implementation of the World Summit on Sustainable Development, http://www.un.org/esa/sustdev/documents/WSSD_POI_PD/English/WSSD_PlanImpl.pdf

World Wildlife Fund and World Conservation Union (WWF & IUCN) (1998) Creating a Sea Change: The WWF/IUCN Marine Policy. World Wildlife Fund and the World Conservation Union

CHAPTER 4

CATCHMENT–COAST INTERACTIONS IN THE ASIA-PACIFIC REGION

SHU GAO
*Ministry of Education Key Laboratory for Coast and Island Development,
Nanjing University, China*

4.1   INTRODUCTION

Catchment-coast interactions have been the focus for several large research programs. The second phase of the Land-Ocean Interactions in the Coastal Zone (LOICZ) project, one of the sub-programs of International Geosphere-Biosphere Programme (IGBP) (LOICZ IPO 2004, Crossland et al. 2005), is investigating this issue, placing emphasis on human induced changes and the resultant societal/economic consequences (with anthropogenic factors related to global climate and physical environment changes being considered against a background of natural variation). In Europe, a project known as EUROCAT (European Catchments), involving 23 institutions and 70 researchers, has been implemented for several years (Salomons 2004). Based upon regional case studies, this project attempts to: (1) identify the impact of catchment changes on the coast; (2) develop integrated models incorporating catchment biophysical and socio-economic aspects; (3) formulate regional environmental change scenarios for the near future; (4) evaluate future coastal changes using the scenarios and the models; and (5) explore the potential application of the research findings by stakeholders and policymakers. Likewise, the National Science Foundation (NSF) of the United States has recently updated the science plan for catchment-ocean—or "source to sink"—research (MARGINS Office 2003). The revised plan calls for investigations into the processes of material transfer from rivers to deep oceans on large temporal scales in response to climate changes and anthropogenic activities.

Compared with the European and American catchments, the problems associated with the Asia-Pacific region are more complex. The Southeast Asia region is the most densely populated in the world, and there are large catchment systems with rivers originating from the highest plateau of the world (i.e. Qinghai-Tibet Plateau). These rivers include many of global importance (e.g. Indus, Brahmaputra-Ganges, Irrawaddy, Mekong, Pearl, Changjiang, and Yellow rivers), together with numerous smaller ones (Anikiyev et al. 1986; Bornhold et al. 1986; Wright et al. 1990; Jorgensen et al. 1993; Allison 1998; Saito et al. 2000, 2001; Ta et al. 2001; Hori et al. 2004; Thanh et al. 2004). The region is also characterised by numerous islands where small rivers are able to carry huge sediment loads. High elevation of the catchment basin, monsoon climate, and intense weathering lead to high water and sediment discharges for both large and small rivers of the region. More than two-thirds of the world's river sediment discharges are from this region (Figure 4.1a; Milliman and Syvitski 1992), and human activities further intensify this pattern (Figure 4.1b). Against the background of global changes in climate,

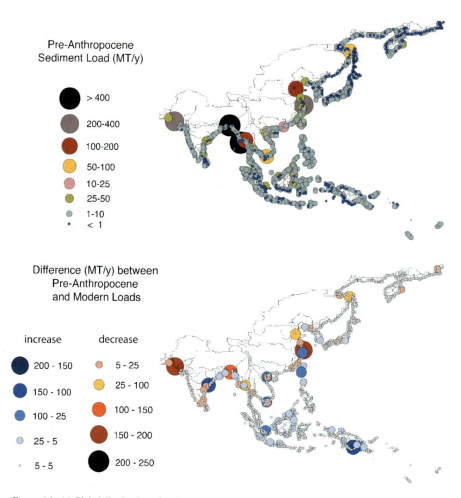

Figure 4.1. (a) Global distribution of prehuman sediment flux, based on model predictions. (b) Differences between prehuman and modern sediment load, based on combined model predictions and observations (Syvitski et al. 2005). Figures courtesy of James PM Syvitski

geomorphology, and ecosystems, natural processes and anthropogenic activities in the catchments of these rivers are important in controlling the environmental characteristics of adjacent coastlines. This modification of coastal environments, in turn, affects the catchment areas. There have been many studies on catchment-coast interaction issues, but knowledge gaps exist with regard to catchment-coast characteristics, the related processes and mechanisms, methods and techniques to improve investigations, and application of scientific output to management practices.

Interactions between the catchment and adjacent coastal areas can be considered in terms of energy and material exchanges and social-economic feedback links among the different locations within the catchment-coast system. Kinetic energy provided by runoff supports material fluxes of freshwater, sediment, nutrients, and pollutants, which are transported from the catchment towards the sea, modifying the physical, environmental, and ecological conditions of the coastal waters. In estuarine waters, there are influences of river flow, tides, waves, and gravitational circulations. Near the river mouth, large harbours and ports are built and exert social and economic pressures on the catchment regions. The patterns of all these interactions are changing in response to changes in the catchment and coastal areas. As far as the Asia-Pacific region is concerned, the most significant changes result from material fluxed from catchments and intense human activities at river mouths.

The purpose of this study is to identify some crucial research gaps associated with the catchment-coast system from the point of view of natural science, which may form important research topics for the future study of the Asia-Pacific region. These issues include both basic and applied aspects and are divided by the following groups: (1) regional characteristics of the catchment-coast systems, in terms of changes in discharges of freshwater, sediment, nutrients, and pollutants; (2) the processes and mechanisms responsible for the discharges and their changes; (3) new methods and techniques that should be developed for future investigations; and (4) applications of the research findings to catchment-coast development and management practices. The proposed research topics represent a component of global change studies.

4.2 REGIONAL CATCHMENT AND COAST CHARACTERISTICS

The catchment and the adjacent coast interact by means of energy and material exchange. Kinetic energy is provided by river flows, tides, and waves. Runoff represents the kinetic energy transferred by rainfall, and the river flow carries solid and dissolved materials towards the continental shelf. At the river mouth, wave and tidal energy propagate into the catchment areas, causing salt, nutrient, pollutant, and sediment exchange between the catchment and the open sea. Biological activities (e.g. fish migration) indirectly transport energy between the two components. An understanding of the energy and material (i.e. water, salt, nutrient, pollutant, and sediment) changes is crucial in characterising the catchment-coast system.

The rivers in the south-eastern Asian region deliver a large percentage of the global freshwater and sediment discharges into the coastal zone. The quantities of water and sediment inputs from some large rivers of the region have been monitored and calculated. The majority of the river inputs to the world's oceans are from this region. The large rivers, including the Indus, Narmada, Krishna, Godavari, Brahmaputra, Ganges, Irrawaddy, Salween, Chao Phraya, Mekong, Red (Song Hong), Pearl, Changjiang, and Yellow rivers are well known for huge quantities of water and sediment discharge, which is of the order of 2.0×10^{10} t a^{-1} (Milliman and Syvitski 1992). Smaller rivers, although they discharge less water than larger rivers, still contribute a significant portion of the region's sediment load. For instance, the total sediment discharge from the small rivers of the region exceeds that from the large rivers (Milliman et al. 1999). This is because a part of the sedimentary materials are trapped by the catchment of large rivers, whilst in small rivers sediment is transported directly to the sea waters.

Human activities within the catchments of these rivers have been rapidly modifying the water and sediment discharges as well as discharges of other materials (i.e. nutrients and contaminants). Since the early-middle 20th century, a large number of dams have been built for the purposes of irrigation, water storage for industrial and domestic uses, power generating, and canalisation/navigation (World Commission on Dams 2000). In the catchment of the Changjiang River alone, more than 48,000 dams were constructed by 2004, according to the Ministry of Hydraulic Engineering of China (Figure 4.2), with the Three Gorges Dam being the largest (Figure 4.3). Such human interference has considerably changed the patterns of water and sediment transport, which has modified monthly discharge distribution patterns.

The Changjiang River, with a catchment area of 1.8×10^6 km^2 and a length of 6,300 km, is the largest river in China. According to statistical analysis of the gauge record at the Datong Station, located 624 km west of the Changjiang River

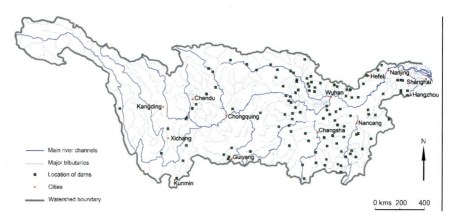

Figure 4.2. Dams (with a reservoir capacity exceeding 10^8 m^3) constructed in the Changjiang River catchment by the end of the 20th century (after Ministry of Hydraulic Engineering 2000)

Figure 4.3. A bird's eye view of the Three Gorges Dam in the middle reaches of the Changjiang River. Photograph courtesy of Mr Bin-cheng Zhu (taken in August 2005)

mouth, the river's long-term average water discharge is of the order of magnitude of 3×10^4 m^3 s^{-1} (or 9×10^{11} m^3 a^{-1}) and the sediment discharge is 15 t s^{-1} (or 5×10^8 t a^{-1}). Traditionally, maximum discharges occur during the summer season, but there are significant annual variations. However, following the construction of more than 48,000 dams within the catchment basin over the last 55 years, the original seasonal patterns of water discharge in terms of both quantity and timing have been modified, although there is no apparent trend in changes to the average water discharge (Figure 4.4).

By itself, the Three Gorges Dam traps a large percentage (up to 70%) of fine-grained sediments from a drainage area of 1.0×10^6 km^2. Since the completion of this dam and many minor ones, the sediment discharge has been reduced to 3×10^8 t a^{-1} or less since the 1990s, a change that implies average suspended sediment concentration has been reduced (Figure 4.4). In the future, the water diversion project to transport freshwater to northern China will eventually reduce the water discharge, and this alternation will have many consequences. For instance, the growth of the Changjiang River delta may be slowed down or even be replaced by coastal erosion in response to decreased sediment discharge; salt intrusion into the lower reaches of the river may be intensified due to decreased freshwater discharge; the catchment geomorphology (including river channel geometry) may change into a non-equilibrium state and trigger readjustment of the river regime; and the ecosystem of the catchment-coast system may be affected.

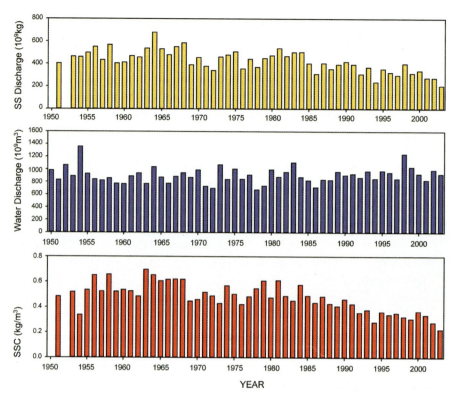

Figure 4.4. Historical records of freshwater and suspended sediment (SS) discharges from the Changjiang River (1950–2003) and associated averaged suspended sediment concentration (SSC) (calculated on the basis of the original data provided by the Changjiang River Hydraulic Engineering Committee)

Reduction of sediment discharge due to dam building is a general feature of the region. An example is the Indus River, where construction of the Korri barrage was started in 1931 some 270 km from the river mouth. During the period from 1931–1980, sediment discharge was reduced to around 20% of the original quantity (Jorgensen et al. 1993), which indicates a significant loss due to trapping in reservoirs. The water discharge also started to decrease in the mid-1950s. Since the catchment area of Indus River (9.6×10^5 km^2) is relatively small compared with the Changjiang River, the effect of damming is more significant.

Increase in nutrient and pollutant discharges resulting from human activities is a recent and highly significant event for the region. The catchment areas of the region are generally densely populated and support an extremely large-scale, rice-based agriculture. Artificial fertilisers and pesticides are used extensively, practices that result in heavy nutrient (nitrogen and phosphorus) loading and POP pollution. In addition, the majority of the region is under development pressure, and chemical industries and other pollutant-generating factories have been established in many developing countries.

Take the Changjiang River, once again, as an example. In contrast to the reduction of water and sediment discharges, the river's discharges of nutrients and pollutants have been increasing in response to rapid social-economic development. Increases in nitrogen and phosphorus inputs are mainly caused by the use of chemical fertilisers and the emission of domestic sewage. Fertiliser use in agriculture has increased by 10% over the past 40 years. As a result, the catchment basin of the river received a nitrogen load in the late 1990s that was four times that of the late 1960s, reaching on the order of 1.0×10^7 t a^{-1} (Yan et al. 2001). Although measures have been taken to control pollution in the river, trends still show an increase in pollutant input, partly due to continued development of the chemical industry along the river banks.

Generally, the present status and future trends regarding material discharges (i.e-water, sediment, nutrients, and contaminants) in the region are only qualitatively understood. In the future, discharges will change in response to changing climate—in terms of both average conditions and occurrence of extreme events—and intense human activities. Even less is known about the consequences of these changes in material discharge on adjacent coastal environments and ecosystems. In the case of the Changjiang River, future reductions in water and sediment discharges combined with enhanced nutrient and pollutant loads may have a number of consequences, including: increase in sediment retention in the middle-lower reaches of the river; intensified saline water intrusion in the estuarine areas; slowing down river delta growth or even coastal recession; decreased amounts of nutrients and pollutants deposited with fine-grained sediments; erosion of the mud deposits on the continental shelf; reduction in the residence time for the dissolved nutrients and pollutants in the adjacent shelf waters; modified nutrient compositions that deviate from the Redfield ratio; changes in the quantity and composition of net primary production in the adjoining marine waters; and changes in the fishing grounds near the Changjiang River mouth.

In order to evaluate these consequences, important issues associated with patterns of water, sediment, nutrient, and pollutant discharges from the Asia-Pacific region should be addressed. In studying regional characteristics of the discharges, emphasis should be placed on the acquisition of quantitative information, including identification of the large, but unknown, amount of pollutants discharged into the rivers and coastal waters everyday. Although the effect of the nutrient and pollutant discharges may be evaluated by a scenario-based modelling approach, there is also a need to obtain measured nutrient and pollutant input data.

Specific research topics for quantifying material fluxes and balance for catchment-coast systems, changes in natural conditions and anthropogenic activities associated with catchment-coast systems, as well as the impact of natural changes and anthropogenic activities on the estuarine and coastal environment are listed below.

4.2.1 Patterns of Catchment Hydrological Cycling

The water balance of a catchment consists of precipitation, evaporation, runoff, and seepage. These parameters are not constant; rather, they evolve with global climate changes and with geomorphological and ecological changes in the catchment.

Rainfall patterns may be modified by large-scale (regional-global) climate changes, deforestation, and land use changes. Intensity of evaporation may differ because the proportions of evaporation through water surface, plants, and soils have been altered. With changes in the vegetation cover, soil erosion, runoff, and seepage patterns will also differ. Thus, the parameters that define catchment hydrological cycling should be analysed to examine the changes of water balance. Freshwater discharges, under the influences of climate change, water use, and dam construction, should be monitored. Further, the seasonal distribution and variability of the freshwater input should be examined.

4.2.2 Sediment Yield and Input to Coastal Waters

Sediment yield is indicated by denudation rate, which is controlled by slope and soil erosion. After entering the river channel, a part of sediment material will be trapped by reservoirs, lakes, and flood plains. Thus, although there is a trend of intensified soil erosion over the region (e.g. Sheng and Liao 1997, Ahmad et al. 1998), the sediment discharge into the sea may not be enhanced. An effect of dam construction is to reduce the occurrence of small flood events and, in many cases, to decrease the peak flow during big floods. In many river basins freshwater diversion projects are completed or on-going, reducing the annual average water discharge. This trend implies that accumulation of fine-grained sediment on flood plains will be reduced. Additionally, material will be trapped by the reservoirs. The net effect is a reduction in the quantity of sediment that is available for transport towards the sea. Measurements and calculations should be undertaken to continuously monitor the sediment discharge and identify changes in existing trends. Once again, the timing, seasonal distribution, and variability of sediment yield should be analysed on the basis of measured data.

4.2.3 River Channel Morphological Evolution

Fluvial geomorphology has been studied by many investigators. Some observations from these studies include: (1) the evolution of the river system (or changes in the hypsometric pattern of the river basin) is controlled by a base of erosion; (2) the cross-sectional area of the river channel is determined by a characteristic flow, together with other factors such as sediment grain size and suspended sediment concentration; and (3) the river system evolves towards an equilibrium status in terms of its longitudinal profile, with fluvial deposits being formed in the middle and lower reaches. After a long period of time (i.e. 10^4–10^7 years) of development, most rivers in the Asia-Pacific region have been approaching their equilibrium. However, climate changes and anthropogenic activities are causing disturbance to these systems. Sea-level rise is changing the rate of erosion, dam construction is modifying river flow and the sediment composition carried by the flow, and changes in the hydrological regime are shifting the river system into disequilibrium. Hence, it is anticipated that adjustments of the river channel morphology, sediment transport capacities, and fluvial deposition patterns will occur.

So far in the Asia-Pacific region, there have been few reports about adjustments in morphology, sediment transport, or fluvial deposition as a result of changes in catchment basins. This scarcity of reports seems to indicate a need to strengthen monitoring of river system evolution following catchment changes. Data sets should be established on the basis of measurements and observations, with regard to cross-sectional areas of the river channel, sediment grain sizes of bed-load and suspended load, current velocities of the river flow, as well as longitudinal profiles of the river bed. On such a basis, morphological changes of the river system can be determined quantitatively.

4.2.4 Accretion and Erosion Patterns in the Estuary and on the Adjacent Coast

Accretion takes place in the estuary and adjacent coastal areas. In some systems, river deltas formed during the Holocene period providing human societies with land resources. For the Asia-Pacific region, large deltas such as those associated with Ganges-Brahmaputra, Mekong, Red, Pearl, Changjiang, and Yellow rivers are well-known examples. The evolution and characteristics of these deltas has been studied extensively by the international scientific community (e.g. Huang et al. 1982; Bornhold et al. 1986; Wright et al. 1990; Xue 1993; Chen, 1998; Saito et al. 2000, 2001). However, knowledge gaps exist regarding quantitative prediction of future evolution of these deltas. After a long period of time, some river deltas may be close to their natural growth limits, which are controlled by regional hydrodynamic, sedimentary, and climatic conditions (Gao 2006). Likewise, changes in sediment supply from the catchment due to human activities may have transformed some estuarine systems from an accretional environment to an eroding environment. At present, estuarine and coastal erosion often becomes a serious problem (Yang et al. 2003).

In order to identify future trends of river delta growth and provide data sets for modelling geomorphological evolution of the estuarine/delta system, measurements of sediment dynamics for some representative estuarine systems are required. In particular, variations in the suspended sediment concentration on various timescales (i.e. the tidal cycle, seasonal, and inter-annual scales) should be analysed with a focus on: long-term measurements over the different sections of the estuarine system; accretion/erosion rates for the estuarine deposits, measured using various methods ranging from repeating levelling operation to ^{210}Pb dating; and the dynamic changes in the deltaic area, monitored using remotely sensed imagery and GIS techniques.

4.2.5 Saline Water Intrusion Patterns in the Estuary

Seawater flows into estuaries due to gravitational circulation or tidal pumping (Dyer 1997). In gravitational circulation the range of saline water intrusion is controlled by the position of the null point (i.e. the upstream limit of seawater in the lower part of the water column), which is affected, in turn, by the freshwater discharge and the bathymetry of the estuarine system. The intensity of tidal pumping is determined by the tidal prism, which is the product of tide range and the basin area of the estuary. Saline water is transported into the estuarine areas by dispersion processes. Seawater enters the river mouth during the flood phase of the tide and is mixed with the estuarine water; during the ebb phase, some of the seawater is retained within the estuary. Eventually the salinity of the estuarine water reaches an equilibrium state determined by the freshwater discharge, tidal range, and the capacity of the estuarine basin. The upstream limit of saline water intrusion will be controlled by the tidal excursion of the saline water mass.

In both gravitational circulation and tidal pumping, the extent of saline water intrusion is associated with freshwater discharge and sediment input. The construction of dams in the catchment modifies both factors. Thus, the timing of water discharge will differ from the original pattern, and the reduction of sediment input may cause morphological changes of the estuary (see above). As a result, the patterns of saline water intrusion and salinity distribution of the estuary and adjacent shelf waters will be changed. Therefore, it is necessary to monitor such changes after the quantity and timing of water/sediment input is modified by natural evolution and social development in the catchment.

4.2.6 Nutrient/Pollutant Emissions and Water Quality Changes

Over the Asia-Pacific region, dense population and rapid economic development have resulted in large emissions of nutrients and pollutants derived from industry, agriculture, and domestic sewage. In the Changjiang River, the observed increase in nitrogen and phosphorus inputs is mainly from chemical fertilisers (Yan et al. 2001). The large scale of the emissions has caused serious environmental problems and is today considered a regional issue. For the purpose of environmental protection and the sustainability of future development, it is important to be able to control the emissions. In the near future it is necessary to investigate the sources and quantities of the emissions, identify the influences on the water quality changes, and establish monitoring and survey schemes for typical catchment-coast systems.

4.2.7 Influences of Material (sediment/nutrient/pollutant) Input on Estuarine and Adjacent Continental Shelf Ecosystems

With development and changes in the catchment, the quantity, budgeting, and timing of material input (in terms of water, sediment, nutrients, and pollutants) in the estuarine and coastal waters will differ considerably from their original patterns. There will be a number of consequences in the future because of such changes: (1) salinity distribution patterns will be different, modifying the ranges of the saline, brackish, and freshwater areas; (2) the suspended sediment concentration may be reduced, with enhanced transparency of the estuarine waters; and (3) nutrient and pollutant loadings will be enhanced in the estuary and coastal waters. Collectively, these changes will lead to modification of the estuarine and shelf ecosystem in terms of the burial and accumulation of nutrients and pollutants, the intensity of photosynthesis in the estuary, primary production and its composition, the range of brackish waters and the distribution of associated species, and the distribution of fishing grounds near the river mouth. For the Asia-Pacific region, these issues should be addressed and monitoring activities that allow the identification of changes should be organised.

4.2.8 Influences of Large Estuarine/Coastal Engineering Schemes

Throughout the region rapid and large-scale economic development, which exceeds levels experienced in the past, has prompted the construction of huge harbours and ports in estuarine and coastal areas. The new Shanghai Harbour as well as the large bridges being built for the harbour and in Hangzhou Bay (both have a length of more than 30 km) are recent examples. These activities are driven by the economic and social development of the Changjiang River basin. In order to meet the transport requirements for the development, such "super" harbours are thought to be a solution. However, the large estuarine and coastal engineering schemes must have extensive and profound influences on the adjacent ecosystem and environment. It is important to be able to have knowledge of the impacts of the large engineering schemes. In particular, disturbance to water circulation patterns, coastal hydro-dynamics, sediment movement patterns, primary production, seabed benthic organisms, and fishery resources should be investigated. This understanding will provide the basis for establishing guidelines for future large engineering schemes.

4.3 PROCESSES AND MECHANISMS ASSOCIATED WITH CATCHMENT-COAST SYSTEMS

Data sets, such as the recorded material input from the catchment to the coast, derived from measurements and observations are useful in analysing patterns of change occurring in the catchment-coast system. However, it is insufficient to use merely the values and/or time series of such data sets for the purpose of predicting the future. In order to enhance our ability to predict future changes in the behaviour of the catchment-coast system, an improved understanding of processes and mechanisms that are responsible for the observed data is required. For instance, it is difficult to predict future freshwater discharges from the Changjiang River from only a time series record of water discharge data; on the other hand, knowing the factors that control the discharge and the way the factors work allows formulation of a better predictive model.

The terms "process" and "mechanism" should not be used in a vague sense. Understanding these terms in a well-defined framework will promote "earth system science" and encourage interdisciplinary research. According to systems theory, "process" means the response of a system to external forcing, and "mechanism" refers to the various combinations of different processes. Alternatively, "process" may be understood as the relationship between a factor and the system under consideration, and "mechanism" means the combined effects of these factors on the system. If all the factors that control catchment-coast system changes and their interrelationships are known, then a modeling approach can be adopted to simulate present and future changes, with the measured data being used as calibration tools. From such a viewpoint, the following research topics about the processes and mechanisms of the catchment-coast interaction may be appropriate for the Asia-Pacific region.

4.3.1 Processes of Catchment Hydrological Changes

Changes of catchment hydrological cycling are taking place in response to climate and sea-level changes and human induced modifications to the catchment. In addition to monitoring patterns of catchment hydrological cycling, the processes of water transport between the various sub-systems (i.e. atmosphere, river channels, soil, and underground passages) on different spatial and temporal scales should be a focus for future studies. Past hydrological cycling studies have been undertaken for normal conditions, where the water balance relationship has remained more or less stable in terms of mean values and amplitudes of fluctuations, and these studies have not considered long-term trends of changes in the driving forces and boundary conditions. However, because the components for water balance (i.e. precipitation, evaporation, runoff, and seepage) are changing due to catchment alterations, both the mean values and amplitudes of fluctuations will become unstable, a change that implies the underlying dominant processes will be different. With a shift in the processes, the mechanisms for the water cycling will become highly variable. Hence, an improved, quantitative understanding of the processes and mechanisms controlling future water cycling in catchment-coast systems is required.

4.3.2 Retention of Sediment within Catchment-Coast Systems

Apart from water, sediment plays an important role in the evolution of catchment-coast systems in terms of morphology/bathymetry, material cycling, and ecosystem behaviour. Retention of sediment in a catchment-coast system refers to a condition under which a part of the sediment will eventually accumulate within the system, although the total amount of the sediment entering and leaving the system may be very large. The efficiency of sediment retention may be expressed by a "Retention Index" (G.M.E. Perrilo, personal information). The factors affecting the index include freshwater discharge, sediment input, bathymetry of the catchment-coast system, and shelf hydrodynamics (wave, tides, and shelf currents). Apparently, there is a need to quantitatively formulate the Retention Index, taking into account the various factors mentioned above. The processes associated with changes in these factors, together with the resultant changes of retention characteristics, should be investigated not only for the whole system, but also for the various parts and sub-systems. Further, the Retention Index may be a scale-related parameter (i.e. given different temporal scales the value will be different). The Retention Index, in relation to shelf processes, spatial distribution patterns, and time-scale related characteristics, is an important research topic for the future.

4.3.3 Sediment Mobility in Estuarine and Coastal Waters with Source-Sink Changes

Sediment mobility is important for the ecosystem in estuarine and coastal waters because it influences the photosynthesis and nutrient cycling processes. Mobility is measured by comparing the entire time period with the time span when the near-bed current is strong enough to reach the threshold for initial sediment motion. For example, in a highly mobile area the seabed sediment may be set in motion more than 80% of the time. The mobility influences the suspended sediment concentration, water transparency, mineral regeneration processes in the water column, and the burial of nutrients and pollutants. The patterns of these phenomena will change with the future changes in freshwater discharge and sediment input that have taken place in the catchment basin. Therefore, it is necessary to re-examine the sediment mobility patterns under new conditions. In particular, resuspension processes versus net accumulation rates should be investigated in conjunction with long-term changes in the tidally averaged suspended sediment concentration.

4.3.4 Estuarine and Coastal Morphodynamic Processes

Sediment discharge from the land causes accretion in the estuarine waters and river delta growth. The relationship between the discharge and the accretion rate or delta growth rate may be established by modelling exercises. In this research field, combined process and geometric models appear to be a useful tool. In quantitative stratigraphy, geometric models with large time-scales have been developed to simulate the formation of sediment sequences and sedimentary basin filling (Paola 2000). However, morphological evolution tends to be associated with smaller temporal scales (i.e. 10^0–10^3 years); for example, during the growth of a river delta, the shoreline position of the delta is controlled by sediment supply and availability, sediment transport and accumulation rates, and the original conditions of the sedimentary environment. Hence, there is a need to incorporate sediment dynamics into the geometric models to simulate the accretion and morphological evolution of estuarine–delta systems.

Long-term changes in sediment supply, in terms of both quantity and composition, should be understood. For the simulation of long-term morphological evolution, sediment supply is an important variable; for large catchment basins, changes in natural conditions (climate, crustal movement, soil erosion, etc.) and anthropogenic activities (land use, dam construction, deforestation, vegetation cover, etc.) can significantly modify the pattern of sediment supply. There is a need to understand variations in sediment supply during the Holocene period for the region. In addition, small time-scale processes for sediment supply changes, which are responses to the various events taking place in the catchment basin (e.g. construction of large dams), should be incorporated.

4.3.5 Changing Mixing and Dispersion Processes for Nutrients and Pollutants

Mixing and dispersion are basic processes in estuarine and coastal waters for material fluxes (Lewis 1997) and the formation of turbidity maxima (Dyer 1997; Gao et al. 2004). Since nutrients and pollutants tend to be present in the form of dissolved matter, or adsorbed by suspended sedimentary particles, their transport and dispersal are controlled by these processes, which, in turn, affect water quality and primary production. Changes in the quantity and timing of freshwater discharge, mainly caused by dam building and increasing nutrient and pollutant emissions from human activities in the catchment basins, modify the hydrodynamics and suspended matter distribution patterns in the estuarine-coastal system. Hence, mixing and dispersion processes in these waters are changed accordingly. These new conditions should be taken into account by examining modifications in the spatial distribution patterns of both flux and concentration of nutrients and pollutants.

4.3.6 Processes for Estuarine and Coastal Ecosystem Evolution

Even before significant changes in the catchment basins were felt, changes in the estuarine and coastal ecosystem have been serious problems for the Asia-Pacific region. These problems include overfishing, disease in sea farming, and harmful algal blooms. Inappropriate use of the sea areas has caused deterioration of the marine environment and resources. Under the changed conditions of material discharges from land (modified water discharge patterns, reduced sediment supply, and enhanced nutrient/pollutant input), new problems are added to the old, unsolved problems. In these systems, the processes associated with the estuarine and coastal ecosystem evolution are much more complex than in an ordinary, natural catchment-coast system. Thus understanding the resultant ecosystem responses requires studies about how the changed material fluxes influence the physical environment (e.g. transparency of the water column, the level dissolved oxygen, and material exchange at the water-sediment interface) and how the changed nutrient loading influences primary production (in terms of quantity and composition), trophic levels, and food web characteristics. In particular, emphasis may be placed on the processes and mechanisms associated with material (i.e. water, sediment, nutrient, and pollutant) input and the ecosystem response, represented by spatial and temporal distributions of primary production, the biomass of dominant species at high trophic levels, and the trend of ecosystem evolution in the estuarine and coastal waters.

4.3.7 Comparison between Large and Small Catchment-Coast Systems

Since large catchment basins produce more water, sediment, nutrient, and pollutant inputs to the coastal waters, there is a tendency to concentrate research activities in large catchment-coast systems, as indicated by the publications in this research field.

However, it is likely that the role played by small basins has been underestimated. In terms of sediment yield and flux, Milliman and Syvitski (1992) and Milliman et al. (1999) have shown that, globally, small rivers deliver more sedimentary materials than large rivers because the number of small rivers is much bigger than the number of large rivers. Additionally, the efficiency of sediment trapping is relatively low in small rivers and there have been fewer dams constructed in small drainage basins. With regard to carbon burial and its influence on ecosystem characteristics, a recent study (Gao and Jia 2004) has shown that a small catchment-coast system may have a large effect. In the Asia-Pacific region, there are numerous small catchment-coast systems. Hence, the overall effect of these systems on sediment, nutrient, and pollutant fluxes should be evaluated against the situations observed for large systems of the region. Comparative studies may be conducted to identify the similarities and differences in the rate of changes in material fluxes and the resultant physical, environmental, and ecosystem changes.

4.4 METHODS AND TECHNIQUES FOR INVESTIGATIONS

Studies of regional characteristics and the involved processes and mechanisms require improved methods and techniques for monitoring and investigations, including survey techniques, modeling and simulation approaches, and geochemical tracing methods. On the modeling side, there are different approaches according to the purpose and the level of the study, including: (1) conceptual modeling to identify the major factors that need to be considered and to gain insight into the scientific problems involved; (2) process modeling to identify the roles played by the various processes and mechanisms; and (3) simulation modeling to take into account both the processes involved and the real boundary/initial conditions, allowing predictions of spatial and temporal changes in a system. The models will not only provide the output of calculations (i.e. numbers), but also offer a useful tool to understand the real world. For the Asia-Pacific region, all three types of modeling studies are required to understand the catchment-coast system changes and future trends in evolution. It should be noted that calibration of the models is necessary, and the data sets obtained by survey and measurement schemes (see above) can serve this purpose.

As a parallel effort, geochemical tracing techniques should be developed, which will be highly beneficial to the study of regional patterns of material sources, the transport and dispersal of the various materials, and the history of regional environmental changes. Using a mixing model (e.g. Owens et al. 2000) the sources of sedimentary material in a catchment basin have been determined quantitatively. Analysis of tracer distribution patterns, such as the use of artificial tracers by a Spatial Integration Method (Madsen 1989) or exotic tracers by a mass conservation model (Gao and Collins 1995) have delineated sediment transport pathways. Finally, tracers in sedimentary records (sampled from cores) have been used as proxies for climate and environmental changes. The application and further improvement of these methods can enhance understanding of the catchment-coast changes of the region not only for sediment, but also for other materials such as nutrients and pollutants.

For research activities to be fruitful, it is important to identify suitable study areas where the phenomena under consideration are significant, thus allowing high quality research to be conducted and enhancing the efficiency of investigations. Because there are numerous large and small catchments in the Southeast Asian region, it is difficult to undertake detailed studies on the material cycling patterns and the related processes/mechanisms on the basis of *in situ* measurements and observations for all of the catchments. For the "source to sink" studies in the USA, factors such as the driving forces of material transfer, the signal of the source areas, characteristics of stratigraphic records, magnitude of the discharges, type of coast and catchment, climate conditions, human activities, as well as research infrastructure and facility availability have been considered in establishing the criteria for the selection of study areas (MARGINS Office 2003). These experiences should inform the Asia-Pacific region. Taking into account regional characteristics, the following criteria may be adopted in the selection of key study areas: climate zones, population density, degree of urbanisation, characteristics of catchment development, river type and size, material discharges, and coastal ecosystem types. Representative catchment-coast systems should be selected for detailed studies to obtain regional patterns. Relevant research topics are listed below.

4.4.1 Process Modeling Approaches to Material Discharges

Sedimentary material discharge from a catchment basin is controlled by temperature (i.e. climate), basin elevation, and catchment area (J. P. M Syvitski, personal communication). If these factors are known, the discharge can be determined using an empirical equation. However, because there are a number of processes involved in the relationship between discharge and the catchment conditions, this approach may not be sufficient for changing temperature (i.e. climate), basin elevation, and catchment area conditions. For instance, climate conditions, including temperature and rainfall, in combination with basin geology affect the weathering of bedrocks and the yield of sediment; the stratigraphy determines the river drainage pattern, which influences the transport and accumulation of sedimentary materials; and the hypsometric patterns of the catchment influence the formation of river terrace deposits and the quantity and composition of the sediment discharged into the coastal waters. Although predictions using the three variables (temperature, basin elevation, and catchment area) are generally correct in terms of order of magnitude, scattering occurs and the method cannot be used to predict the discharges influenced by human activities. This situation is also the case for other materials (i.e. water, nutrients, and pollutants). Therefore, it is important to consider the processes of both natural and anthropogenic changes so that the relative importance of the various processes can be determined. The approach of process modeling may be adopted to determine the role played by the factors such as rock types and strata in the basin, vegetation cover, dam building, chemical fertiliser uses, and industrial/domestic sewage discharges in the changes of material fluxes.

4.4.2 Modeling River Channel and Delta Evolution

New techniques have been developed in coastal morphodynamics to simulate the geomorphological evolution of coastal systems (e.g. De Vriend et al. 1993). These techniques are based on the calculation of sediment movement and accumulation/erosion. The key issue is the determination of sediment transport rate, which can be calculated with enhanced accuracy thanks to progress in sediment dynamics. Thus, morphological models can be established to investigate the evolution of river channels (in terms of cross-sectional area, width, and depth) and delta growth patterns under the changed conditions of water and sediment discharge patterns. Such studies have shown that new equilibrium conditions will be established for the river channel (Raudkivi 1990) and the river delta may be sensitive to changes of sediment supply (Gao 2006). In order to improve spatial resolution, two-dimensional models are required for the simulation of morphological evolution, in addition to time-dependent (i.e. 0-dimensional) and one-dimensional models. The development and applicability of such models depends on calibration using ground-truthed data sets. In the Asia-Pacific region, long-term monitoring data are available for some important river system (e.g. the Changjiang River). It would be very useful to establish a sufficient number of morphological models for the region that are well calibrated.

4.4.3 Geochemical Tracing Method to Define Material Retention

Mixing models for closed systems are often applied to obtain quantitative information on sediment sources (e.g. Owens et al. 2000). However, for an open system like an estuarine-coast system, such models may not be sufficient because the quantity and composition of the materials exported from the river mouth towards the shelf have to be considered. Based on the assumption that the areas for sediment deposition consist of an estuarine-coast sink and an open shelf sink, with the sources being from both the catchment and the ocean sides, mixing models for open systems may be established. Such an effort will enable us to use geochemical tracers to define material retention in the estuarine-coast system. There may be potential to apply the model not only to sediments, but also to other materials, such as nutrients and pollutants in particulate forms or adsorbed by sedimentary particles.

4.4.4 Forward Modeling of the Formation of Sedimentary Records to Identify the Evolution History of the Catchment-Coast System

Inverse methods are often used in the analysis of sedimentary records (e.g. Bhandari et al. 2005). It is a common practice to re-sample from cores, and various types of analyses are then carried out for the sub-samples. The data sets obtained are used to formulate proxies of environmental changes in order to reconstruct the history of the evolution of the catchment-coast system. A limitation of this approach is that the continuity of the sedimentary record cannot be evaluated fully and, therefore, the resolution of the time series from core analysis may not be sufficiently high to identify recent estuarine and coastal changes. Thus, it is desirable to develop "forward modeling" techniques to simulate the formation of sedimentary sequence on the basis of sediment dynamics and morphodynamic calculations. In this way, the nature of the sequence can be understood, which is important in the selection of appropriate records for analysis. Further, some crucial environmental information that cannot be derived with the inverse method, such as deposition rates on small time scales, sediment mobility during the accumulation and preservation potential of the record, can be obtained by forward modeling. This method may become an important tool for the study of the evolution of the catchment-coast system in the near future.

4.4.5 Databases for the Simulation of Catchment-Coast System Behaviour

Thanks to progress in the techniques of data collection and *in situ* monitoring, large data sets will accumulate at a high rate. Every year, data sets from field measurements and laboratory analyses (e.g. hydrodynamic, sediment grain size, geochemical, biological, and remote sensing data) are added. These data are normally used for describing the regional characteristics and calibrating models, but powerful tools like GIS must be used to deal with the large amount of data. Hence, it is necessary to establish databases (including multi-layer databases) for the region.

4.4.6 Methods to Generate Appropriate Scenarios for Future Management

For future development, planning, and management of the Asia-Pacific region, scenarios of environmental change, trends of social and economic growth, climate conditions, ecosystem evolution, etc., must be formulated. Due to an inability to sufficiently predict future conditions, uncertainties exist in many scenarios that are currently being adopted. Such uncertainties enhance the cost of management and planning. Therefore, new or improved methods should be developed to generate more realistic, appropriate scenarios that incorporate historical records, present trends, and future predictions on the basis of process and mechanism studies.

4.4.7 Artificial Intelligence Tools for Integrated Catchment-Coast Management

Many tools are already available to support management practices, including databases to store scientific data sets and other types of information (e.g. legislation documents), regional plans for socioeconomic development and land use status, and GIS software for data processing. In addition to these tools, artificial intelligence management models should be developed in the future to enhance the quality and the efficiency of management. The model should be both intelligent and robust. Being "intelligent" means that, given sufficient information, decisions will be suggested by the model in the form of model output; likewise, if the information available is insufficient to provide a decision, the required information will be specified by the model. Being "robust" implies that any decision made from running the model will not violate any regulations and/or management guidelines, and, if there are shortcomings in the regulations and/or guidelines, the mistakes can be identified and remedy measures suggested. Such new tools will be beneficial to integrated catchment-coast management practices.

4.5 APPLICATIONS TO CATCHMENT-COAST DEVELOPMENT AND MANAGEMENT

The purpose of the research activities outlined above is to support integrated, effective environmental resource management (Mitchell 2002) of the catchment-coast system. In the past, the collection of scientific data was often considered mainly of interest to scientists, and, at its best, as background information for management. As a result, in developing countries it has been a general practice that management activities are separated from research activities. To improve this situation, interdisciplinary efforts are required so that researchers with different knowledge backgrounds (i.e. natural and social sciences) are able to work together with managers, policymakers, and stakeholders to formulate management models and tools. The following are examples for investigations into the application of scientific results for management practice.

4.5.1 Regional Planning for Future Catchment-Coast Development

It can be expected that powerful management tools (e.g. multi-layer databases, GIS, and artificial intelligence management models) will be made available to the manager in the near future. At the same time, improved scenarios of future changes (e.g. environmental change, trends in social and economic growth, climate conditions, and ecosystem evolution) will be formulated by future scientific research. In order to realise the goal of sustainability in regional development, scientific data, other information, and predictions of future changes, should be synthesised within the framework of regional planning. Because of the large population and extremely high pressures exerted by human activities on the catchment-coast environment, appropriate, science-based planning is essential, and efforts should be made to enhance the quality of planning.

4.5.2 Protection of Coastal Wetland and Coral Reef Ecosystems

Coastal wetlands and coral reefs are widely distributed along the Asia-Pacific coastlines. These systems provide important goods and ecosystem services (Raffaelli and Hawkins 1996, Mitsch and Gosselink 2000, Kathiresan and Bingham 2001). However, the increasing influences of nutrients and pollutants derived from catchment areas threaten the future of the wetlands and coral reefs (e.g. Richardson 1997, Coles and Brown 2003). Environmental and ecosystem degradation in these systems has been observed, and the protection of these features has become a task of many agencies and organisations.

For the protection of the coastal wetlands and coral reefs of the region, research in marine ecosystem dynamics (McLusky and Elloitt 2004, Neumann 2004) must be strengthened. Coastal ecosystems differ from their land or open sea counterparts in that the environment is subjected to both land-based factors (e.g. inter-tidal landforms, river discharges, and rainfall) and marine-based factors (e.g. waves, tides, and salinity). In investigating material cycling patterns in these systems, the increasing input of nutrients and pollutants from the catchment should be taken into account. On this basis, changes in primary production, food webs, trophic level structures, dominant species, and mechanisms of recruitment can be evaluated. Furthermore, proxies of ecosystem health may be formulated. These studies will be helpful to the establishment of protection measures for these ecosystems.

4.5.3 Managing Estuarine and Coastal Environmental Changes

In response to catchment development activities, represented by changes in material fluxes and intense use of coastal land, the estuaries over the region have been modified significantly, as indicated by the geomorphology, water quality, and ecosystem changes. Thus, it is necessary to enhance our management ability for the estuarine systems. Unfavourable modifications to the estuarine shoreline should be avoided, pollution and saline water intrusion should be reduced to enhance water quality in estuarine areas, and biodiversity and ecosystem health should be maintained. All these tasks depend upon an appropriate management framework.

4.5.4 Potential for Coastal Land Reclamation

River deltas, together with the adjacent coastline, prograde towards the sea because of sediment supply from land. This process generates new land on many deltas, which generates precious resources for reclamation. On the Changjiang River delta, for example, future urban and industrial development depends to a large extent on the newly reclaimed land from the deltaic wetlands. However, the growth of the delta is constrained by sediment supply, coastal ground subsidence, and sea-level change (Yang et al. 2003, Gao 2006). Therefore, there is a need to estimate the potential increase in coastal land for the future so that appropriate measures can be taken to plan for reclamation schemes and maintain a balance between land requirements and coastal wetland conservation/protection.

Using knowledge of future sediment discharge and techniques of morphodynamic modeling of shoreline changes, accretion rates and the growth pattern for delta systems may be determined. Information on future sediment sources (from land, offshore areas, or nearby coastal recession), in terms of quantity and composition, can be obtained from catchment-coast interaction studies; sediment retention can be estimated using estuarine hydrodynamic and sediment dynamic data; dispersal of sedimentary material towards the open sea, after leaving the river mouth, can be calculated using modeling techniques; net deposition rates on the delta, together with the grain size composition of the deposited material, can be delineated on the basis of deltaic sedimentary facies analysis; and, finally, the accretion status and the sediment availability can be compared to derive the rate of coastal land area changes. Based upon this methodology, dynamic conditions of the deltaic area can be determined and future land area changes can be predicted.

4.5.5 Managing Aggregate Resources in the River Channel and Coastal Areas

Sand and gravel mining for road construction and the building industry is an intense activity in the region. This activity is often excessive. For example, in the Changjiang River channel and on the inner shelf adjacent to the river mouth, illegal mining has continued for a long time. Every year a large amount of sand is extracted and sold to the building market, which causes instability of the river bank or the seabed and is destructive to benthic biological communities and fish spawning grounds. For reasonable mining activities, environmental impact assessments (EIA) (Merritt et al. 2004) must be carried out, including an evaluation of the effect of aggregate mining on the river morphological stability and marine ecosystem. Since the accretion/erosion patterns have been modified by the change in river sediment discharge, the EIA should evaluate new sediment supply conditions using morphological dynamic modelling to determine potential affects on the river channel and delta evolution.

4.5.6 Flood Defence

Over the Asia-Pacific region, estuarine and coastal flooding is often caused by typhoon action, and severe flooding events tend to be related to astronomical tides and high freshwater discharges from the drainage basin, which also occur in conjunction with typhoons. For flood defence, algorithms of flooding water level with various frequencies of occurrence have been developed for "normal" conditions, meaning those that do not include changes in climate and sea level, water and sediment discharges, or geomorphological characteristics. However, these algorithms are no longer applicable when these conditions are varied. Because of climate and sea-level changes, the sea may be "stormier" than the past, with larger amplitudes of typhoon-induced waves and storm surges. Likewise, changes in water and sediment discharges from land, together with the effects of coastal engineering schemes, influence the evolution of the estuarine and coastal geomorphology. Therefore, new standards for flood defence must be established to replace the outdated standards. For instance in a tidal inlet system, water and sediment discharges from the drainage basin may be modified by the construction of dams, a part of shoreline may be reclaimed for urban development, and the entrance channel may be regulated to facilitate navigation; all these changes may result in new flooding patterns in the tidal basin. Thus, models should be established to calculate the flooding water levels under modified conditions so that updated flood prevention standards can be established.

4.5.7 Water Quality Improvement and Ecosystem Health

In the estuary and adjacent coastal waters, water quality and ecosystem health are crucial issues related to coastal development. The water quality should be assessed using the information on pollutant input and saline water intrusion associated with catchment changes. At the same time, effort should be made to synthesise nutrient, pollutant, and fine-grained sediment fluxes and to formulate ecosystem health indices. On such a basis, the data sets obtained and analysed can be combined to evaluate future ecosystem characteristics, in terms of degree of eutrophication, dissolved oxygen content, photosynthesis, biological community structures, recruitment and reproduction, and habitat conditions. These evaluations provide necessary information for the sustainability of coastal development and its management in the future.

4.5.8 Sustainable Utilisation of Biological Resources

Areas near large river mouths tend to have high biological production. For example, the fishing grounds near the Changjiang River mouth are one of the largest of the region. Here, primary production is on the order of 300–600 g C m^{-2} a^{-1}, and these areas are well known for fish, shrimp, crabs, and shellfish products (Anonymous 1987). However, the fishery resources have been threatened by the combined effect of overfishing, eutrophication due to enhanced nutrient loading, and harmful algal blooms. Furthermore, mariculture, or sea farming, is an intensive human activity throughout the region that leads to environment deterioration. For the purpose of sustainable utilisation of biological resources, the potential for fisheries and sea farming in the region should be re-analysed to consider changed nutrient conditions, and new planning and guidelines should be formulated.

4.5.9 Prospect of Regional Catchment-Coast Development

Population growth will continue to be a trend of the Asia-Pacific region; hence, further catchment-coast development in the future can be expected. In the past, the management of the catchment and the coast tended to be considered separately (e.g. Han et al. 1988; Becu et al. 2003). For the future, the land use patterns, river valley and coastal urbanisation, industry, agriculture, tourism, and other human activities of the catchment-coast system of the region should be examined in an integrated way. Additionally, the various management options should be evaluated on a regional scale to optimise future management practices.

4.6   SUMMARY

In the Asia-Pacific region, the interactions between the catchment and the adjacent coastal areas are characterised by rapid changes in freshwater, sediment, nutrient, and pollutant fluxes from land, and by extremely high pressure exerted from intense human activity in the catchment and estuarine/coastal environments. The patterns of the various interactions are changing in response to changes in the catchment and coastal areas.

For the purpose of integrated management for the catchment-coast systems of the region, basic and applied research should be undertaken to: define the general characteristics of catchment-coast interactions and their variations; understand the underlying processes and mechanisms so that insight can be gained to predict future changes; develop new research methodologies that enhance our ability to generate more realistic scenarios for future management; and analyse the implications of scientific data and information derived from basic research in terms of resource conservation, environmental protection, and integrated catchment-coast management.

For the studies of regional characteristics, catchment-coast changes and their impact on the estuarine and coastal environments will be a focus, including: patterns of catchment hydrological cycling; sediment yield and input to coastal waters; river channel morphological evolution; accretion and erosion patterns in the estuary and on the adjacent coast; saline water intrusion patterns in the estuary; nutrient/pollutant emissions and water quality changes; influences of sediment, nutrient, and pollutant inputs on estuarine and adjoining continental shelf ecosystems; and influences of large estuarine/coastal engineering schemes.

Further studies on the basic processes and mechanisms that are responsible for the changes will be beneficial to the quantitative prediction of future changes. These processes and mechanisms include: the processes of catchment hydrological changes, retention of sediment within catchment-coast systems, sediment mobility in estuarine and coastal waters with source-sink changes, estuarine and coastal morphodynamic processes, changing mixing and dispersion processes for nutrients and pollutants, processes for estuarine and coastal ecosystem evolution, and comparison between large and small catchment-coast systems. For these studies, the following methods and techniques should be developed: process modeling approaches to material discharges; modeling river channel and delta evolution; geochemical tracing methods to define material retention; forward modeling of the formation of sedimentary records to identify the evolutionary history of the catchment-coast system; databases for the simulation of catchment-coast system behaviour; methods to generate appropriate scenarios of catchment-coast changes; and artificial intelligence tools for integrated catchment-coast management.

The findings obtained from the above studies can be applied to catchment-coast development and management practices in the future. Many research topics can be formulated in this respect, with the examples including: regional planning for future catchment-coast development, protection of coastal wetland and coral reef ecosystems, managing estuarine and coastal environmental changes, evaluation of the potential for coastal land reclamation, managing aggregate resources in the river channel and coastal areas, flood defence, improvement to water quality and ecosystem health, sustainable utilisation of biological resources, and regional management of future catchment-coast development.

ACKNOWLEDGEMENTS

The writing of the article was originated from an APN Synthesis Forum, held during the Sixth International Conference on the Environmental Management of Enclosed Coastal Seas (Bangkok, Thailand, 18th–22nd November, 2003). This study is supported financially by the natural Science Foundation of China (Grant Number 40231010). The Changjiang River Hydraulic Engineering Committee provided the data of water and sediment discharges. The author wishes to thank Professor James PM Syvitski for providing Figure 4.1, Mr Bin-cheng Zhu for providing Figure 4.3, and Dr Ya Ping Wang and Mr Niu Zhan-sheng for their help in the preparation of Figures 4.2 and 4.4.

REFERENCES

Ahmad T, Khanna PP, Chakrapani GJ, Balakrishnan S, 1998. Geochemical characteristics of water and sediment of the Indus River, Trans-Himalaya, India: Constraints on weathering and erosion. Journal of Southeast Asian Earth Sciences, 16, 333–346.

Allison MA, 1998. Historical changes in the Ganges–Brahmaputra delta front. Journal of Coastal Research, 14, 1269–1275.

Anikiyev VV, Zaytsev OV, Hieu TT, Savilyeva IT, Starodubtsev Ye G, Shumilin Ye N, 1986. Variation in the space-time distribution of suspended matter in the coastal zone of the Mekong River. Oceanology, 26(6), 725–729.

Anonymous (ed.), 1987. The Three Gorges Project on the Changjiang River: Implications on Ecology and Environment. Chinese Science Press, Beijing, 1126pp (in Chinese).

Becu N, Perez P, Walker A, Barreteau O, Page C Le, 2003. Agent based simulation of a small catchment water management in northern Thailand: description of the CATCHSCAPE model. Ecological Modelling, 170, 319–331.

Bhandari S, Maurya DM, Chamyal LS, 2005. Late Pleistocene alluvial plain sedimentation in Lower Narmada Valley, Western India: palaeoenvironmental implications. Journal of Asian Earth Sciences, 24, 433–444.

Bornhold BD, Yang ZS, Keller GH, Prior DB, Wiseman WJ, Wang Q, Wright LD, Xu WD, Zhuang ZY, 1986. Sedimentary framework of the modern Huanghe (Yellow River) Delta. Geo-Marine Letters, 6, 77–83.

Chen XQ, 1998. Changjiang (Yangtze) River delta, China. Journal of Coastal Research, 14, 838–858.

Coles SL, Brown BE, 2003. Coral bleaching – capacity for acclimatization and adaptation. Advances in Marine Biology, 46, 183–223.

Crossland CJ, Kremer HH, Lindeboon HJ, Crossland JIM, Tissier MDA Le (eds), 2005. Coastal Fluxes in the Anthropocene. Springer-Verlag, Berlin, 231pp.

De Vriend HJ, Capobianco M, Chesher T, De Swart HE, Latteux B, Stive MJF, 1993. Approaches to long-term modeling of coastal morphology: A review. Coastal Engineering, 21, 225–269.

Dyer KR, 1997. Estuaries: A physical introduction (2nd edn). John Wiley, Chichester, 195pp.

Gao JH, Gao S, Cheng Y, Dong LX, Zhang J, 2004. Suspended sediment movement and the formation of turbidity maxima in Yalu River estuary. Journal Coastal Research, Special Issue, 43, 134–146.

Gao S, 2006. Modeling the limit of the Changjiang River delta growth. Geomorphology. In press.

Gao S, Collins M, 1995. Net transport direction of sands in a tidal inlet, using foraminiferal tests as natural tracers. Estuarine, Coastal and Shelf Science, 40, 681–697.

Gao S, Jia JJ, 2004. Accumulation of fine-grained sediment and organic carbon in a small tidal basin: Yuehu, Shandong Peninsula, China. Regional Environmental Change, 4, 63–69.

Han MK, Zhao SS, Ge LQ, 1988. China's coastal environment, utilization and management. In: Borgese EM, Ginsberg N, Morgan JR (eds), Ocean Yearbook 7. The University of Chicago Press, Chicago, 223–240.

Hori K, Tanabe S, Saito Y, Haruyama S, Nguyen V, Kitamura A, 2004. Delta initiation and Holocene sea-level change: Example from the Song Hong (Red River) delta, Vietnam. Sedimentary Geology, 164, 237–249.

Huang ZG, Li PR, Zhang Z, Li K, Qiao P, 1982. The Pearl River delta. Science Press, Guangzhou (China) (In Chinese).

Jorgensen DW, Harvey MD, Schumm SA, 1993 Morphology and dynamics of the Indus River: Implications for the Mohen Jo Daro site. In: Shroder JF (ed.), Himalaya to the Sea: Geology, Geomorphology and the Quaternary. Routledge, London, 288–326.

Kathiresan K, Bingham BL, 2001. Biology of mangroves and mangrove ecosystems. Advances in Marine Biology, 40, 83–251.

Lewis R, 1997. Dispersion in Estuaries and Coastal Waters. Wiley, Chichester, 312pp.

LOICZ IPO, 2004. Land–Ocean Interaction in the Coastal Zone (LOICZ) Science Plan and Implementation Strategy Structure. The Netherlands, 72pp.

Madsen OS, 1989. Transport determination by tracer. A: Tracer theory. In: Seymour RJ (ed.), Nearshore Sediment Transport. Plenum Press, New York, 103–114.

MARGINS Office, 2003. NSF MARGINS Program Science Plan 2004. Columbia University, New York.

McLusky DS, Elloitt M, 2004. The Estuarine Ecosystem: Ecology, Threats and Management. University Press, Oxford, 214pp.

Merritt WS, Croke BFW, Jakeman AJ, Letcher RA, Perez P, 2004. A biophysical toolbox for assessment and management of land and water resources in rural catchments in Northern Thailand. Ecological Modelling, 171, 279–300.

Milliman J, Farnsworth KL, Albertin CS, 1999. Flux and fate of fluvial sediments leaving large islands in the East Indies. Journal of Sea Research, 41, 97–107.

Milliman J, Syvistski JPM, 1992. Geomorphic/tectonic control of sediment discharge to the ocean: The importance of small mountainous rivers. Journal of Geology, 100, 525–544.

Ministry of Hydraulic Engineering (ed.), 2000. The flooding of 1998 in the Changjiang catchment and hydraulic monitoring and forecasting. China Hydraulic Press, Beijing, 200pp (in Chinese).

Mitchell B, 2002. Resource and Environmental Management (2nd edn). Pearson/Prentice Hall, Singapore, 367pp.

Mitsch WJ, Gosselink JG, 2000. Wetlands (3rd edn). John Wiley, New York, 920pp.

Neumann FW, 2004. Introduction to the Modeling of Marine Ecosystems. Elsevier, 330pp.

Owens PN, Walling DE, Leeks GJL, 2000. Tracing fluvial suspended sediment sources in the catchment of the River Tweed, Scotland, using composite fingerprints and a numerical mixing model. In: Foster IDL (ed.), Tracers in Geomorphology, Wiley, Chichester, 291–308.

Paola C, 2000. Quantitative models of sedimentary basin filling. Sedimentology, 47(Suppl. 1), 121–178.

Raffaelli D, Hawkins S, 1996. Intertidal ecology. Chapman & Hall, London, 356pp.

Raudkivi AJ, 1990. Loose Boundary Hydraulics (3rd edn). Pergamon Press, Oxford, 538pp.

Richardson K, 1997. Harmful or exceptional phytoplankton blooms in the marine ecosystem. Advances in Marine Biology, 31, 301–385.

Saito Y, Wei HL, Zhou YQ, Nishimura A, Sato Y, Yosuta Y, 2000. Delta progradation and chenier formation in the Huanghe (Yellow River) delta, China. Journal of Asian Earth Sciences, 18, 489–497.

Saito Y, Yang ZS, Hori K, 2001. The Huanghe (Yellow River) and Changjiang (Yangtze River) deltas: A review on their characteristics, evolution and sediment discharge during the Holocene. Geomorphology, 41, 219–231.

Salomons W, 2004. European catchments: Catchment changes and their impact on the coast. Institute for Environmental Studies, Amsterdam, 57pp.

Sheng JA, Liao AZ, 1997. Erosion Control in South China. Catena, 29, 211–221.

Syvitski PM, Vorosmarty CJ, Kettner AJ, Green P, 2005. Impact of humans on the flux of terrestrial sediment to the global coastal ocean. Science, 308, 376–380.

Ta TKO, Nguyen VL, Tateishi M, Kobayashi I, Saito Y, 2001. Sedimentary facies, diatom and foraminifer assemblages in a late Pleistocene–Holocene incised-valley sequence from the Mekong River Delta, Bentre Province, Southern Vietnam: the BT2 core. Journal of Asian Earth Sciences, 20, 83–94.

Thanh TD, Saito Y, Huy DV, Nguyen VL, Ta TKO, Tateishi M, 2004. Regimes of human and climate impacts on coastal changes in Vietnam. Regional Environmental Change, 4, 49–62.

World Commission on Dams, 2000. Dams and Development: A New Framework for Decision-making. Earthscan, London, 404pp.

Wright LD, Wiseman WJ, Yang ZS, Bornhold BD, Keller GH, Prior DB, Suhayda JN, 1990. Processes of marine dispersal and deposition of suspended silts off the modern mouth of the Huanghe (Yellow River). Continental Shelf Research, 10, 1–40.

Xue C, 1993. Historical changes in the Yellow River delta, China. Marine Geology, 113, 321–329.

Yan WJ, Zhang S, Wang JH, 2001. Trend of nitrogen input from the Changjaing drainage basin over the last 30 years in relation to DIN in the Changjiang water. In: Hu DX, Han WY, Zhang S (eds), Land–Ocean Interactions at Changjiang and Pearl Rivers and their Adjoining Coastal Waters. China Ocean Press, Beijing, 17–26 (in Chinese).

Yang SL, Belkin IM, Belkina AI, Zhao QY, Zhu J, Ding PX, 2003. Delta response to decline in sediment supply from the Yangtze River: Evidence of the recent four decades and expectations for the next half-century. Estuarine, Coastal and Shelf Science, 57, 689–699.

CHAPTER 5

COASTAL EVOLUTION IN THE ASIA-PACIFIC REGION

PATRICK D. NUNN[1] AND ROSELYN KUMAR[2]
[1]*Department of Geography, The University of the South Pacific, Fiji*
[2]*Institute of Applied Sciences, The University of the South Pacific, Fiji*

5.1 INTRODUCTION

As in most parts of the world, the proportion of coastal lands (relative to non-coastal lands) varies immensely throughout the Asia-Pacific region. For example, in some of the most distant parts of this vast region such as the tropical Indian Ocean or Pacific Islands, every piece of land is coastal in the sense that it is affected directly by coastal processes. Yet for the largest land areas in the region, coasts by any definition comprise only a small proportion of the total land area. Such statements may be misleading, however, because, in terms of their importance to humans as locations for settlement, economic activities, and food production, coasts are generally more valuable than most other land areas of comparable size in the Asia-Pacific region. At the same time, coasts are more vulnerable to change than other land types, whose degree of natural resilience is commonly greater.

As elsewhere in the world, the positions and the characters of Asia-Pacific coasts have changed through time. These changes have sometimes brought about profound alterations to the lifestyles of coastal-dwelling humans in the region, yet also presented new opportunities for their descendants. In the same way, it is clear that changes within the past 100 years — a time of unprecedented increases in human population pressure on most parts of the Asia-Pacific coastal zone — have been more rapid than at most earlier times, causing widespread disruption to human lifestyles and posing significant challenges for the next hundred years; challenges this book is trying to help solve.

Coastal changes can occur at a variety of scales, but it is useful, when assessing coastal history, to separate local from regional changes. Local changes may be unique and present problems that arise from a distinct set of circumstances, which may have little relevance to what is happening in other areas. There is little scope to ponder such local changes in a book that considers the entire Asia-Pacific coastal zone. Of far greater importance are regional changes, usually arising from climatic or environmental drivers that have similar effects across large regions. By studying such changes, even at a local scale, it is often possible to identify commonalties that are relevant to many other geographical situations.

In the following subsections (5.1.1–5.1.2), the distribution and nature of coasts in the Asia-Pacific region is discussed. There follows a section in which we look at the principal natural drivers of coastal change in this region (section 5.2). Readers should note that a discussion of humans – important as agents of coastal change in only comparatively recent times – is given in Chapter 6. Next is a summary of the history of coasts in the Asia-Pacific region (section 5.3). We then review existing research on the Asia-Pacific coastal zone by theme and by region, giving particular emphasis to research gaps (sections 5.4–5.6).

5.1.1 Asia-Pacific Coasts Discussed in this Chapter

For the purposes of this book, the vast Asia-Pacific region extends from the Pakistan/Iranian coastal border in the west to the Russian coast just north of Japan and includes the island archipelagos of Southeast Asia, Melanesia and Remote Oceania as far east as French Polynesia and adjacent island groups (Figure 1.2). However, for ease of discussion this chapter focuses on a more limited geography. It covers coasts that span ten countries on the Asian mainland (Bangladesh, Burma, Thailand, Malaysia, Singapore, Cambodia, Vietnam, South Korea, North Korea and China) and 24 on islands (Indonesia, Brunei, Philippines, Taiwan, Japan, Papua New Guinea, Solomon Islands, Vanuatu, New Caledonia, Nauru, Palau, Federated States of Micronesia, Northern Marianas Islands, Marshall Islands, Kiribati, Tuvalu, Fiji, Tonga, Wallis and Futuna, Samoa, Tokelau, Niue, Cook Islands and French Polynesia).

For both convenience of discussion and pragmatic reasons involving climate and the nature of particular characteristics (see following subsection), the coasts of this region are divided into three groups. The *Southeast Asia group* comprises seven continental countries (Bangladesh, Burma, Thailand, Malaysia, Singapore, Cambodia, and Vietnam), the *East Asia group* three continental countries (South Korea, North Korea and China), and the *Asia-Pacific Islands group* the 24 island countries mentioned above.

The proportion of coasts in these countries varies (Table 5.1). In China, for example, a country of 9.5 million km^2, coasts account for less than 1% of the land area; whereas in Indonesia, a country of some 1.9 million km^2, coasts comprise about 3% of the land area, and in archipelagic French Polynesia the figure is close to 70%. The degree of importance of the coast varies between countries of this region. In both China and Indonesia, for instance, the largest cities are along the coast, but in China there is understandably a greater proportion of the total

Table 5.1. Characteristics of coasts in the Asia-Pacific region

Group/Nation	Land area (km^2)	Coastline length (km)	Insularity index (coastline length/land area) × 100	Population density (persons per km^2)	GDP per capita (US$)
Southeast Asia group					
Bangladesh	133,910	580	0	1034	1720
Cambodia	176,520	443	0	74	1556
Malaysia	328,550	4675	1	70	8591
Burma	657,740	1930	0	65	1733
Singapore	682.7	193	28	6751	24,389
Thailand	511,770	3219	1	126	6936
Vietnam	325,360	3444	1	251	2251
Means			4.4		6739
Totals	2,134,533	14,484		172	
East Asia group					
China	9,326,410	14,500	0	138	4653
North Korea	120,410	2495	2	187	991
South Korea	98,190	2413	2	492	19,497
Means			1.3		8380
Totals	9,545,010	19,408		142	
Asia-Pacific islands					
Brunei	5,270	161	3	68	18,151
Cook Islands	240	120	50	88	4998
Federated States of Micronesia	702	6112	871	154	—
Fiji	18,270	1129	6	48	5551
French Polynesia	3,660	2525	69	72	4959
Indonesia	1,826,440	54,716	3	129	3040
Japan	374,744	29,751	8	339	28,699
Kiribati	811	1143	141	122	802
Marshall Islands	181.3	370.4	204	311	2038
Nauru	21	30	143	599	4773
New Caledonia	18,575	2254	12	11	14,231
Niue	260	64	25	8	3543
Northern Marianas Islands	477	1482	311	168	—
Palau	458	1519	332	43	—
Papua New Guinea	452,860	5152	1	12	2051
Philippines	298,170	36,289	12	284	4487
Samoa	2,934	403	14	61	5612
Solomon Islands	27,540	5313	19	18	1571
Taiwan	32,260	1566.3	5	701	17,962
Tokelau	10	101	1010	142	1058
Tonga	718	419	58	151	2182
Tuvalu	26	24	92	435	1079
Vanuatu	12,200	2528	21	16	2823
Wallis and Futuna	274	129	47	57	1907
Means			144.0		5480
Totals	3,077,101	153,300.7		155	
Overall means			6866.5		1827
Grand totals	14,756,644	187,193		149	

Source: CIA World Factbook, December 2003, via in early May 2004 www.nationmaster.com.

population living inland and a significant amount of non-coastal economic activity. In island groups like French Polynesia, most people inhabit the coast because inland areas are often steep with high relief; lifestyle options in inland areas are few and generally unattractive.

Furthermore, in most island countries of this region, trade across the ocean was always important because there was no alternative. Most sizeable settlements are therefore coastal. In contrast, overland communication was far more common in the larger continental countries with sea trade less important. Thus in China, for examples, two of the four largest cities (Shanghai and Tianjin) are coastal while the other two (Beijing and Chongqing) are inland. In some continental countries of this region, cities were established originally at the ocean mouth of a large river, but, as the river delta has extended seaward, now find themselves inland; examples include Bangkok (Thailand) and Tokyo (Japan).

5.1.2 Nature of Asia-Pacific Coasts

The nature of modern Asia-Pacific coasts is controlled by a number of region-wide factors, such as Holocene climate and sea-level changes and 20th century human settlement, as well as local and/or sub-regional factors. All of the coasts in this part of the world were drowned during the postglacial sea-level rise of 120–130 m that affected almost every part of the world's coastline between about 15,000 and 6,000 years ago. This drowning is shown today by the embayed nature of the region's coast – at both large- and small-scales – and by the formation of buried valleys in the lowland parts of the region's larger river valleys. The history of human settlement in the Asia-Pacific region is discussed in Chapter 6, but the 20th century and the first few years of the 21st century have clearly involved increased population densities along Asia-Pacific coasts, which increase various demands on coastal areas and, in most parts, result in a pastiche of human-environment interactions that are unsustainable in the longer-term. Climate is a major control on the nature of modern Asia-Pacific coasts. Other controls are considered primarily as "drivers" of coastal change and are therefore considered in section 5.2.

All coasts in the *Southeast Asia group* (defined above) are tropical. Those in the north of the region have a marked summer monsoon, and many inland areas are dry in winter. Good examples are Yangon (Rangoon) that receives 250 cm of precipitation yearly, almost all between May and October, and Singapore that receives 240 cm yearly evenly distributed throughout the year. These patterns give rise to rainforests in the south and to tropical grasslands and deciduous forest farther north. Many low-lying coasts in this region are fringed with mangrove forests. Coral reefs grow in many places, but are constrained by the high turbidity and low salinity of many nearshore waters.

In the *East Asia group*, there are major zonal contrasts in coastal environment associated mostly with climate. In the northern part of the coast of China (including North Korea and South Korea), the climate is temperate and supports mostly grassland. Most of the southern China coast is subtropical and covered with grasslands and patches of forests. In winter, cold dry winds blow off China (the winter monsoon), while in summer warm humid air from the southeast blows toward land bringing rain (the summer monsoon).

Since the *Asia-Pacific Islands group* covers such an enormous area, there is more variation in the nature of modern coasts attributable to differences in climate. These features vary from the vast mangrove-fringed aggrading delta at the mouth of the Fly River on high New Guinea Island (politically part of both Indonesia and Papua New Guinea) (Harris et al. 1993), to the islands (*motu*) of countries like Kiribati, the Marshall Islands, Tokelau and Tuvalu, where most of the habitable land is largely made up of reef detritus deposited onto the surfaces of broad living coral reefs (Nunn 1994). The nature of coasts in this group is determined at a local level, mostly by the volume of sediment arriving on a particular coast and by the dominance of particular sources of sediment. On high islands, terrigenous sediment dominates, but is often confined to building lowlands around the mouths of large rivers. In contrast, on reef-fringed islands, calcareous sediment is often the main sediment type to reach island shorelines and tends to build them out uniformly, particularly along windward coasts.

5.2 DRIVERS OF COASTAL CHANGE IN THE ASIA-PACIFIC REGION

This section looks at the principal natural drivers of coastal change in the Asia-Pacific region, including geotectonic and geomorphological processes, climate change, and sea-level rise. Geotectonic drivers involve geological and tectonic processes, many of which have been influencing the region's coasts for millions of years. Geomorphological drivers are processes of landscape change that operate on a variety of timescales and bring about changes that are often significant to the biotas of particular areas (including humans). Climate change and sea-level change are isolated from other drivers because of their overwhelming importance in fashioning the modern landscapes of the Asia-Pacific coast and generating some of the most important challenges for its future sustainability. Humans are also significant as drivers of coastal change, but are considered in Chapter 6.

5.2.1 Geotectonic Influences on Coastal Change

In a general sense, the modern coasts of the Asian continent and many of the offshore islands owe their characters to the geological and tectonic processes by which they evolved. Thus the Gulf of Thailand began forming in the early Tertiary as a result of the collision of India with Eurasia (Bustin and Chonchawalit 1995), eventually creating not only the ~3,000 km long gulf coastline, but a broad lowland basin now drained by the Chao Phraya River system. As a result of its low-lying character and its dominantly alluvial composition, the lowlands at the head of the Gulf of Thailand are now threatened by rising sea level and its associated effects (Jarupongsakul 2003).

In East Asia, the continental craton (ancient core), which has existed since the Palaeozoic, became significantly larger during the Mesozoic and Cenozoic eras (like other

Table 5.2. Selected rates of uplift and subsidence along Asia-Pacific coasts

Country	Area	Uplift or subsidence rate (mm a^{-1})	Uplift or subsidence magnitude (m)	Source of information
1. Aseismic uplift and subsidence				
Bangladesh	Dhaka-Sundarbans	−0.6 to −5.5		Huq et al. 1995
Bangladesh	Surma Basin	>−20		Huq et al. 1995
French Polynesia	Anaa, Tuamotus	0.1		Pirazzoli et al. 1987
French Polynesia	Moruroa, Gambier Islands	−0.12		Labeyrie et al. 1969
Hong Kong		0.6 to 1.0		Yim, 1995
Indonesia	Alor Island	1.0 to 1.2		Hantoro et al. 1994
Indonesia	Sumba Island	0.5		Pirazzoli et al. 1993
Japan	Hateruma, Ryukyus	0.1 to 0.3		Pirazzoli and Kawana, 1986
Japan	Honshu (general)	−1.0 to −1.2		Tjia, 1970
Marshall Islands	Enewetak	−0.1 to −0.2		Yonekura, 1983
New Caledonia	Maré, Loyalty Islands	0.7		Dubois et al. 1974
Niue		0.13 to 0.16		Nunn and Britton, 2004
Papua New Guinea	Huon Peninsula	1.0 to 3.0		Goudie, 1995
Timor	various	0.03 to 0.5		Chappell and Veeh, 1978
Vanuatu	Torres Islands	0.7 to 0.9		Taylor et al. 1985
2. Coseismic uplift and subsidence (abrupt)				
Fiji	Beqa		1.32 (Holocene)	Nunn, 1990
Japan	Honshu		3.0–6.0 (AD 1703)	Kayanne and Yoshikawa, 1986
Japan	Honshu		−0.45 (AD 1964)	Rothé, 1969
Japan	Okinawa, Ryukyus		2.5 (late Holocene)	Kawana and Pirazzoli, 1985
Solomon Islands	Guadalcanal		1.5 (AD 1961)	Grover, 1965
Vanuatu	Malakula		1.2 (AD 1965)	Taylor et al. 1980
Vanuatu	Santo		−0.15–0.20 (AD 1971)	Prevot and Chatelain, 1983
3. Rates of ground subsidence arising from groundwater extraction				
China	Shanghai		−1.64 (1921–1992)	Wang, 1998
China	Tanggu Harbour		−1.1 (1975–1991)	Wang, 1998
China	Tianjin		−0.83 (1975–1991)	Wang, 1998
Indonesia	Jakarta (northern)		−0.3 to −0.7 (1980–1995)	McBeth, 1995
Japan	Tokyo		−4.0 (1892–1972)	Wang, 1998
Japan	Osaka		>−2.8 (1935–1973)	Wang, 1998

Coastal Evolution in the Asia-Pacific Region 99

parts of the Pacific Rim) as a result of the accretion of terranes (fragments of ancient oceanic and/or continental lithosphere). This process created areas of folding and, because of associated extension, areas of rifting. Many of the major river valleys in East Asia now follow downfolded areas or lines of extensional faulting. Indeed, because of the geological history of this area a small number of very large river systems has developed, as opposed to a large number of smaller river systems. As a result, coastal sedimentation has been more localized than it might have otherwise been, and this situation has, in turn, influenced the modern spatial vulnerability of the East Asia coastal zone.

Most islands of Southeast Asia, Papua New Guinea, and the island arcs from the Solomon Islands to New Zealand owe their origins to processes operating along convergent plate boundaries. Many of these islands are still highly tectonically active today; typically the rates of these processes (Table 5.2) are an order of magnitude less than the rate at which sea level is expected to rise in the next century.

In the central Pacific, distant from plate boundaries, islands are uplifted, sometimes they even emerge above sea level for the first time, as a result of passing across intraplate swells or hotspots (Nunn 1994). But for most islands away from plate boundaries, subsidence is the most common condition. For example, the Vanuabalavu Island group in Fiji is sinking and, as a result, shoreline erosion is much more pronounced than it is on nearby stable islands (Figure 5.1; Nunn et al. 2002).

5.2.2 Geomorphological Influences on Coastal Change

The character of Asian coasts is also an outcome of the geomorphological processes that have been operating — sometimes for millions of years — in particular places. In China, for example, the valleys of the Huanghe (Yellow River)

Figure 5.1. The west-facing coast of southern Vanuabalavu Island in eastern Fiji. This island is subsiding so shoreline erosion – explicable elsewhere in Fiji largely by sea-level rise – is more severe here

and Changjiang (Yangtze River), along with adjacent lowlands, represent almost the only place along the entire Pacific Rim that is not bordered by a range of mountains. There are many implications of this geomorphology. Winds linked to the monsoons are unobstructed, and the loess carried from distant inland areas finds its way far out into the Pacific Ocean (Duce et al. 1980, Inoue and Naruse 1991). Sediments carried by these vast rivers have built large deltas and coastal lowlands; on the continental shelf of the East China Sea, evidence of Pleistocene river deltas has been discovered, showing that these rivers have been an important conduit for sediment and water transport to adjoining marginal seas for a long time (Liu et al. 2000). In fact, the first humans to occupy the coast of the Pacific Ocean, perhaps 40 million years ago, probably did so because they had followed these valleys from the continent's interior.

Yet significant geomorphological changes also occur on short timescales. Shifts in the outlets of large river channels in the vast deltas of East and Southeast Asia are normal, yet can bring about changes in the nature of shoreline processes along long stretches of coast. For example, in 1855, the main outlet of the Huanghe in China shifted northward into the Bohai Sea. This shift left a large part of the North China coast with no terrigenous sediment supply, resulting in shoreline retreat in areas that had previously been characterized by progradation. Shoreline erosion occurred between 1855 and 1980 at an average rate of 170 m a^{-1}, with the loss of 1,400 km^2 of land (Li and Li 1981).

Islands in the Asia-Pacific region understandably have fewer large rivers than mainland Asia. However, where they exist, such rivers have built large deltas, which were targeted by the earlier settlers and have remained popular locations for settlement ever since. On smaller islands, geomorphological processes have been critical in forming habitable land (Dickinson 2003), and understanding these processes remains important for the purpose of sustaining human habitability on these islands (Sherwood and Howorth 1996).

5.2.3 Influences of Climate and Sea-Level Change

It is important to remember that the essential character of almost every part of the world's coastline has changed within the past 18,000 years as a result of the net sea-level rise associated with postglacial warming and land-ice melt. It is salutary to reflect, when we are anxious about future sea-level rise, that for about 90% of the past 150,000 years or so, the ocean surface has been below its present level. The corollary is that humans have been living with, and adapting to, sea-level rise since they colonized Asia-Pacific coasts in large numbers approximately 12,000 years ago.

Asia-Pacific coasts have all been drowned since the Last Glacial Maximum around 18,000 years ago. This drowning has taken many forms. Lowland coasts that hosted large rivers during the Last Glaciation have seen the valleys of these rivers inundated and filled with sediment. Cliffed coasts may have experienced little change until the cliffs were overtopped, but then inundation would have been rapid. But most importantly, shorelines throughout the Asia-Pacific region receded (moved landward) as sea level rose. Connections between land masses were severed. Some landmasses even disappeared beneath the rising sea. One of the best-known examples involves the flooding of the Sunda Shelf that transformed the southeastern extremity of the Asian continent into an archipelago.

In most parts of the Asia-Pacific region, sea level is thought to have reached a maximum during the middle Holocene (6,000–3,000 years Before Present [BP]), perhaps 1–4 m above its present level. Since that time, sea level has generally fallen, exposing both aggradational coastal flats and erosional benches that had developed below mean sea level on many coasts during the middle Holocene (Umitsu 1991, Somboon and Thiramongkol 1992, Dickinson 2001, Yim and Huang 2002, Boyd and Lam 2004). Erosional benches became the foundations for coastal platforms around many Pacific Islands, and their permanent exposure above sea level is likely to have been an important factor in the chronology of human colonization of many islands (Dickinson 2003). Around the mouths of large rivers, late Holocene sea-level fall exposed large areas of fertile alluvial lowland that attracted large numbers of agriculturalists. The Zhujiang (Pearl River) Delta in south China is dotted with thousands of hills that were islands at the time of the middle Holocene sea-level maximum and acted as foci for alluvial sediment accumulation (Wu 2002).

Within the past 200 years or so, sea level has been rising throughout the Asia-Pacific region causing inundation and erosion. The latter process, attributable to the Bruun Effect, is manifested throughout the region by beach erosion (Sato and Mimura 1997). Good examples come from Malaysia (Teh 1997), Japan (Nishioka and Harasawa 1998), and Fiji (Mimura and Nunn 1998).

Coastal waters also warmed during postglacial times to such an extent that broad coral reefs became established. It is difficult to overestimate the importance of coral reefs in the maintenance of coasts in the tropical Asia-Pacific region. In the tropical part of the Asia-Pacific region around 4,500 years ago, at the time of the sea-level maximum, there were far fewer coral reefs at the ocean surface than there are today. The reason for this difference was that, in many parts of the region, sea level had been rising faster than coral reefs could grow upwards. So although there were a few areas where reef surfaces were able to "keep up" with sea-level rise, in most places it was only when sea level began falling in the late Holocene that reef surfaces were able to "catch up" with sea level. In some places, reefs had to "give up" trying to grow upwards as sea level rose (Neumann and MacIntyre 1985, Nunn 1999).

Thus in open-ocean locations where oceanographic conditions favoured coral-reef growth, like the equatorial atolls of western Kiribati (the Gilbert group), reefs successfully "kept up" with postglacial sea-level rise. But this success appears to be exceptional rather than normal. For example, there are a number of drowned atolls west of Savai'i Island in Samoa that are interpreted as reefs which "gave up" trying to grow up at the same rate as postglacial sea-level rose (Nunn 1994). In most parts of the region, coral reefs appear to have only caught up with sea level 1,000 years or more after it reached its Holocene maximum level. This delay meant that, at the time of the Holocene maximum level, waves with larger amplitude than at present reached the coast; in Hopley's (1984) phrase, there was a "high-energy window" open. Considerable erosion of many Asia-Pacific coasts occurred during the time (typically 6,000–3,000 years ago) this high-energy window remained open, but erosion was commonly replaced by progradation and nearshore (lagoonal) sedimentation once reefs caught up with sea level during the late Holocene.

Another important and widespread ecosystem in the tropical Asia-Pacific region is mangroves that form broad, continuous intertidal forests between the land and ocean along many shallow, sediment-floored coasts. Being unable to adapt to rapidly rising sea level, most mangrove forests in this region probably began to develop only when postglacial sea-level rise slowed around 7,000–6,000 BP; examples come from northern Australia (Chappell 1993, Mulrennan and Woodroffe 1998). But slower sea-level rise, followed by stabilization of sea level in most of the region around 5,000–4,000 BP, was not the only condition needed for the development of mangrove forests. A shallow, sediment-covered sea floor that mangrove roots could easily penetrate was also required, and this condition was often not met until after the Holocene sea-level maximum (perhaps 4,300 BP in most of the region) when many reefs reached sea level and nearshore lagoons developed. An unusual example comes from the Sepik-Ramu rivers in Papua New Guinea where tectonic changes after about 3,500 BP enclosed a shallow inland sea that subsequently became colonized with mangroves (Chappell 1993).

5.3 COASTAL HISTORY IN THE ASIA-PACIFIC REGION

This section describes coastal history in the Asia-Pacific region with an emphasis on island coasts because they exhibit more variability than continental coasts. Each subregion has elements of a distinctive coastal history, but equally there is considerable overlap. Between continental and insular Southeast Asia, for example, there is a close similarity in long-term coastal evolution because of the geotectonic connections. Yet, once the latter area became insular (archipelagic) rather than continental, coastline lengths increased abruptly and coasts acquired characteristics that their continental counterparts lacked. Foremost among these characteristics, in many places, were an increase in the contribution of marine-derived sediment supply to shorelines and a corresponding drop in terrigenous sediment supply. This balance is quite a different situation to that prevailing along most continental coasts where the contribution of terrigenous sediments transported to the shoreline by the large-volume rivers (which islands generally lack) dominates in many places.

It is the relative importance of terrigenous and marine sediment inputs to shorelines which justifies the fundamental division between continent and island in this chapter. Owing to their vast catchment areas, the largest continental rivers in the Asia-Pacific region are generally capable of transporting far more sediment (and freshwater) than their insular counterparts. The distinction is not always easy to make because of complications arising from climate and relief. For example, in New Guinea, classified here as an island, tectonic instability, steep slopes, and high rainfall combine to produce erosion rates that are among the highest in the world; for instance, ground-surface lowering in the Fly River catchment, calculated from suspended sediment yields, is 3–4 mm/year (Pickup et al. 1980).

In general though, the insularity index shown in Table 5.1 can be used as a crude guide to the relative importance of coastal sediment inputs from different sources. On Palau, for example, the insularity index of 332 immediately suggests an archipelagic nation, lacking islands of a size capable of supporting large rivers and, therefore, likely to have coasts dominated by marine sediment buildup. In Indonesia, with an insularity index of just 3, it can be inferred that the islands here are larger than those of Palau and therefore capable of generating considerably more terrigenous sediment, which may end up along island shores. But for China, Cambodia, and Bangladesh, the insularity index of 0 (actually each country's index is less than 0.5) clearly indicates they have large rivers carrying huge amounts of terrigenous sediment to the coast where, particularly around river mouths, it greatly outweighs the contribution of marine sediment.

The South Asia continental margin and many of the larger, continental islands offshore are in active tectonic locations, marked by contemporary seismicity and volcanism. The principal influences on this are the continuing collision of India with Eurasia and the northwards movement of the Australian continent towards Southeast Asia. The net effects are the occurrences of zones of extension and zones of compression throughout much of this sub-region. Many areas here are rising comparatively rapidly, a process, which largely through mass movements on unstable slopes, causes high rates of denudation and fluvial sediment inputs to the ocean; a good example comes from the central Rift Valley of Sumatra (Flenley and Butler 2001).

The coast of continental East Asia represents the trailing edge of the modern Eurasian Plate. It is generally classified as a passive margin, formed initially more than 750 million years ago with the breakup of the supercontinent Rodinia, but it has not always been thus. During the early part of the Mesozoic Era, the edge of the continental craton was hundreds, in places even thousands, of kilometres landward of its present position. Terranes – fragments of lithosphere – became accreted onto the edge of the craton creating fold belts. The most spectacular event in the Mesozoic history of the region was the collision of the Yangtze (or South China) microcontinent with the Sino-Korea (or North China) microcontinent, already joined to the craton. The accretion of the Yangtze microcontinent caused the anticlockwise rotation of the Sino-Korea terrane and a consolidation of the cratonic core (Metcalfe 1993). Terrane accretion may also account for the development of the marginal seas of East Asia (Hall 1996); although, some of these — such as the South China Sea — are bordered by active island arcs along their ocean-facing sides.

The Pacific Islands are generally smaller and formed more recently than the islands of South and East Asia. For both these reasons, the coastal histories of Pacific Islands are shorter and often less complex than elsewhere in the region.

5.3.1 Continental-island Southeast Asia

The continental islands of Southeast Asia are part of the adjacent Asian continent and many coasts are comparatively stable. Circular islands, like Borneo, have larger rivers that carry more sediment to their coastal zones compared to more elongate islands with convolute coastlines, like Sulawesi. There is also a contrast in this area between island coasts that face and/or are close to the Southeast Asian continent and those which face away from it and/or are farther away. The former coasts receive more terrigenous sediment inputs, while the latter receive less terrigenous sediment and more marine sediments, especially along reef-fringed coasts.

A large part of continental-island Southeast Asia (and Japan) is actually rising as a result of complex geotectonic processes associated both with the region's long-term geotectonic evolution and with the effects of underplating along the convergent plate margins of the Western Pacific and Sunda. As a result, reliable measurements indicate that relative sea level has actually been falling within this area over the past few decades, rather than rising as elsewhere in the Asia-Pacific region (Figure 5.2).

The coastlines of Borneo and Sulawesi have not changed much since the Pliocene; the Mahakam River on Borneo has been building its delta since that time. In contrast, some of the terranes ("arms") that account for the unusual shape of Sulawesi became accreted onto the main island only during the Quaternary (the last 1.8 million years), and the pattern of coastal evolution here must be considered within the context of closed and closing seaways. Terrane accretion, together

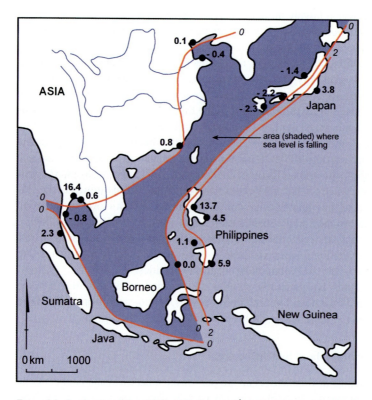

Figure 5.2. Sea-level variation (1950–1991, in mm a^{-1}) in East Asia and Southeast Asia (after Yanagi and Akaki, 1993). Largely as a result of tectonic influences, sea level appears to have fallen within a large part of this region

with volcanism and uplift due to nearby subduction, have been responsible for uncommonly high rates of terrigenous sediment input to the Sulawesi coast at various times during the Quaternary (Moss and Wilson 1998).

5.3.2 Asia-Pacific Islands along Convergent Plate Boundaries

The second group — the Asia-Pacific islands along convergent plate boundaries — include the Sunda Arc (Sumatra, Java, Timor), most of Japan, the Philippines, Papua New Guinea, Solomon Islands, Vanuatu, Fiji, Tonga, and parts of smaller island groups in the northwest Pacific Ocean. All these islands originated, in part, either as non-volcanic islands thrust upwards on buoyant plates at convergent plate boundaries or as volcanic islands formed from magma resulting from the melting of a downthrust plate. Although in some parts of these island groups, convergence has ceased and islands have either stabilized or become volcanically inert, in the other parts they remain tectonically and/or volcanically active. Some of the world's most active volcanoes occur on these islands and influence the nature of adjacent coastlines.

Most of the tectonically active islands in this region are rising rather than subsiding (see Table 5.2). Some of the best studied examples come from the Sunda Arc (Pirazzoli et al. 1993, Hantoro et al. 1994) and the Ryukyus Arc (Kawana and Pirazzoli 1985, Pirazzoli and Kawana 1986). Many such island groups experience both subsidence and uplift, often simultaneously in different parts, which explains why this region has been characterized over the past few decades by both sea-level rise and sea-level fall.

Two examples of uplifted coral-reef limestone islands are shown in Figure 5.3. Both formed along convergent plate boundaries, and both are essentially made from emerged fringing coral reefs of different ages that occupy distinct elevation ranges. In both examples, the highest reef terrace formed about 210,000 years ago during Oxygen-Isotope Stage 7, while lower terraces formed mostly during Stage 5. Raising the sea level up one or more terrace levels serves to reconstruct the former coastlines of these islands as it would have occurred during these interglacial stages.

The island Niue — a one-island state in the Central Pacific — originated in an intraplate location (see below), but emerged above sea level only when it began to ride up the flexure in the lithosphere on the eastern side of the Tonga-Kermadec Trench, 275–300 km to the west. It therefore owes its existence as an island to tectonic processes associated with this convergent plate boundary. The coast of Niue is cliffed almost everywhere and there is only a narrow ephemeral reef fringing it, both consequences of the island continuing to rise at rates of 0.13–0.16 mm a^{-1} (Nunn and Britton 2004).

Many islands along convergent plate boundaries in the Asia-Pacific region combine both volcanic and non-volcanic lithologies. This situation is true of larger islands like Honshu in Japan, Java in Indonesia, the island New Guinea (politically part of both Indonesia and Papua New Guinea), and Vanua Levu island in Fiji. The coasts of these composite islands reflect their particular origins with (younger) limestone coasts typically upstanding and (older, more degraded) volcanic coasts subdued. Since these

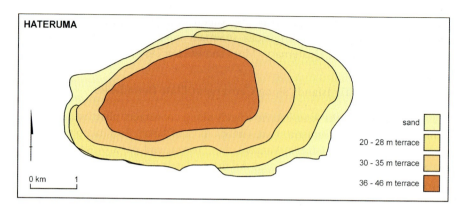

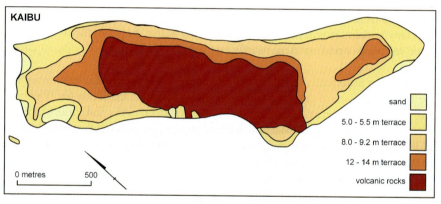

Figure 5.3. Examples of coastal changes on two smaller reef-fringed uplifted islands in the Asia-Pacific region. The locations of emerged coral reefs mark the coastlines during Quaternary interglacials (Stage 9, Stage 7, Stage 5e) compared to the modern coast. The upper map shows Hateruma Island in the Ryukyu Islands of Japan (after Ota and Hori, 1980; Omura, 1984), the lower map shows Kaibu Island in the Lau Islands of Fiji (after Nunn and Omura, 2003; Fiji Mineral Resources Department, 1984)

composite islands tend to be older than the most recent islands forming along the associated convergent plate boundaries, they also tend to be less tectonically and volcanically active. For this reason their coastlines usually exhibit the effects of Holocene sea-level changes more than convergent-boundary processes. Many composite island coasts have complex offshore coral-reef systems that also reflect their long history of vertical movements intertwined with sea-level changes.

The complex coastal history registered in such composite islands is exemplified by the islands of the Lau group of eastern Fiji. Formerly part of the Tonga-Kermadec arc and formed originally as a result of proximal subduction, the Lau Ridge (from which the islands rise) separated from the Tonga Ridge about five million years ago. At the time of separation, the Lau Islands were entirely volcanic. As separation progressed and the islands were gradually removed from the effects of nearby subduction, they began to subside; some sunk so deeply that coral-reef limestone grew over the top of them, while others sunk only slightly so that reef grew only around their sides. Then, around the start of the Quaternary Period (1.8 million years ago), the Lau Islands began rising once again as a detached slab of partly subducted lithosphere beneath them began heating up and expanding (Nunn 1998). The reef-fringed volcanic coasts and atoll coasts of the Lau Islands began emerging and were transformed into high limestone coasts. Within the later part of the Quaternary, uplift slowed down, eventually ceased, and was replaced, once more, by subsidence. Lau island coasts today exhibit signs of late Holocene sea-level changes superimposed on subsiding coasts (Nunn et al. 2002; see Figure 5.1).

5.3.3 Pacific Intraplate Islands

Fewer islands form in intraplate than in any other areas, and in the Pacific Ocean the lack of islands in intraplate areas accounts for their sobriquet "Remote Oceania". The reason so few islands form in these areas is because of the general absence of processes capable of creating islands. The most common of the island-forming processes in intraplate areas is the "hotspot hypothesis", whereby the movement of a plate across a fixed hotspot (a place where magma leaks onto the ocean floor from the asthenosphere many kilometres below) produces over long time periods a line of volcanic islands. Only one of these islands — that lying above the hotspot — is generally active. All others are volcanically extinct and subsiding as the moving lithospheric plate carries them into deeper water.

The coasts of most active hotspot, subaerial volcanic islands, such as Mehetia in French Polynesia, are indistinguishable from those of most other volcanic coasts. The coasts of the older volcanic islands in hotspot island chains are sometimes more distinctive, registering the effects of long-term, often quite rapid, subsidence; good examples are found in the steep-sided Marquesas Islands in French Polynesia (Okal et al. 2002).

As subsidence of volcanic islands in a hotspot chain continues in warm tropical waters, so coral reefs will grow up, around, and eventually above these islands giving rise to atoll (or ring) reefs. Owing to a late-Holocene fall of sea level in the tropical Pacific, most such atoll reefs have emerged by 1–2 m in the past 3,000 years or so, creating a base for islands that were subsequently formed from wave-carried reef detritus. In countries like Kiribati, the Marshall Islands, and Tuvalu, most atoll islands (named *motu*), although young in age and superficial in composition, rest on reefs that rise from the flanks of sunken volcanoes, which formed as part of one or more parallel hotspot island chains.

The long-term coastal history of atoll islands covers almost the entire gamut of possibilities, beginning with exposed volcanic island coasts, followed by reef-fringed volcanic island coasts, followed, in turn, by submergence and atoll development, and finally, by the formation of superficial islands (*motu*) that are among the island-coastal environments most vulnerable to future sea-level rise. On many atoll islands, shoreline erosion associated with recent sea-level rise has succeeded in exposing their low emerged foundations of fossil reef. There is concern that as sea level continues to rise in the future, shoreline erosion will become abruptly more rapid as this reef-rock foundation is overtopped and its cover of unconsolidated sediments begins to be eroded (Dickinson 1999).

5.4 KEY THEMATIC AREAS OF RESEARCH

Long-term coastal evolution is studied in the Asia-Pacific region for much the same reasons as it is studied elsewhere in the world, namely to help identify both global trends and global variations of significance in the course of Quaternary sea-level changes and — by inference — climate changes.

Tectonically active coasts have been the subject of considerable study, in particular along the island arcs of the western rim of the Pacific Basin (Ota and Kaizuka 1991). Studies of Japan (Ota and Omura 1991, Kaizuka 1992), Taiwan (Liew et al. 1990, Hsieh et al. 1994), Papua New Guinea (Chappell 1974, 1983; Ota 1994), and the Sunda Arc (Vita-Finzi 1995, Hantoro et al. 1994) exemplify this theme. Such coasts are of particular interest in the context of coastal-associated hazards, such as abrupt coseismic uplift or subsidence and locally generated tsunami.

A relatively new thematic area of research is the use of archaeological data to determine former shoreline positions (and to understand nature–society interactions along contemporary coasts). Case study 5.1 below explains how late-Holocene coastal history in the tropical west Pacific Islands can be traced by knowledge of where their first inhabitants lived. Similar work has been carried out in Thailand (Boyd et al. 1996) and Japan (Koike 1986).

Coastal planning on the basis of future sea-level rise is another popular area of research. Driven by the estimates of 21st century sea-level rise issued by the Intergovernmental Panel on Climate Change (IPCC), studies looking at how much

---

CASE STUDY 5.1   HISTORY OF MODERN COASTAL PLAINS ON TROPICAL WEST PACIFIC ISLANDS TRACKED BY THEIR HUMAN OCCUPATION

A fruitful line of study in recent years has been the identification of shoreline position and coastal configuration some 3,000 years ago in the West Pacific island archipelagoes of Fiji, New Caledonia, Samoa, Tonga, and Vanuatu through the study of their first human occupants. These people, known informally as the Lapita people, remained identifiable through the manufacture of uniquely decorated pottery for periods of 300–600 years. They subsisted largely on fringing-reef and shallow-water marine foods and consequently settled in the lee of beach dunes (Kirch 1997). Many Lapita settlement sites are now well inland of the modern shoreline, and can be used as precise markers of shoreline position some 3,000 years ago (Nunn 2005).

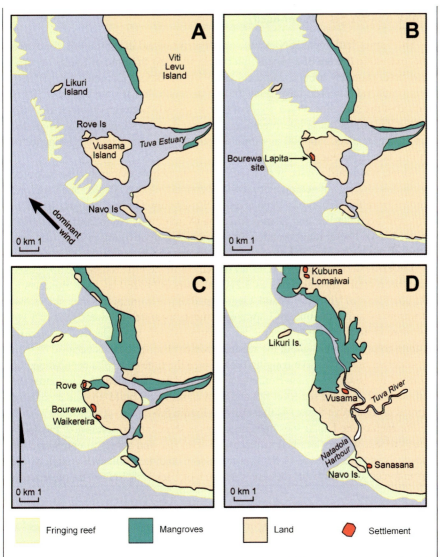

Figure 5.4. Palaeogeographic maps of the Bourewa Lapita site, established as much as 3210 cal yr BP (1260 BCE) on an island offshore Viti Levu Island in Fiji (figure adapted from Nunn, 2005; palaeosea level data from Nunn and Peltier, 2001).
(A) 4,000 years ago. Sea level was 2 m above its present level, no people lived in Fiji. Four bedrock islands existed off the main island where the Tuva River met the sea. The extent of coral reefs was less than today.
(B) 3,000 years ago. Human settlement established at Bourewa on offshore Vusama Island. Sea level 1.5 m higher than today. Coral reefs growing landwards. Mangrove forest spreading across shallow areas.
(C) 2,500 years ago. Human settlement overflows to Rove and Waikereira on offshore Vusama Island. Sea level 90 cm above present. Broad coral reefs. Mangrove filling coastal embayments.
(D) Modern situation. Former Vusama Island now joined to the main island.

> **CASE STUDY 5.1 (Continued)**
>
> The Lapita settlement at Naitabale on Moturiki Island in central Fiji, for example, existed in the lee of a broad high beach ridge. At the time of Lapita occupation here, probably around 2,950 cal yr BP (1000 Before Common Era [BCE]), this beach ridge would have been the active beach, but it now lies 250–300 m inland of the shoreline with a series of beach ridges in between (Nunn et al. 2003, 2007). This situation is interpreted as meaning that the shoreline has prograded seaward here a net 250–300 m within the past 3,000 years under the influence of falling late Holocene sea level.
>
> Along windward coasts in this region, Lapita people evidently also favoured tombolos — strips of sedimentary lowland connecting two islands. Such sites are known from the Bismarck Archipelago of Papua New Guinea, Naigani Island in Fiji, and from the Ha'apai Islands in Tonga (Kirch 2001, Nunn 2005, Dickinson et al. 1994). The gradual broadening of tombolos, by infilling in their lees, and other changes can also be linked to late Holocene sea-level fall.
>
> The earliest-known human (Lapita) settlement in the Fiji archipelago appears to have been established as much as 3,210 cal years BP (1260 BCE) at Bourewa in the southwest of Viti Levu Island. Geomorphological work in this area shows that it was an island at the time of Lapita settlement, separated by water gaps of 1 km or more from the mainland. Smaller offshore islands were clearly favoured by Lapita colonizers throughout the western tropical Pacific Islands because of their generally more productive fringing reefs. Since the time of Lapita occupation, the coast of the island on which Bourewa was located has extended seaward under the influence of both net sea-level fall and alluvial sediment build-up (and associated spread of mangrove forest) as illustrated in Figure 5.4.
>
> For each of the Lapita sites mentioned above, shoreline erosion is now the dominant process. This is ascribed to recent sea-level rise and serves to emphasize the point that long-term sea-level changes are seldom unidirectional, even though we are often forced to suppose that they are when we look back at them from the present day.

inundation, for example, might occur along particular coasts by particular dates have become widespread. Such studies are required of country signatories to key international agreements.

5.5 KEY REGIONAL AREAS OF RESEARCH

Some of the most detailed work on coastal evolution for the purpose of understanding Quaternary sea-level change has been carried out along rapidly-rising coasts, where shorelines formed below present sea level may have emerged (see Table 5.2). An important site in China is the Qingfeng section, from which sea-level behaviour for a large part of the East Asian coast has been inferred (Zhao et al. 1995).

There has been considerable research on pre-Holocene coastal evolution in certain (parts of) active island arcs in the western Pacific. These areas include Japan and Papua New Guinea. In Japan, long-term coastal evolution has become an increasingly popular (yet largely academic) field as recognition has grown about the importance of pre-Holocene studies for understanding the present state of the nation's coasts (Ota and Kaizuka 1991, Ota and Omura 1991) Although most parts of Papua New Guinea have received very little attention from scientists interested in pre-Holocene coastal evolution, the Huon Peninsula, where the long-term uplift rate can be assumed to be monotonic and unidirectional, has become a touchstone for the understanding of late Quaternary sea-level changes (Chappell 1974, 1983; Ota 1994).

The coastal evolution of East Asia has received a lot of attention in recent years, largely because of its vulnerability, occasioned both by its low-lying, largely unconsolidated character, and by its high population density. Attention has focused on the Huanghe Delta (Saito et al. 2000, Yi et al. 2003), the Changjiang Delta (Hori et al. 2002, Wang et al. 2005), and the areas in between. One of these — the Jiangsu coastal plain — is located between the modern Changjiang mouth and the former Huanghe mouth (open between AD 1128 and AD 1855; Ren and Shi 1986). The 3,000 km^2 area is covered with postglacial sediments, both terrigenous and marine, ranging from 10 m to >30 m in thickness. During the early Holocene (12–7 kilo anno Before Present [ka BP]), the area was influenced largely by Changjiang sediment, but during the middle Holocene (7–3 ka BP), occasional incursions of the Huanghe across the area led to rapid shoreline progradation. The record for the late Holocene is more precisely dated and shows that coastal processes were controlled largely by the position of the Huanghe. Thus during the period AD 1128–1855, there was rapid progradation, but since then the Jiangsu coastal plain has been eroding rapidly because of the lack of terrigenous sediment arising from the Changjiang and Huanghe entering the East China Sea and Bohai Sea respectively (Ren and Shi 1986, Li et al. 2001).

5.6 GAPS IN RESEARCH

Among the continental countries of the Asian continent within the region of study, coastal history is probably least well known in Burma and India, with still much more research needed in Cambodia, Korea, Thailand, and Vietnam before the understanding can reach the level it has achieved elsewhere in the region. Perhaps unsurprisingly, very little is known about long-term coastal history in North Korea, and there has been comparatively little in South Korea (Park and Yi 1995).

Among the islands of the Asia-Pacific region, some of those where coastal history is least well known are the large islands of Southeast Asia, such as Borneo, Java, Sulawesi, and Sumatra. While research has focused on the more geologically conspicuous parts of these islands, comparatively little systematic study of coastal history has been made, a comment that applies also to most of the smaller islands in Southeast Asia and to the Philippines.

There has been a marked imbalance in research targeting the pre-Holocene evolution of the coasts of active island arcs in the western Pacific. For example, hardly any research has been undertaken in the Solomon Islands where, along with many other Pacific Island groups in such locations (such as Tonga and Vanuatu), many of the key data were reported more than 30 years ago.

In general terms, marine geologists and terrestrial geologists in the Asia-Pacific region would benefit from sharing and understanding one another's results more than they have done. Many geological surveys of land masses are not continued offshore, while marine geological surveys still typically show islands as blank areas. Additionally, there is a need to increase research on megadeltas of the region (see Chapter 10).

REFERENCES

Boyd, W.E. and Lam, D.D. 2004. Holocene elevated sea levels on the north coast of Vietnam. Australian Geographical Studies, 42, 77–88.

Boyd, W.E., Higham, C.F.W. and Thosarat, R. 1996. The Holocene palaeogeography of the southeast margin of the Bangkok Plain, Thailand, and its archaeological implications. Asian Perspectives, 35, 193–207.

Bustin, R.M. and Chonchawalit, A. 1995. Formation and tectonic evolution of the Pattani Basin, Gulf of Thailand. International Geology Review, 37, 866–892.

Chappell, J. 1974. Geology of coral terraces, Huon Peninsula, New Guinea: A study of Quaternary tectonic movements and sea-level changes. Geological Society of America, Bulletin, 85, 553–570.

Chappell, J. 1983. A revised sea-level record for the last 300,000 years from Papua New Guinea. Search, 14, 99–101.

Chappell, J. 1993. Contrasting Holocene sedimentary geologies of lower Daly River, Northern Australia, and lower Sepik-Ramu, Papua New Guinea. Sedimentary Geology, 83(3–4), 339–358.

Chappell, J. and Veeh, H.H. 1978. Late Quaternary tectonic movements and sea-level changes at Timor and Atauro Island. Geological Society of America, Bulletin, 89, 356–368.

Dickinson, W.R. 1999. Holocene sea-level record on Funafuti and potential impact of global warming on central Pacific atolls. Quaternary Research, 51, 124–132.

Dickinson, W.R. 2001. Paleoshoreline record of relative Holocene sea levels on Pacific Islands. Earth-Science Reviews, 5, 191–234.

Dickinson, W.R. 2003. Impact of mid-Holocene hydro-isostatic highstand in regional sea level on habitability of islands in Pacific Oceania. Journal of Coastal Research, 19, 489–502.

Dickinson, W.R., Burley, D.V. and Shutler, R. 1994. Impact of hydro-isostatic Holocene sea-level change on the geologic context of island archaeological sites, northern Ha'apai Group, Kingdom of Tonga. Geoarchaeology, 9, 85–111.

Dubois, J., Launay, J. and Recy, J. 1974. Uplift movements in New Caledonia-Loyalty Islands area and their plate tectonics interpretation. Tectonophysics, 24, 133–150.

Duce, R.A., Unni, C.K., Ray, B.J., Prospero, J.M. and Merrill, J.T. 1980. Long-range atmospheric transport of soil dust from Asia to the tropical North Pacific: Temporal variability. Science, 209, 1522–1524.

Fiji Mineral Resources Department, 1984. Geology of Nukutolu, Vatuvara, Yacata and Kaibu, Naitauba, Kibibo, Malima, Wailagilala, Katafaga, Vekai, Cicia, Mago, Kanacea, Tuvuca, Tavanuku-i-vanua and Yaroua. 1:25,000 geological map (MRD 785).

Flenley, J.R. and Butler, K. 2001. Evidence for continued disturbance of upland rain forest in Sumatra for the last 7000 years of an 11,000 year record. Palaeogeography, Palaeoclimatology, Palaeoecology, 171, 289–305.

Goudie, A.S. 1995. The Changing Earth: Rates of Geomorphological Processes. Oxford: Blackwell, 302 p.

Grover, J.C. 1965. Seismological and volcanological studies in the British Solomon Islands to 1961. British Solomon Islands Geological Record, 2, 183–188.

Hall, R. 1996. Reconstructing Cenozoic SE Asia, In: Hall R. and Blundell, D. (eds.), Tectonic Evolution of Southeast Asia, Geol. Soc. Spec. Publ, 106, 153–184.

Hantoro, W.S., Pirazzoli, P.A., Jouannic, C., Faure, H., Hoang, C.T., Radtke, U., Causse, C., Borel Best, M., Lafont, R., Bieda, S. and Lambeck, K. 1994. Quaternary uplifted coral reef terraces on Alor Island, East Indonesia. Coral Reefs, 13, 215–223.

Harris, P.T., Baker, E.K., Cole, A.R. and Short, S.A. 1993. A preliminary study of sedimentation in the tidally dominated Fly River Delta, Gulf of Papua. Continental Shelf Research, 13, 441–472.

Hopley, D. 1984. The Holocene. "high energy window" on the central Great Barrier Reef. In: Thom, B.G. (ed.), Coastal Geomorphology in Australia. London: Academic Press, 135–150.

Hori, K., Saito, Y., Zhao, Q. and Wang, P. (2002) Evolution of the coastal depositional systems of the Changjiang (Yangtze) River in response to Late Pleistocene-Holocene sea-level changes. Journal of Sedimentary Research, 72(6), 884–897.

Hori, K., Tanabe, S., Saito, Y., Haruyama, S., Nguyen, V. and Kitamura, A. (2004) Delta initiation and Holocene sea-level change: example from the Song Hong (Red River) delta, Vietnam. Sedimentary Geology, 164(3–4), 237–249.

Hsieh, M.-L., Lai, C.K. and Liew, P.-M. 1994. Holocene climatic river terraces in an active tectonic-uplifting area, middle part of the coastal range, eastern Taiwan. Journal of the Geological Society of China, 37, 97–114.

Huq, S., Ali, S.I. and Rahman, A.A. 1995. Sea-level rise and Bangladesh: A preliminary analysis. Journal of Coastal Research, Special Issue 14, 44–53.

Inoue, K. and Naruse, T. 1991. Accumulation of Asian long-range eolian dust in Japan and Korea from the late Pleistocene to the Holocene. Catena, Supplement 20, 25–42.

Jarupongsakul, T. 2003. Effect of sea-level rise and responsible coastal zone management for the low-lying areas of Bangkok metropolis (Abstract). In: EMECS 2003, 6th International Conference on the Environmental Management of Enclosed Coastal Seas, 18–21 November, 2003, Bangkok, p 69.

Kaizuka, S. 1992. Coastal evolution at a rapidly-uplifting volcanic island: Iwo-Jima, Western Pacific Ocean. Quaternary International, 15/16, 7–16.

Kawana, T. and Pirazzoli, P.A. 1985. Holocene coastline changes and seismic uplift in Okinawa island, the Ryukyus, Japan. Zeitschrift für Geomorphologie, Supplementband 57, 11–31.

Kayanne, H. and Yoshikawa, T. 1986. Comparative study between present and emerged erosional landforms on the southeast coast of Boso Peninsula, central Japan. Geographical Review of Japan, 59, 18–36.

Kirch, P.V. 1997. The Lapita Peoples. Oxford: Blackwell, 353 p.

Kirch, P.V. (ed.) 2001. Lapita and its Transformations in Near Oceania: archaeological investigations in the Mussau Islands, Papua New Guinea, 1985–88. Contribution No. 59, Archaeological Research Facility, University of California at Berkeley.

Koike, H. 1986. Prehistoric hunting pressure and paleobiomass: An environmental reconstruction and archaeozoological analysis of a Jomon shellmound area. In: Akazawa, T. and Aikens, C.M. (eds.), Prehistoric Hunter-Gatherers in Japan: New Research Methods. Bulletin no. 27. Tokyo: The University Museum, The University of Tokyo, pp. 27–53.

Labeyrie, J., Lalou, C. and Delibrias, G. 1969. Etudes des transgressions marines sur l'atoll de Mururoa par la datation des differents niveaux de corail. Cahiers du Pacifique, 3, 59–68.

Li, C. and Li, B. 1981. Studies on the formation of Subei sand cays. Oceanoloca et Limnologica Sinica, 12, 321–331.

Li, C.X., Zhang, J.Q., Fan, D.D. and Deng, B. 2001. Holocene regression and the tidal radial sand ridge system formation in the Jiangsu coastal zone, east China. Marine Geology, 173, 97–120.

Liew, P.-M., Hsieh, M.-L. and Lai, C.K. 1990. Tectonic significance of Holocene marine terraces in the Coastal Range, eastern Taiwan. Tectonophysics, 183, 121–127.

Liu, Z., Berge, S., Saito, Y., Lericolais, G. and Marsset, T. 2000. Quaternary seismic stratigraphy and paleoenvironments on the continental shelf of the East China Sea. Journal of Asian Earth Sciences, 18, 441–452.

McBeth, J. 1995. Water peril: Indonesia's urbanization may precipitate a water crisis. Far Eastern Economic Review, 1st June 1995, 61.

Metcalfe, I. 1993. Southeast Asian terranes: Gondwanaland origins and evolution. In: Findlay, R.H., Unrug, R., Banks, M.R. and Veevers, J.J. (eds.), Gondwana Eight: Assembly, Evolution and Dispersal. Rotterdam: Balkema, 181–200.

Mimura, N. and Nunn, P.D. 1998. Trends of beach erosion and shoreline protection in rural Fiji. Journal of Coastal Research, 14, 37–46.

Moss, S.J. and Wilson, M.E.J. 1998. Biogeographic implications of the tertiary palaeogeographic evolution of Sulawesi and Borneo. In: Hall, R. and Holloway, J.D. (eds.). Biogeography and Geological Evolution of SE Asia. Leiden, The Netherlands: Backhuys, pp. 133–163.

Mulrennan, M.E. and Woodroffe, C.D. 1998. Holocene development of the lower Mary River plains, Northern Territory, Australia. Holocene, 8(5), 565–579.

Neumann, A.C. and MacIntyre, I. 1985. Reef response to sea-level rise: Keep-up, catch-up or give-up. In: Proceedings of the 5th International Coral Reef Congress, 3, 105–110.

Nishioka, S. and Harasawa, H. (eds.) 1998: Global Warming: The Potential Impact on Japan. Tokyo, Japan: Springer-Verlag, 244 pp.

Nunn, P.D. 1990. Coastal geomorphology of Beqa and Yanuca islands, South Pacific Ocean, and its significance for the geomorphology of the Vatulele-Beqa Ridge. Pacific Science, 44, 348–365.

Nunn, P.D. 1994. Oceanic Islands. Oxford: Blackwell.

Nunn, P.D. 1998. Late Quaternary tectonic change on the islands of the northern Lau-Colville Ridge, Southwest Pacific. In: Stewart, I.S. and Vita-Finzi, C. (eds.) Coastal Tectonics. Geological Society, London, Special Publications, 146, 269–278.

Nunn, P.D. 1999. Environmental Change in the Pacific Basin: Chronologies, Causes, Consequences. London: Wiley.

Nunn, P.D. 2005. Reconstructing tropical paleoshorelines using archaeological data: Examples from the Fiji Archipelago, southwest Pacific. Journal of Coastal Research, Special Issue 42, 15–25.

Nunn, P.D. and Britton, J.M.R. 2004. The long-term evolution of Niue Island. In: Terry, J. and Murray, W. (eds.). Niue Island: Geographical Perspectives on the Rock of Polynesia. Paris: INSULA (International Scientific Council for Island Development), 31–74.

Nunn, P.D. and Omura, A. 2003. Quaternary shorelines of Kaibu Island, southwest Pacific Ocean: Implications for Last Interglacial sea-level history and uplift of the Lau-Colville Ridge. Quaternary Australasia, 21, 33–38.

Nunn, P.D. and Peltier, W.R. 2001. Far-field test of the ICE-4G (VM2) model of global isostatic response to deglaciation: Empirical and theoretical Holocene sea-level reconstructions for the Fiji Islands, Southwest Pacific. Quaternary Research, 55, 203–214.

Nunn, P.D., Matararaba, S. and Kumar, R. 2003. An early Lapita site on Moturiki Island, central Fiji: Preliminary report. Domodomo, 16, 7–14.

Nunn, P.D., Ollier, C.D., Hope, G.S., Rodda, P., Omura, A. and Peltier, W.R. 2002. Late Quaternary - sea-level and tectonic changes in northeast Fiji. Marine Geology, 187, 299–311.

Nunn, P.D., Ishimura, T., Dickinson, W.R., Katayama, K., Thomas, F., Kumar, R., Matararaba, S., Davidson, J. and Worthy, T. 2007. The Lapita occupation at Naitabale, Moturiki Island, central Fiji. Asian Perspectives forthcoming.

Okal, E.A., Fryer, G.J., Borrero, J.C. and Ruscher, C. 2002. The landslide and local tsunami of 13 September 1999 at Fatu Hiva (Marquesas Islands, French Polynesia). Bulletin de la Société Géologique du France, 173, 359–367.

Omura, A. 1984. Uranium-series ages of the Riukiu Limestone on Hateruma Island, southwestern Ryukyus. Transactions and Proceedings of the Palaeontological Society of Japan, 135, 415–426.

Ota, Y. (ed.). 1994. Study on coral reef terraces of the Huon Peninsula, Papua New Guinea: establishment of Quaternary sea level and tectonic history. Yokohama National University, Japan.

Ota, Y. and Hori, N. 1980. Late Quaternary tectonic movements of the Ryukyu Islands, Japan. The Quaternary Research (Japan), 18, 221–240.

Ota, Y. and Kaizuka, S. 1991. Tectonic geomorphology at active plate boundaries–examples from the Pacific Rim. Zeitschrift fur Geomorphologie, Suppl-Bd 82, 119–146.

Ota, Y. and Omura, A. 1991. Late Quaternary shorelines in the Japanese islands. The Quaternary Research (Japan), 30, 175–186.

Park, Y.A. and Yi, H.-I. 1995. Late Quaternary climatic changes and sea-level history along the Korean coasts. Journal of Coastal Research, Special Issue, 17, 163–168.

Pickup, G., Higgins, R.J. and Warner, R.F. 1980. Erosion and sediment yield in the Fly River drainage basins, Papua New Guinea. Publications of the International Association of Hydrological Sciences, 132, 438–456.

Pirazzoli, P.A. and Kawana, T. 1986. Détermination de mouvements crustaux quaternaires d'après la déformation des anciens rivages dans les îles Ryukyu, Japon. Revue de Géologie Dynamique et de Géographie Physique, 27, 269–278.

Pirazzoli, P.A., Koba, T., Montaggioni, L.F. and Person, A. 1987. Anaa (Tuamotus) – an incipient rising atoll? Marine Geology, 82, 261–269.

Pirazzoli, P.A., Radtke, U., Hantoro., Jouannic, C., Hoang, C.T., Causse, C. and Borel Best, M. 1993. A one million-year-long sequence of marine terraces on Sumba Island, Indonesia. Marine Geology, 109, 221–236.

Prevot, R. and Chatelain, J.L. 1983. Sismicité et risque sismique à Vanuatu. Nouméa: ORSTOM, Rapport 5–83.

Ren, M.E. and Shi, Y.L. 1986. Sediment discharge of the Yellow River (China) and its effect on the sedimentation of the Bohai and Yellow Sea. Continental Shelf Research, 6, 785–810.

Rothé, J.P. 1969. The Seismicity of the Earth. Paris: UNESCO, 336 p.

Saito, Y., Wei, H., Zhou, Y., Nishimura, A., Sato, Y. and Yokota, S. 2000. Delta progradation and chenier formation in the Huanghe (Yellow River) Delta, China. Journal of Asian Earth Sciences, 18, 489–497.

Sato, Y. and N. Mimura, 1997. Environmental problems and current management issues in the coastal zones of south and southeast Asian developing countries. Journal of Global Environmental Engineering, 3, 163–181.

Sherwood, A. and Howorth, R. 1996. Coasts of Pacific Islands. SOPAC Miscellaneous Report 222.

Somboon, J.R.P. and Thiramongkol, N. 1992. Holocene highstand shoreline of the Chao Phraya delta, Thailand. Journal of Southeastern Asian Earth Sciences, 7, 53–60.

Tanabe, S., Ta, T.K.O., Nguyen, V.L., Tateishi, M., Kobayashi, I. and Saito, Y. 2003. Delta evolution model inferred from the Holocene Mekong Delta, Southern Vietnam. In: Sidi, F.H., Nummedal, D., Imbert, P., Darman, H. and Posamentier, H.W. (eds.), Tropical Deltas of Southeast Asia – Sedimentology, Stratigraphy, and Petroleum Geology, SEPM Special Publication no. 76, 175–188.

Taylor, F.W., Isacks, B.L., Jouannic, C., Bloom, A.L. and Dubois, J. 1980. Coseismic and Quaternary vertical tectonic movements, Santo and Malekula islands, New Hebrides island arc. Journal of Geophysical Research, 85, 5367–5381.

Taylor, F.W., Jouannic, C. and Bloom, A.L. 1985. Quaternary uplift of the Torres islands, northern New Hebrides frontal arc: Comparison with Santo and Malekula islands, central New Hebrides frontal arc. Journal of Geology, 93, 419–438.

Teh, T.T. 1997. Sea-level Rise Implications for Coastal and Island Resorts: Climate Change in Malaysia. Universiti Putra, Selangor Darul Ehsan, Malaysia, pp. 83–102.

Tjia, H.D. 1970. Rates of diastrophic movement during the Quaternary in Indonesia. Geologie en Mijnbouw, 49, 335–338.

Umitsu, M. 1991. Holocene sea-level changes and coastal evolution in Japan. The Quaternary Research (Japan), 30, 187–196.

Vita-Finzi, C. 1995. Pulses of emergence in the outer-arc ridge of the Sunda Arc. Journal of Coastal Research, Special Issue, 17, 279–281.

Wang, Y. 1998. Sea-level changes, human impacts and coastal responses in China. Journal of Coastal Research, 14, 31–36.

Wang, Z.H., Saito, Y., Hori, K., Kitamura, A. and Chen, Z.Y. 2005. Yangtze offshore, China: Highly laminated sediments from the transition zone between subaqueous delta and the continental shelf. Estuarine, Coastal and Shelf Science, 62(1–2), 161–168.

Wu, C. 2002. In: Hong, G.H., Kremer, H.H., Pacyna, J., Chen, C.-T.A., Behrendt, H., Salomons W. and Crossland, J.I.M. (eds.). East Asia Basins: LOICZ Global Assessment and Synthesis of River Catchment–Coastal Sea Interaction and Human Dimensions. LOICZ Reports & Studies No. 26 (LOICZ IPO, Texel, The Netherlands), pp 112–117.

Yanagi, T. and Akaki, T. 1993. Variations of the mean sea level around Japan. Umi no Kenkyu, 2, 423–430 [in Japanese].

Yi, S., Saito, Y., Oshima, H., Zhou, Y. and Wei, H. 2003. Holocene environmental history inferred from pollen assemblages in the Huanghe (Yellow River) delta, China: Climatic change and human impact. Quaternary Science Reviews, 22, 609–628.

Yim, W.W.-S. 1995. Implications of sea-level rise for Victoria Harbour, Hong Kong. Journal of Coastal Research, Special Issue, 14, 167–189.

Yim, W.W.-S. and Huang, G. 2002. Middle Holocene higher sea-level indicators from the South China coast. Marine Geology, 182, 225–230.

Yonekura, N. 1983. Late Quaternary vertical crustal movements in and around the Pacific as deduced from former shoreline data. In: Geodynamics of the Western Pacific: Indonesian Region. Washington: American Geophysical Union, 41–50.

Zhao, X., Tang, L., Wang, S. and Shen, C. 1995. Holocene climatic and sea-level changs in Qingfeng section, Jianhu County, Jiangsu Province: A typical example along the coastal areas of China. Journal of Coastal Research, Special Issue, 17, 155–162.

CHAPTER 6

HUMAN RESPONSES TO COASTAL CHANGE IN THE ASIA-PACIFIC REGION

PATRICK D. NUNN[1], CHARLES T. KEALLY[2], CAROLINE KING[3], JAYA WIJAYA[4], AND RENATO CRUZ[5]

[1] *Department of Geography, The University of the South Pacific, Fiji*
[2] *Department of Comparative Culture, Sophia University, Japan*
[3] *United Nations University, International Network on Water, Environment and Health, McMaster University, Canada*
[4] *Department of Marine Affairs and Fisheries, Indonesia*
[5] *Manila Bay Environmental Management Project, Department of Environment and Natural Resources, Philippines*

6.1 HUMAN-ENVIRONMENT INTERACTIONS IN ASIA-PACIFIC COASTAL REGIONS

As coasts have changed in the past – for whatever reason – this shift has commonly elicited a response from local inhabitants. Clearly the nature of the human response depends on both the nature and the magnitude of the change. Swifter and more profound changes, such as tsunami or tropical-cyclone (typhoon) effects, will generally have more immediate impacts on affected coastal populations. In contrast, a slower change of temperature, sea level, or even urbanization, for example, may produce only slow or delayed change as a human system's natural resilience absorbs short-term stresses.

The introductory section of Chapter 6 looks at the overall picture of human-environment interactions in the Asia-Pacific region. The section moves from the context of modern interaction – by far the most visible imprint on most coasts in this region – in section 6.1.1, through a discussion of early (section 6.1.2), late Holocene (section 6.1.3), and modern (section 6.1.4) interactions, to a brief consideration of future interactions (section 6.1.5). This chapter then goes on to describe the *changes* that took place along Asia-Pacific coasts within the period of human occupation (section 6.2) and to examine how humans *responded* to coastal changes (section 6.3). Section 6.4 highlights thematic and geographic areas that have been the focus of research and identifies research gaps. Finally, section 6.5 suggests research priorities for the future.

6.1.1 Constraints and Opportunities for Responding to Coastal Change along Contemporary Asia-Pacific Coasts

The degree to which human societies can rise to the challenges posed by environmental change is related to contextual factors that constrain or enable adaptation. In the Asia-Pacific region, the ability to adapt to coastal change will increasingly become a significant issue because large populations of people inhabit coastal areas. It is notable that one quarter of the world's 75 largest cities are located on or close to Asia-Pacific coasts (ESCAP 1995). Effectively responding to this important issue requires prioritisation of investments both geographically and thematically, based on an understanding of vulnerability and historical strategies for navigating change. This section explores how coastline length, overall land area, population density, and Gross Domestic Product (GDP) are useful measures for identifying the comparative vulnerability of contemporary Asia-Pacific countries to undesirable coastal changes (cf. Table 5.1). Subsequent sections review historical human responses to coastal change.

An understanding of the nature of human interaction with coasts in the Asia-Pacific region is helped by both the coastline length and the insularity index columns in Table 5.1. Coastline length is a crude, yet absolute, measure of coastal vulnerability, but it must be weighed against total land area, which can be seen as a rough measure of a country's resilience to undesired coastal change. In other words, coastlines are especially vulnerable to change due to their location at the intersection of the hydrosphere, the atmosphere, and the lithosphere. If a nation has ample land inland of the coast for relocation of vulnerable populations and activities, then that nation is more resilient to undesired coastal change than one without such an area inland. Thus, China's huge land area (more than 9 million km^2) offsets, at least in theory, some of the vulnerability associated with the vast length of the country's coast (14,500 km). Contrast continental China with archipelagic Palau. Palau's coastline length of only 1,519 km might suggest that the country's vulnerability to coastal change is lower than that of other nations within the region. However, since the total land area of Palau is just 458 km^2, it becomes clear that Palau is potentially far more vulnerable to the effects of undesired coastal change than China, irrespective of the significant differences in the coastline lengths of the two countries.

The insularity index in Table 5.1 allows this relationship to be specified more readily for comparative purposes. It is likewise a crude measure of a country's vulnerability to undesired coastal changes. Thus China has a value of less than 0.5 (insignificant for comparative purposes) while Palau has a value of 332, a reflection of both vulnerability and the archipelagic nature of the country. Thus in countries with a high insularity index, it is likely that almost all parts are effectively coastal. Such an index could better represent a nation's degree of vulnerability if it also accounted for other natural features that confer resilience to coastal change, like topography, lithology (rock type), and the presence or absence of offshore reefs.

Another measure of vulnerability is population density (Table 5.1). Countries with a high population density are generally more vulnerable than those with lower population densities. Thus the atoll nation of Tuvalu with a population density of 435 persons per km^2 has potentially more of a problem than the neighbouring atoll nation of Kiribati with a population density of 122 and a comparable insularity index. However, in most cases it is not enough to look at population density in isolation. For example, on the basis of population density alone (even when considered in conjunction with insularity index), Singapore might appear more vulnerable to coastal change than Bangladesh, which it is not. One reason for Singapore's greater resilience lies in its per capita GDP and understanding that Singapore apparently has the wherewithal to fund optimal solutions to undesired coastal changes, whereas Bangladesh probably does not.

The ability of a nation to implement the most appropriate solutions to problems of the coastline obviously depends on many factors, but as a crude comparative measure, the GDP per capita (Table 5.1) can be used. This approach assumes that the most appropriate solutions require expensive investments. In reality, there are many examples of solutions being found at low cost through the cooperation of coastal peoples, although there are legitimate questions about the sustainability of some such solutions. The comparative extent to which GDP enables countries to successfully respond to coastal change must be weighed against the size of potential challenges, as measured by the insularity index. Thus Singapore, with the second highest per capita GDP, has more of a problem (insularity index = 28) than some other continental countries in the Southeast Asia group (insularity indices = ≤ 1) even though these all have a considerably lower per capita GDP.

In terms of assistance required from external sources for coping with coastal problems, a good illustration is provided by Taiwan and Tokelau. Taiwan, with a reasonably high per capita GDP and an insularity index of 5, has a problem, but probably one it could – on the basis of the limited data in Table 5.1 – be expected to deal with without outside assistance. But Tokelau, with one of the lowest per capita GDPs in the region and by far the highest insularity index, clearly faces challenges that require outside assistance.

If we are to understand the modern context of coastal vulnerability and to make decisions about appropriate responses to particular problems, it is helpful to understand the history of human interactions with Asia-Pacific coasts.

6.1.2 Early Human Interactions with the Coast in the Asia-Pacific Region

Humans in the Asia-Pacific region have not always been attracted to the coast. There is abundant evidence that the first anatomically modern humans to see the Pacific Ocean regarded it as a barrier rather than an opportunity. These humans, having become accustomed over tens of thousands of years to hunting and gathering foods in inland areas, did not abruptly change their diet upon encountering the ocean. Even 30,000 years ago – perhaps 10,000 years after first seeing the Pacific coast – people in East Asian coastal regions apparently still favoured terrestrial foods (Ahn 1993). This lack of interest in Pacific coasts could be indulged while people still followed a nomadic lifestyle; but once they became more sedentary, as they did following the first domestications of plants and animals in East Asia, so they became less able to both ignore the opportunities provided by coastal living and avoid the challenges posed by coastal changes.

The first modern humans to see the Pacific coast were probably those who entered continental Southeast Asia or followed the broad East Asian river valleys of the Huanghe (Yellow River) and Changjiang (Yangtze) downstream about 40,000–50,000 years ago. Although they were adept at using fire, the nomadic lifestyles and low numbers of these people were such that any environmental impacts they may have had were minor and impossible to detect beneath the overprint of later human impacts.

In the history of humanity as a whole, two significant events occurred in the Asia-Pacific region. First, perhaps 40,000–50,000 years ago, humans succeeded in crossing ocean gaps at least 70 km wide from Southeast Asia (Sunda) to Australasia (Sahul). Besides the achievement of making such a crossing (Birdsell 1977), this event was significant because it introduced humans into a "naïve" environment where plants and animals had evolved over millions of years in the absence of large predators such as humans. Humans no longer had to compete with other large predators for the same food sources.

In Australasia, sufficient food was generally simple to find although significant differences in lifestyle subsequently developed between New Guinea and Australia. The nomadic lifestyle of native Australians, so well suited to the size of the continent and the vagaries of its climate, meant that there was little human impact on most of the Australian coastal zone until the arrival of Europeans. In New Guinea, as today, lowland areas were avoided by people whenever possible, perhaps because of the presence of malarial mosquitoes, and most of the early inhabitants lived in higher, cooler areas where they deliberately manipulated forest composition to boost food production as much as 35,000 years ago (Pavlides and Gosden 1994).

Second, the appearance of agriculture in lowland East Asia about 11,000 years ago may be the earliest such event anywhere in the world, although, agriculture appeared independently within a thousand years (or less) at the other centres in Central America and Mesopotamia (Mannion 1999). It is likely that agriculture in East Asia developed (as in the other centres) as a necessity following the start of the Younger Dryas cooling period (approximately 11,000–10,000 years ago) when insufficient food was able to be acquired for the resident population by hunting and gathering (Pringle 1998). It is likely that the vast alluvial lowlands of the Huanghe and Changjiang, then far greater in extent than at present owing to the lower sea level 11,000 years ago, became home to hosts of nascent agriculturalists during the Younger Dryas and for millennia afterwards. Once the Younger Dryas ended, and more climatically equable conditions returned to this region, agriculturalists began to produce surpluses, setting the stage for the development of cities and more complex societies by around 7,000 years ago (Yan 1993, Nunn 1999). The dependence of these cities on agriculture rather than either ocean-food production or sea trade meant that they were all located inland rather than along the coast.

The development of agriculture in lowland East Asia meant a reduced dependence on hunting and gathering as a means of food acquisition, and consequently the appearance of more sedentary societies. In areas of coastal lowland, sedentism is a more vulnerable lifestyle option than nomadism because river floods and/or storm surges can sweep through an area destroying the food resource base abruptly. During the early Holocene period (approximately 10,000–6,000 years ago) in lowland East Asia, this vulnerability was exacerbated by the rising sea level. It has become clear only quite recently that early Holocene (postglacial) sea level rise did not proceed monotonically, but included comparatively short-lived bursts of rapid sea level rise linked perhaps to the breaching of dams holding back large volumes of glacial meltwater, particularly in North America (Blanchon and Shaw 1995, Locker et al. 1996).

Figure 6.1 shows a graph of postglacial sea level rise depicting, in particular, these CREs (catastrophic rise events) and a theoretical model of associated settlement pattern changes in an unspecified part of the contemporary Huanghe-Changjiang coastal lowlands. Although there are three CREs shown during the period of postglacial sea level rise, it is considered that population densities in coastal-lowland East Asia were great enough only during the last CRE (CRE-3) for significant disruption to human settlement pattern to occur. Human settlements in Maps A and B are shown by filled squares each having its own sphere of influence (primarily for food gathering). Overlaps between spheres of influence at an earlier time would have led to a decrease in hunting and gathering and a concomitant increase in agricultural production.

It is envisaged (Map A) that around 8,000 years ago, the slowing rate of sea level rise and the gradually increasing river discharges (attributable to increasing precipitation and inland forest clearance) would have led to significant shoreline progradation around the mouths of the largest rivers. Human settlements would have spread out across these new coastal lowlands. Yet when sea level rose 6.5 metres in less than 140 years during CRE-3, these settlements would have been inundated (Map B) and their inhabitants moved inland. Here population densities were already high and also increasing because of the effects of CRE-3. The response of certain human groups was to increase their dependence on marine foods (represented by the extension of spheres of influence across the ocean in Map B), while other groups set out across the ocean on intentional voyages of colonization. It is contended that one or more of these enforced oceanic migration events around 7,400 years ago ultimately led to the colonization of tropical west Pacific Islands about 3,500–3,000 years ago by a distinctive group known as the Lapita people (see Case study 5.1), whose ancestry can be traced using genetics and linguistics back to the region of southern China and Taiwan (Green 1997, Young 2004).

It is ironic that our knowledge of human-coastal interactions in the Asia-Pacific region (and elsewhere in the world) is itself limited by coastal change – the fact that most of the information we require is below the ocean surface. We have only

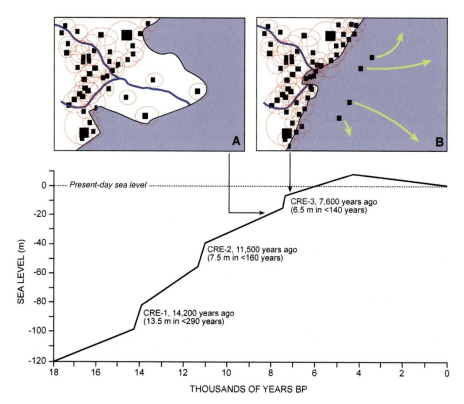

Figure 6.1. The general pattern of sea level rise in the Asia-Pacific region since the Last Glacial Maximum, which was around 18,000 years ago, shows three catastrophic (sea level) rise events (CREs) as determined from studies by Blanchon and Shaw (1995). (A) This map shows a plausible situation in an Asian continental delta around 8,000 years ago. Sedentary groups of agriculturalists, represented by black squares, and their spheres of influence are pictured along the back of a newly formed deltaic lowland. Even though sea level is rising in this area, the amount of terrigenous sediment entering the coastal zone offsets the inundation and erosion associated with this sea-level rise and builds a delta outwards. Pioneering groups of agriculturalists are starting to occupy this delta. (B) Following CRE-3, the recently formed delta is flooded and its inhabitants are displaced, together with many of the people located along the pre-delta shoreline. Overlaps between spheres of influence cause coastal communities to depend more on marine food resources, and leads some to migrate out of the area across the sea. It is this event that may have eventually led to the colonization of Pacific Islands

a very restricted portion of the evidence we need to understand the historical context of contemporary human-environment interactions along Asia-Pacific coasts. Yet it is clear that humans have not always regarded the coast as most regard it today – a desirable place to occupy. The following section describes how this attitude gradually became entrenched during the last few thousand years of the late Holocene.

6.1.3 Late Holocene Human Interaction with the Coast in the Asia-Pacific Region

Today we hear a lot about how sea level, which has been rising for around 200 years, will continue to rise during the 21st century and probably beyond. But sea-level rise is a relatively new phenomenon. Around 6,000 years ago sea level reached close to its present level in most parts of the Asia-Pacific region and subsequently exceeded its present level by between 1–4 metres before falling in the late Holocene (3,000–0 years ago) to its present level (Umitsu 1991, Somboon and Thiramongkol 1992, Dickinson 2001, Yim and Huang 2002, Boyd and Lam 2004). It is therefore really only for the late Holocene that we begin to find enough information to gain a complete picture of human-coastal interactions, and one must be cautious not to overemphasize the importance of events during this period just because, in contrast to earlier times, they are more readily visible and more easily intelligible. However, the existing evidence suggests that the effects of falling sea level during the late Holocene drove many human-coastal interactions.

In Chapter 5 (section 5.2.3), the changes wrought to many tropical coasts in the Asia-Pacific region by the absence of offshore coral reefs during the early and middle Holocene (10,000–3,000 years ago) were outlined. It was further explained that, when reefs finally caught up with sea level in the later Holocene, this occurrence brought about a fundamental change in the nature of many such coasts (see Figure 6.3). Whereas before reefs caught up with sea level, erosion had been the norm along many coasts, after reefs caught up and created lagoons, nearshore areas became shallower, shorelines stabilized, and in many places extended seawards. Not only did many coastal areas become more stable, but they also became richer in marine foods. These food resources would have come from lagoons, reef surfaces, and from the mangrove forests which extended both laterally and seawards along the shallowing soft-sediment coasts in many tropical parts of the Asia-Pacific region around 3,000 years ago. For many Asia-Pacific peoples, this change presented opportunities for new, easier lifestyles centred around marine food consumption and, as a direct consequence, coastal habitation.

So great, perhaps, was the change in lifestyle from inland to coastal during the late Holocene that the desire to find unexploited areas of rich marine foods may also have been a factor that drove people to colonize the islands far from the western borders of the Pacific Basin at this time. For tropical Pacific Islands, it has been argued that initial colonization from the western margins of the region was a deliberate process by people, perhaps cognizant of the recently increased attractiveness of distant island coasts (Nunn 1988), and that their eastward colonization continued in stages, the timing of which can be explained by variations in the timing of the total exposure of coastal flats by sea level fall (Dickinson 2003).

In general, as sea level fell during the late Holocene throughout the Asia-Pacific region, coasts were increasingly favoured for settlement because of the bounty of marine foods – comparatively easy to obtain – that they offered. But this situation did not continue everywhere, and the most important causes of disruption of the human occupation of coasts in pre-modern times appear to have been extraneous and region-wide, rather than internal to particular societies and localized. Sea-level change continued to be the major cause of disruption of coastal settlements just as it is likely to be in the future.

During the "AD 1300 Event" (see Case study 6.3 below), it was sea level fall that, by lowering water tables on coastal plains and exposing (and killing) the most productive parts of coral reefs, led to a massive and sudden reduction in the food resource base for coastal dwellers in some parts of the Asia-Pacific region. Many responded by violent competition for the remaining food resources, particularly on smaller, more remote islands where these resources were visibly limited, and by abandoning undefendable coastal settlements for the safety of mountain-top villages, caves, and smaller uninhabited offshore islands (Nunn 2000a, 2003). Most such coastal areas were only re-occupied some centuries later, around the same time new people arrived. An understanding of this situation provides the backdrop to modern human interactions with the coast in the Asia-Pacific region, where new demands on coastal environments have overprinted what was, in many places, only a fairly recently established settlement pattern.

6.1.4 Modern Human Interaction with the Coast in the Asia-Pacific Region

Modern human demands on the coasts of the Asia-Pacific region are quite different in both type and magnitude compared to those of just a few hundred years ago. In the past, most people who occupied the coast subsisted from the foods they found there, and whenever this strategy proved impossible, coastal dwellers would move elsewhere. Today, while subsistence dwellers live in many parts of the Asia-Pacific coastal zone, its ability to sustain human life has been compromised by the huge numbers of people who have flocked to live there. The growth of some of the world's largest cities has seen decreased productivity of the immediate coastal zone, increased demand for food from the peripheral coastal zones, and changed demands on the physical fabric of the coastal zone, which range from infrastructure development and building, to commercial agriculture, logging, mining and aquaculture. As discussed in section 6.3.4 below, the effects of these local stressors on the coastal environment can only be properly understood by also looking at the regional/global stressors, such as global warming and sea-level rise, that also impact all Asia-Pacific coasts.

The political development of the Asia-Pacific region over the past century has resulted in large, in some cases, archipelagic nations as well as a considerable number of smaller island states. Despite their administrative and demographic differences, all countries in the Asia-Pacific region (defined in Chapter 5) are vulnerable to a range of coastal hazards (Solomon and Forbes 1999; Huang 1997), as follows:
- tidal waves;
- flooding;
- shoreline erosion;
- wind damage during tropical cyclones;

- accelerated sea-level rise;
- increased saline intrusion;
- sediment transport processes;
- reef degradation due to storm damage, coral bleaching, disease and predation;
- sedimentation on the reef caused by storm runoff (exacerbated by clear-cut logging), reef blasting and mining;
- beach mining; and
- pollution and physical damage by heavy recreational use.

Sea-level rise during the past 100 years or so has posed considerable problems for many Asia-Pacific coasts, particularly for those that are low-lying and vulnerable to storm surges. The coastal-river lowlands of Bangladesh, for example, have been progressively inundated and subjected to storm surges reaching increasingly farther inland (Huq et al. 1995, Nunn 2004a). Through the Bruun Effect, recent sea-level rise has also caused erosion of most soft-sediment shoreline in the region; in Peninsular Malaysia, for example, 73% of the coast is classified as eroding (Midun and Lee 1995). Beach erosion is a problem throughout the region (Sato and Mimura 1997).

The islands of the Pacific Ocean are generally smaller and more remote than the Asian islands along its western continental margins, and these attributes accentuate their vulnerability to coastal change. During the last 100 years or so, many Pacific Island coasts have experienced significant land losses with serious consequences for their inhabitants (Kaluwin and Smith 1997). This situation has served to highlight the vulnerability of low atoll islands (such as those in Kiribati, Marshall Islands, Tokelau and Tuvalu; Roy and Connell 1991) but has also, paradoxically, diverted attention away from the problem in higher-island nations which are erroneously, yet popularly, perceived as being less threatened by future sea level rise (Nunn 2004b).

Yet the most manifest impacts of this sea-level rise have been on low-lying coasts, especially those where problems associated with sea-level rise have been exacerbated by development. Examples include coastal areas where the magnitude and frequency of river flooding has been increasing because of land-use changes (especially logging) in the upper parts of catchments, and coastal areas where groundwater extraction has caused coastal lowlands to subside thereby making larger areas more vulnerable to inundation and storm surges.

Improved understanding of coastal history and coastal process regimes during the 20th century have led to answers about questions concerning shoreline protection from undesired changes. Of particular concern are the shorelines of the many densely populated parts of the Asia-Pacific region where calls to halt undesired changes have become progressively louder. The problems are typified by Indonesia (Case study 6.1 below).

CASE STUDY 6.1 INDONESIA

Indonesia is a nation of 17,504 islands in which ocean comprises more than 60% of its total territory of 5.9 million km^2. As an archipelagic state and the world's fourth most populous country, Indonesia's coastal and marine

CASE STUDY 6.1 (Continued)

environment play a key role in terms of habitation, living and non-living resources, and other services. More than 60% of Indonesia's 210 million people live in the coastal zone which is less than 40% of the country's total area. More than eight million people depend on marine-based economic activities. The most pressing issues associated with the development and utilization of coastal and marine resources are deterioration of coastal habitats and their uncontrolled exploitation.

Economic growth, increasing (coastal) population, technology transformation, and urbanization are the principal drivers of recent change in Indonesia's coastal and marine environment. For two decades, from 1975 to 1995, Indonesia's rapid economic growth of 8% annually triggered excessive and unsustainable coastal resource use. The combination of high economic growth with other factors, such as rapid urbanization and population growth, exacerbated the vulnerability of the coastal zone and marine environment. The origins of the pressures on Indonesia's coastal zone and marine environment lie in past land-based development strategies focused on economic growth not environmental sustainability, uncoordinated development in coastal zones, and the institutional complexity that frustrated appropriate top-down responses. The result is an alarming degree of disturbance to coastal and marine resources that inhibits sustainable development in the future. Of particular concern is overfishing (especially of certain species) in some parts of Indonesia's fishing grounds, physical and chemical damage to coral reefs, the dwindling areas of mangrove forest, and coastal and marine environmental pollution.

The total area of coral reef in Indonesia is approximately 85,200 km^2, of which 40% is severely damaged, 24% considerably damaged, and only 6% in good condition. The degraded condition of coral-reef ecosystems in Indonesia is comparable to other countries in Southeast Asia.

Again, as elsewhere in this region, mangrove forests have been a conspicuous casualty of increased demands on the coastal zone of Indonesia within the past few decades. The mangrove forests of Kalimantan (Borneo) decreased in area by 81% in 21 years, from 443,450 hectares in 1980 to only 80,308 hectares in 2001. In Sumatra, the area of mangrove forest declined by 63% in the same period and covered an area of 285,598 hectares in 2001 (Spalding et al. 1997).

The importance of properly functioning institutions for regulating human impacts on coastal and marine environments in Indonesia cannot be overstressed. While many modern problems can be attributed to the institutional chaos of past times, the beginning of the "decentralization era" in 1999 has not yet brought about the widespread improvements in coastal and marine environments that were hoped for. If anything, decentralization has compounded the problems these environments face because of the uncoordinated and piecemeal response of the decentralized institutions (Gellert 1998). One critical difficulty is that terrestrial environmental management has been separated from coastal and marine environmental management, a difficulty that must be overcome if institutions are to play a role in stopping poor environmental practices and ensuring future sustainability of Indonesia's coasts.

CASE STUDY 6.1 (Continued)

Figure 6.2. Yos Sudarso Bay in Jayapura–Papua is one of the eastern parts of Indonesia where the coastal and marine environment is still in good condition (photo: Jaya Wijaya)

Finally there is a geographic contrast within Indonesia that needs to be understood. Western parts of Indonesia generally have an alarming rate of coastal and marine resources exploitation, which is linked to the driving forces such as population density, urbanization rate, and economic growth. Less affected is the eastern part of Indonesia (Figure 6.2) where the coastal and marine environments have a great opportunity to develop in ways that are economically productive yet environmentally sustainable and socially acceptable. Whether or not this happens depends on the institutions and strategies that are created and implemented.

6.1.5 Future Human Interaction with the Coast in the Asia-Pacific Region

The future of the Asia-Pacific region will clearly depend on many factors, both physical and human, but for most occupants of the region's low-lying coasts, it is sea-level rise and its direct consequences (such as shoreline erosion, groundwater salinization, increased mobility of nearshore sediments) that pose the greatest threats to their continued occupation at present (or higher) population densities. Bleak prognoses abound for low-lying Pacific Island nations, considered to be, perhaps, the most vulnerable to total erasure by 21st century sea-level rise (Roy and Connell 1991), but these potential consequences are clearly not the only–or, as measured by the numbers of affected people, the most severe–examples in the region. In Bangladesh, for example, predicted 21st century sea-level rise will cause the inland movement of the shoreline by many kilometres, causing hundreds of thousands of people to be displaced from the homes they currently occupy in the low-lying delta of the Ganges-Brahmaputra-Meghna (Huq et al. 1995). But that is not necessarily the greatest concern, for much of this part of Bangladesh is already vulnerable to storm surges amplified by being funnelled into the Bay of Bengal. These almost annual events can reach 160 km inland; an estimated 200,000 people drowned in the 1970 storm surge (Flierl and Robinson 1972). As sea level rises this century, so the inland reach of storm surges will increase.

A similar situation is found in China, the country with the longest coast in the region, despite around 70% of its length having a mountainous rocky character. There are eight coastal plains in China (including Taiwan) that are both considered highly vulnerable to change and host the region's largest cities (Figure 7.1). A key concern for the future is urban populations (also see Chapter 7 on Megacities). About 35% of the people living in the Asia-Pacific region are urban (compared to 43% globally). Thirteen of the world's 25 largest cities are in this region. By 2015, a projected 903 million people are expected to live in Asian cities with populations of more than 1 million (WRI/UNEP/UNDP/WB 1996).

However, just as in the past, future changes will elicit many and diverse responses from humans. Because solutions to projected changes may seem beyond reach today, does not mean that they will remain so, or even that novel solutions will not be found. It is with this end in mind that we discuss later in this chapter the principal gaps in research that exist and the priorities for future research into the issues confronting the coastal zone of the Asia-Pacific region.

6.2   COASTAL CHANGES WITHIN THE PERIOD OF SIGNIFICANT HUMAN OCCUPATION OF THE ASIA-PACIFIC COAST

As discussed above, it was the sea-level fall during the late Holocene that drew many humans to occupy the Asia-Pacific coast. In this section, the coastal (natural rather than human) changes that took place within the past 5,000 years in this region are discussed. A fuller description of coastal history (without an emphasis on humans) is given in Chapter 5.

In most parts of the Asia-Pacific region, the ocean surface (sea level) fell during the past 3,000 years from a maximum during the middle Holocene, variously 3,000–6,000 years ago (see above). The magnitude of this maximum varied between 1 and 4 m above present mean sea level, and so, therefore, did the coastal response to its subsequent fall in particular parts of the region.

A good example comes from the archaeological site of Nong Nor in the southeast part of the Bangkok Plain in Thailand (Boyd et al. 1996). The site was occupied first by people during the middle Holocene while sea level was still rising up to its Holocene maximum, about 4,600 BP here. The site was drowned and abandoned around this time, and it was subsequently reoccupied during the late Holocene when sea level was falling and it became dry land once more. In the early period of occupation, the site was on the side of a river estuary that provided a great diversity of foods (including shellfish, estuarine and open-ocean fish) to its inhabitants. In the later period of occupation, the site had become incorporated into the Bang Pakong river floodplain, an environment that offered fewer lifestyle options to its inhabitants, who probably focused on growing wetland rice.

In the tropical Pacific Islands, coastal flats suitable for human settlement and associated fringing-reef flats, essential to human sustenance, came into existence at different times in different island groups because of variations in the course of late Holocene sea level fall (Dickinson 2003). A detailed example of the interaction between changing coastal landscapes and human occupation was given for the Lapita era in Fiji in Chapter 5 (Case study 5.1); the earliest inhabitants of these islands appeared to favour smaller islands offshore over the larger ones (cf. Figure 5.4).

As an illustration of coastal change within the period of significant human occupation of the Asia-Pacific region, the example of Japan is given (Case study 6.2). Japan is a good example because it has both been part of the contiguous Asian continent and an island archipelago, and it is affected by tectonic movements that obscure and complicate an understanding of the history of coastal change compared to elsewhere in the region.

CASE STUDY 6.2 JAPAN

Japan consists of four main islands – Hokkaido, Honshu, Shikoku and Kyushu – and 6,848 smaller to very small islands (Statistics Bureau 2004). There are two major chains of islands stretching away from the main four islands. The Izu-Ogasawara chain extends from the Izu Peninsula-Tokyo Bay coast of eastern Honshu almost to the Mariana islands. The Nansei Islands extend southwest from the southern end of Kyushu almost to Taiwan.

The west of the archipelago is bounded by the Sea of Japan and the East China Sea; the eastern side is washed by the Pacific Ocean. The warm Kuroshio (Black Current) flows along the eastern coasts, its Tsushima Current branch flowing into the Sea of Japan. The cold Oyashio current flows along the northern reaches of the islands until it meets the Kuroshio off the shores of northeastern Honshu.

The total coastline of Japan is about 33,889 km long (Statistics Bureau 2004) and is highly variable in nature. This coastline has expanded and contracted with the rise and fall of the sea level over the past 35,000–50,000 years of human settlement (see Chapter 5). The rise and fall of the sea has also caused changes in the food resources available in coastal waters and tidal flats of the islands. Tectonic movements have added further changes to the coastline. Changes in the climate over the past 50,000 years have led to temporal variations in terrestrial and marine biodiversity.

Any study of human responses to coastal change in Japan must consider the richness and diversity of both terrestrial and marine (especially nearshore) food resources. It is quite possible that humans, given a rich and diverse food base

CASE STUDY 6.2   (Continued)

inland, would choose to live inland rather than face the uncertainties associated with dwelling on or near the coast – higher winds, tsunami, rough seas during typhoons, shifting tides, and generally dangerous waters.

Japanese prehistory and history are divided into 8 broad periods – Palaeolithic, Jomon, Yayoi, Kofun, Ancient, Medieval, Early Modern and Modern. The Palaeolithic begins controversially somewhere between about 35 ka BP and 50 ka BP, with the possibility of earlier settlement of the islands still under discussion (Sato 2001). The oldest pottery in Japan – the diagnostic of the Incipient Jomon culture (see Case study 6.5) – is dated ca. 13 ka BP (ca. 16 ka cal yr BP), but pottery is not common until ca. 9.5 ka BP (ca. 11.2 ka cal yr BP) (Keally et al. 2003). The farming-based Yayoi culture began in the 10th century BC (calibrated radiocarbon) (Harunari et al. 2004: 73), and ended ca. AD 250. The transition from Protohistoric to Historic in Japan spans the Asuka (AD 552–646), Hakuho (AD 646–710) and Nara (AD 710–794) phases of the late Kofun and early Ancient periods. The Ancient Period is generally dated AD 710–1185, the Medieval Period AD 1185–1573, the Early Modern AD 1573–1967, and the Modern Period from AD 1868 to the present.

The sea level is thought to have fallen below its present level from shortly after the Last Interglacial ca. 120–130 ka BP (see Chapter 5). Estimates for the maximum depression of the sea level at the Last Glacial Maximum (20–17 ka cal yr BP) range from -80 m to -140 m, with -100 m usually taken as the best overall estimate (Naruse 1981; Umitsu 1991). The sea level then rose to about -20 m, but dropped again to -40 m ca. 11–10 ka cal yr BP (the Younger Dryas–see Section 6.1.2 above). The sea level then rose slowly from ca. 10 ka cal yr BP to about 8.0 ka cal yr BP. Between 8.0 ka cal yr BP and 6.5 ka cal yr BP the sea level rose rapidly at about 13 mm/yr (Naruse 1981) to 15–20 mm/yr (Umitsu 1991). The maximum sea level and maximum sea transgression, occurred ca. 6.5–5.0 ka cal yr BP, reaching 2–5 m higher than present. This event was followed by one or more minor regressions to 2–3 m below present sea level between the Middle Jomon Period and the Yayoi Period (Umitsu 1991). Later minor fluctuations in the sea level are also reported (Umitsu 1991).

The Japanese islands are in constant motion because of their location along a major convergent plate boundary. Raised beaches are one of the most conspicuous manifestations of uplift relevant to the islands' human history. For example, Kikuchi (2001) found a Late Pleistocene beach dated ca. 52 ka BP is now 74 m above present sea level, although at the time it was the beach it would have been about 28 m below the present sea level. This raised beach is on the southern tip of the Boso Peninsula in Chiba Prefecture just east of Tokyo. The northern end of that same peninsula has been sinking over time.

The exposure and drowning of former land connections (land bridges) between Japan and the Asian mainland is an important aspect of coastal change relevant to human history. During most periods of lowered sea level, the Japanese main islands Honshu, Shikoku, and Kyushu are joined by land into a single island islands Honshu, Shikoku, and Ktushu are joined by land into a single island called Hondo. Geologists and palaeontologists today seem to agree that Hondo was connected to the continent three times in the Pleistocene: ca. 1.0 Ma, ca. 630 ka, and ca. 430 ka (the latter two dates sometimes appear as 500 ka and 300 ka) (Dobson and Kawamura 1998, Kawamura 1998, 2001). Hokkaido was connected to Honshu via land at 630 ka and 430 ka, but not at the Last Glacial Maximum. However, at the Last Glacial Maximum, a number of large mammals did cross the narrowed Tsugaru Strait from Hokkaido to Honshu, probably via an "ice bridge" that formed at least part of the year. But to the north, Hokkaido was connected to the continent via Sakhalin Island from ca. 100 ka to ca. 10 ka cal yr BP. Kawamura (1998) says hat the northern Nansei Islands (Okinawa and northward) were not connected to either Hondo or the Asian continent during the Pleistocene. But the southern Nansei Islands, the Miyako group, were connected to Taiwan and the continent in the Middle and Late Pleistocene.

The rise and fall of sea level, combined in many places with tectonic movements, opened and closed landbridges, but they transformed Japan's coasts in other ways. At the time of the Last Glacial Maximum, Japan was a much larger landmass than it is today, but it possibly had a simpler, and shorter, coastline. In contrast, at the maximum Holocene transgression, the sea invaded much of the country's coastal lowlands, making Japan a smaller country, but one with a more complex and possibly longer coastline than it has today. The regression that followed the maximum transgression expanded Japan's coastal lowlands both by removing the water that had inundated low coasts and river floodplains and by depositing large quantities of alluvium in many coastal regions.

Esaka (1967) shows the maximum extent of the Holocene transgression in the river floodplains in the eastern part of the Kanto Plain near Tokyo. On the Ara River, the sea reached about 45 km inland from the northern end of the present Tokyo Bay. On the Edo River, the sea invaded 60 km, possibly 70 km, inland. And on the Tone and Kinu rivers the transgression reached as much as 100 km up the present river valleys from the Pacific coast. No exact number is given, but Esaka's map suggests that as much as 30% of the present land area of the Kanto Plain was under water 5,500 years ago, and the coastline was more convoluted and longer than it is today.

As the sea began to recede after the maximum transgression, deltas and beaches began to expand. The literature discusses a number of these in detail; here are two examples.

The Nobi Plain at Ise Bay is a delta about 45 km long and 30 km wide. At the Last Glacial Maximum, this area was exposed. During the Holocene transgression, sediments were deposited in the bay that formed where the plain is today (Yamaguchi et al. 2003). As the sea began to recede after ca. 5.9 ka cal yr BP, the present delta began to expand. The oldest and deepest sediments (deeper than -35 m at 9.5 ka cal yr BP) on the Nobi Plain are freshwater deposits. The

CASE STUDY 6.2 (Continued)

sediments from −31 m to −21 m (9.0–5.0 ka cal yr BP) are marine deposits, but the fresh water sediments gradually increase after that depth and age. At 5.9 ka cal yr BP, the delta front was inland 32 km and extending at a rate of 6 m/yr. At 4.2 ka cal yr BP this rate increased to 10 m/yr. At the end of the Jomon Period and the beginning of the Yayoi Period, ca. 2.8 ka cal yr BP, the delta front was extending at a rate of only 5 m/yr, but it was still 9 km inland from today's coast. The average annual expansion rate of the delta front during the last 3,000 years of the Jomon period extended that delta front about 275–280 m in the 35-year lifetime of a Jomon individual. This would not have gone unnoticed by the Jomon people.

The Kujukuri coastal plain is located on the northeastern Pacific shore of the Boso Peninsula in Chiba Prefecture, just east of Tokyo (Matsuda et al. 2001). This plain is about 60 km long and about 10 km wide. This entire area was flooded at the maximum Holocene transgression. As the ocean began to recede, natural levees and sand dunes developed in bands that are today increasingly younger from inland to coastal areas. There are wetlands – lagoons and floodplains – among the levees and dunes. The beach front prograded at an average rate of 1.4–1.6 m/yr, varying from 3.5 m/yr to 0.8 m/yr. It is likely that any human sites on this plain before the transgression were obliterated as the seas invaded. During the last three millennia of the Jomon Period, individuals would have seen the beach (shoreline) move inland more than 50 m in their lifetime.

Similar studies have been conducted in other regions of Japan. For example, the Niigata Kosakyu Gurupu (Niigata Ancient Dune Research Group, 1974) studied the development of the large coastal plain on the Sea of Japan coast in Niigata Prefecture. The width of the plain varies but averages around 20 km. Judging from the archaeological remains found on the dune ridges, the Group concluded that the oldest ridges were formed in the Early and Middle Jomon Periods, the middle ridges in the Kofun Period, and the outer ridges in the Medieval Muromachi Period.

Soeda and Akamatsu (2001) studied changes on a smaller and more recent scale at Lake Saroma in eastern Hokkaido. The lake is separated from the sea by a narrow ridge of sand dunes. Minor rise or fall of the sea level caused changes in the salinity of the lake in both the late 10th to 11th centuries and the late 14th to late 16th centuries.

Within the past 5,000 years, changes in precipitation and in sea level have produced changes along tropical Pacific Island coasts (Figure 6.3). The middle Holocene in this part of the region was significantly wetter than the later Holocene, which had consequences for natural inputs of terrigenous sediment into the coastal zone. More importantly, the ocean surface fell from just over 2 m above its present mean level to its current level between about 4,000 and 1,000 years ago (Nunn 1995, Nunn and Peltier 2001). Within this period, many coral reefs, whose upward growth had lagged behind rising sea level during the earlier Holocene, finally reached the ocean surface and began to grow laterally. The dynamics of tropical Pacific Island coasts changed with the arrival of coral reefs at the ocean surface, both because their erosion began providing marine (biogenic) sediments to island coasts and because, by creating enclosed lagoons off island coasts, most sediment which entered the coastal zone remained within it – a contrast to the situation that prevailed before lagoon creation and enclosure. Thus, as seen in Figure 6.3, the contribution of terrigenous sediments to island shorelines decreased during the last 5,000 years as the climate became drier and marine sediment inputs more important.

Last-millennium (but pre-AD 1900) coastal changes in much of the Asia-Pacific region have proved difficult to isolate. This is both because of a lack of diagnostic dates and because of the massive disturbances to many coasts, particularly in the continental part of the region, within the past 100 years or so. Nevertheless it has been possible to identify sea-level changes within the last millennium for the Pacific Islands region (Nunn 2000c). The Medieval Warm Period (approximately AD 750–1250), or Little Climate Optimum, was a time when sea level was rising slowly (together with temperature and aridity). Most islanders occupied island coasts and depended on marine foods gathered from shore flats and nearshore reefs. A fall of sea level of typically 70–80 cm during the subsequent "AD 1300 Event" (approximately AD 1250–1350) led to the exposure of the most productive parts of these areas, causing a huge drop in the food resource base and creating a widespread societal crisis (see Case study 6.3). Sea level remained low during the ensuing Little Ice Age (approximately AD 1350–1800) and began rising only about 200 years ago. Twentieth-century rates of sea-level rise in this region averaged 1.0–1.5 mm a^{-1} (Wyrtki 1990), apparently slightly less than the global average of 1.5–2.0 mm a^{-1} (Miller and Douglas 2004).

Recent changes to Asia-Pacific coasts – defined as those that occurred since AD 1900 – were faster than most earlier changes and also had more profound and far-reaching effects on coastal-dwelling humans. The reasons are twofold. First, the sea level has been rising reasonably uniformly within this period causing progressively more coastal inundation, amplified during storms, and erosion of sandy shorelines throughout the region (Sato and Mimura 1997). Second, population densities along many Asia-Pacific coasts, especially urban coasts, have been increasing rapidly throughout this period (Figure 6.4) placing greater demands than ever before on already vulnerable coastal environments.

Twentieth century sea-level rise caused inundation along almost every coast in the Asia-Pacific region, exceptions being those very few coasts which appear to have risen faster than sea-level. Along subsiding coasts, the rate of relative sea level rise has been much greater (Table 6.1).

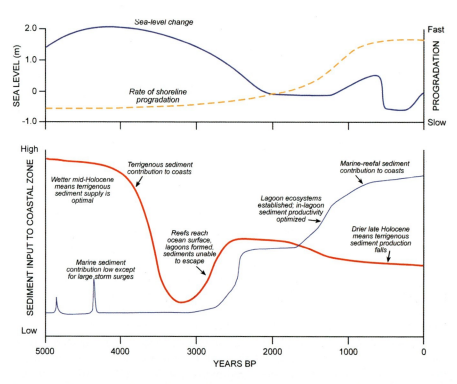

Figure 6.3. Changes in sea level and associated nearshore coastal processes over the past 5,000 years in the tropical western Pacific Islands (after Nunn 2005). Upper – Sea level changes in Fiji over the past 5,000 years (from Nunn and Peltier 2001). The suggested rate of shoreline progradation is also shown. Lower – Likely variations in terrigenous and marine sources of sediment to island coasts in Fiji over the last 5,000 years associated with changes in climate and reef-surface level

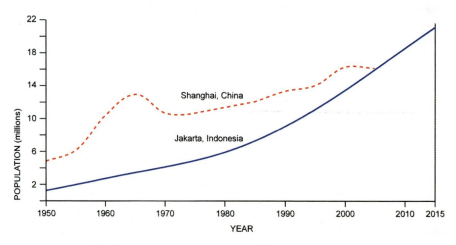

Figure 6.4. Recent and future population increases in Jakarta and Shanghai

Table 6.1. Selected rates of relative 20th century sea-level rise for the Asia-Pacific region

Location	Rate (mm a^{-1})	Duration	Source of information
global	1.0–2.0	1901–2001	IPCC WG1, 2001
global	1.5–2.0	1901–2001	Miller and Douglas, 2004
global	3.1	1992–1999	Nerem, 1999
ASIA			
China (general)	1.5–2.0		ESD-CSA, 1994
Korea (east)	0.57	long-term	Cho, 2003
Korea (south)	3.13	long-term	Cho, 2003
Korea (west)	2.64	long-term	Cho, 2003
Korea (east)	4.6	1992–2001	Cho, 2003
Korea (south)	4.8	1992–2001	Cho, 2003
Japan (all)	2.2	1955–1989	Uda et al., 1992
Japan (southwest)	4.0	1985–2000	Nakano et al., 2002
Japan (north)	2.7–15.3	1980–2000	Nakano et al., 2002
PACIFIC ISLANDS			
Honolulu, Hawaii	1.5	1901–1988	Wyrtki, 1990
Kwajalein, Marshall Islands	0.9	1937–1988	Wyrtki, 1990
Truk, Federated States of Micronesia	0.6	1947–1988	Wyrtki, 1990
Tutuila, American Samoa	1.4	1947–1988	Wyrtki, 1990
Viti Levu, Fiji (Naloto)	2.5	1919–1989	Nunn, 1990

6.3 INSTITUTIONAL RESPONSES TO COASTAL CHANGE IN THE ASIA-PACIFIC REGION

During the last 100 years, awareness of the effects of global environmental changes on the countries of Asia and the Pacific has grown. The countries of this region have progressively strengthened their capacity to respond to these issues at the international level and have harnessed regional cooperation in order to respond to shared environmental challenges.

6.3.1 Responding to Coastal Hazards

In many cases, human behaviour has been found to affect the severity of coastal hazards, for example through the emission of greenhouse gases and the destruction of forests (Arthurton 1998, Valencia 1981). The construction or floatation of settlements in vulnerable areas has also increased exposure to these hazards in the region. Thus some commentators have argued that coastal communities should prioritize mitigation rather than response to coastal change.

At a workshop on coastal area development and management held in Manila in 1979, James Mitchell recommended that the countries of Asia and the Pacific should adopt natural hazard management strategies to protect their coastal zones from the disasters brought on by the convergence of tectonic plates and the effects of tropical typhoons. Almost twenty years later, a growing awareness of global warming and its effects entered the problem definition. Most of the mitigation measures then proposed by Arthurton (1998) remained quite similar to the recommendations of Mitchell (1981) before the emergence of the issue of global warming (Table 6.2). Some reservations concerning the suitability of the engineering solutions had emerged, as coastal managers in the region learned from their experiences, and a greater emphasis on the use of scientific data and quantitative assessment methods was added.

An useful illustration of the problem of adaptation to climate-change vulnerability in Asia and the Pacific was offered by Yim (1996). For Hong Kong, this author found some difficulty with engineered mitigation measures, including raising the level of land on new reclamation areas and the construction of seawalls to protect against storm surge flooding and future sea-level rise, since these sometimes increased rainwater flood-risks. Rather he recommended improvements to drainage, land-use planning, and construction methods for the Hong Kong urban area. Yim also highlighted the importance of enhanced monitoring activities to detect changes and identify areas at risk.

In conducting his vulnerability and adaptation assessment for Hong Kong, Yim was able to draw upon research dating back to the early part of the century and beyond: instrumental temperature and precipitation records compiled at the Royal Observatory's Nathan Road Station since 1884 (with a gap during the Second World War), tide gauge information since 1954, and typhoon data collected since 1946, together with damage records for typhoons, landslides, floods, and droughts. Therefore, the empirical foundations for the response had been laid even earlier than the emergence of the current understanding of the problem of global change.

In general, whilst the larger states of the region face vast challenges in the protection of their lengthy coastlines, they have been able to build on their existing institutions, and capitalize on considerable economic benefits obtained from activities in the coastal areas to create the foundations for scientifically based coastal management (although see Case Study 6.1). In general, the Pacific Island States have faced greater institutional challenges, often having had to build environmental management systems almost from scratch (Table 6.3).

Table 6.2. Recommendations for natural hazard management

From Mitchell, 1979	From Arthurton, 1998
• land-use planning;	• long-term strategic planning, involving relocation and capital protective works;
• emergency preparedness programs;	• emergency planning;
• early warning systems; and	• effective warning networks; and
• engineering of protective structures (seawalls, dikes, levees, typhoon shelters, breakwaters).	• predictive capacity, data collection, monitoring, modeling, assessment of socio-economic costs and benefits of mitigation.

Table 6.3. Change in staffing levels of environment units in selected Pacific Island countries

Country	1990	1995
Kiribati	1	2
Fiji	3	7
Palau	3	6
Solomon Islands	4	5
Tokelau	1	5 (includes coordinators for each of the 3 atolls)
Tuvalu	0	1
Vanuatu	3	8
Western Samoa	5	12

Source: (Miles et al. 1995).

6.3.2 Evolution of Coastal Governance

Christie and White (1997) identify a series of phases in the development of coastal management in tropical countries, beginning with a Precolonial period, followed by the institution of centralized management by the colonialists. A similar development was presented by the World Resources Institute in their Reefs at Risk study (WRI 2001), observing that whilst centralized frameworks of natural resource management were introduced in many parts of Southeast Asia by colonial powers, many Southeast Asian nations instituted their own centralized styles of management even after they had gained independence.

Early coastal governance was driven by territorial and economic interests that emerged at the 1958 UN Conference on the Law of the Sea in Geneva. According to Hance (1995) and Hong and Chang (1997), the development of coastal management law and administration was influenced by a number of global phenomena, as follows:
- The development of advanced underwater systems for mineral exploitation on and beyond the continental shelf.
- The development of a system of property rights for fisheries management.
- Expansion of marine-navigation management systems based on global standards.
- Establishment of a permanent global ocean-atmosphere monitoring system.

The nascent institutions responsible for coastal management responded by creating a series of legal and other measures intended to protect the rights of particular nations to the economic benefits of coastal development and exploitation of the coastal shelf. The Chinese State Oceanic Administration was created in the 1960s. It is responsible for marine environmental research and monitoring and offshore oil development. In Indonesia, the establishment of institutions and activities to deal with coastal problems also began in the 1960s (Ongkosongo 1979).

Since the 1970s, the rapid growth of economies in the northeast Asian region has been accompanied by the establishment of a supra-national transport network (Hong and Chang 1997). In China from the 1970s, the government accorded development priority to "Special Economic Zones" in coastal areas. By 1992, such areas represented 13% of the country's land area and produced 54% of GNP. These developments demonstrated the value of the coastal zone and justified the resources for research to be carried out there. Whilst environmental protection was not always the primary objective of these development-driven changes in coastal management provisions, many important additions were made to coastal management systems during this period.

During the 1970s, scientific research institutes were established by many of the larger countries in the region. The Korea Ocean Research and Development Institute was established in 1973. In 1977, the National Environmental Protection Council of the Philippines (NEPC) was instructed to organize and coordinate an inter-agency task force to conduct research on vulnerable ecosystems including the country's coastal areas, and a Coastal Zone Management Committee was created (Zamora 1979). In Indonesia, the National Committee on Marine Research was established in 1978 to coordinate marine research activities including coastal studies (Ongkosongo 1979). In Thailand, the establishment of an Office of Coastal Land Development was approved by the National Cabinet in 1977 (Ruangchotivit 1979).

The confirmation by the IPPC in 1988 of the forecasted effects of climate change, the universal acceptance of Agenda 21 of the United Nations Conference on Environment and Development (UNCED) in 1992, and the entry into force of the United Nations Convention on the Law of the Sea in November 1994, led to institutional reforms to achieve integrated coastal management in many parts of the region. Within the past decade or so, specialist agencies have been created to handle coastal management issues. In Indonesia, the Environmental Impact Management Agency, a non-departmental Government Agency, developed a marine and coastal pollution control program to overcome negative impacts on the coastal and marine environment (UNEP 1997, Indonesia 1995). The Korean government established the Ministry of Maritime Affairs and Fisheries on 8 August 1996, integrating the ocean-related functions previously distributed amongst ten government authorities (Hong and Chang 1997). The Maritime Institute of Malaysia was created in 1996.

At the beginning of the 1990s, Paw and Chua (1991) documented the growing awareness of the potential effects on coastlines in Asia and the Pacific of an intensifying greenhouse effect that could cause profound climate changes and accelerate sea-level rise. They proposed strategies for the mitigation of sea-level rise impact including zoning and land-use management, erosion and flood control, water management, reinforcement of existing coastal structures, and waste management.

During this period, the internationally recognized theoretical discipline of integrated coastal management became established through international and regional efforts, and the development of advanced management techniques, including the use of geographical information systems and formal Environmental Impact Assessment, became incorporated into ICM practices.

Programmes for regional cooperation in coastal management introduced during the 1990s were also important developments. They include the Association of Southeast Asian Nations (ASEAN) Strategic Plan of Action on the Environment, the South Asia Co-operative Environment Programme (SACEP) Strategy and Programme (SPR-1, 1992–96), the UNEP Regional Seas Programmes, and the 2001 Asia-Pacific Economic Cooperation (APEC) Action Plan on Sustainability of the Marine Environment. The Asia Pacific Network (APN) has offered essential support to the development of coastal management research activities throughout the region.

6.4 UNDERSTANDING HUMAN RESPONSES TO COASTAL CHANGE: FOCAL AREAS AND GAPS IN RESEARCH

The nature of coastal change in the Asia-Pacific region was summarized in section 6.2 and was the subject of more detailed treatment in Chapter 5. The present section specifically looks at what is known about how humans in this region have responded to the changes experienced along its coasts. It should be noted that human impacts on Asia-Pacific coasts since AD 1900 were generally much greater than for earlier times and, in some places, have outweighed – in both magnitude and effects – the natural drivers of coastal change.

This section is organized into key thematic areas of research, key geographical areas of research, and a discussion of major challenges involved in responding to coastal change and the research gaps (thematic and geographical) that remain.

6.4.1 Key Thematic Areas of Research

Research into the responses of coastal dwellers in the Asia-Pacific region to coastal change has been mostly targeted at specific sites; commonalities between responses across the region have generally attracted less attention than local-area studies. The regional or subregional themes discussed in this sub-section begin with Case study 6.3, which explains the Pacific-wide responses of coastal people to a rapid sea-level fall around AD 1300. This is followed by accounts of two key coastal management issues in the Asia-Pacific region, namely the construction of artificial structures and the clearance (or conversion) of the mangrove forests that were present along many coasts in the region during the middle Holocene (see above). Two associated themes, coastal reclamation and future sea-level rise, are then discussed. The issue of tsunami impacts in Asia-Pacific countries is discussed separately.

CASE STUDY 6.3   RESPONSES OF PACIFIC PEOPLES TO THE AD 1300 EVENT

Around AD 1300, throughout the tropical Pacific Islands and continental Rim, there was a rapid cooling and sea level fall marking the transition from the generally warm dry Medieval Warm Period (Little Climatic Optimum) and the cool dry Little Ice Age. This so-called "AD 1300 Event" depleted food resources for many coastal-lowland dwellers to such an extent that massive and enduring societal disruption followed (Nunn 1999, 2000a, 2003).

During the Medieval Warm Period, most people living in the tropical Pacific, particularly on islands, occupied coasts and depended in large part on marine foods, especially those that were readily collectable from fringing reefs or lagoon-edge environments. The sea-level fall during the AD 1300 Event may have been as much as 80 cm in places, and this drop exposed the most productive, surface parts of coral reefs killing off the associated biota. This sea-level fall also decreased lagoonal circulation, making lagoonal environments more turbid and therefore less ecologically productive. The sea-level fall also lowered water tables along coastal plains depriving crops, which had been grown successfully in such areas during the Medieval Warm Period, of sufficient water. There is also evidence for an increase in storminess in some parts of this region, which had major impacts on water-conservatory infrastructure that had been created to cope with the aridity of the later part of the Medieval Warm Period.

The abrupt reduction in marine foods and in the supply of lowland crops led to a sharp reduction in available food for coastal dwellers in many parts of the Asia-Pacific region. This situation led to a crisis in many places as people were forced to compete for the remaining food resources. The associated conflict is most visible in the archaeological records from the Pacific Islands. On Easter Island (Rapanui), for example, the open lowland settlements used during the Medieval Warm Period were abandoned around AD 1300 in favour of caves, rockshelters, and offshore islands. Conflict became widespread; the famous stone statues (*moai*) were toppled and smashed, and the island's obsidian was mined for the first time to manufacture spearheads (*mataa*) (Bahn and Flenley 1992). In Palau, people abandoned the main islands after the AD 1300 Event in favour of the aptly named Rock Islands, where newly established clifftop settlements were readily defendable (Lucking and Parmentier 1990). Around this time, people abandoned their large coastal settlements on Viti Levu Island in the Fiji group and moved inland, establishing smaller settlements on mountain tops and beginning a period of conflict that lasted several hundred years (Field 2004, Kumar et al. 2006).

The AD 1300 Event provides an excellent example of how extraneous changes to environments can drive enduring societal change, which is a perspective that has been unpopular for a long time, but is beginning to find broad acceptance (deMenocal 2001, Haug et al. 2003, Nunn 2003).

In the past, especially in the times before some coastal dwellers built houses intended to "last", the most common response to coastal flooding or a similar calamity would possibly have been to move elsewhere and then rebuild. Yet where humans have invested a lot of money and/or effort in developing a particular coastal location, especially perhaps over a long period of time, it is natural to try to protect that location from the vagaries of nature. In many cases, responsible authorities have responded to pressures for coastal protection by investing in artificial structures. These range from precisely engineered seawalls (Figure 6.5A) to a variety of "ad hoc", often informal solutions (Figure 6.5B; see also Figure 6.8A). The limited success of many early versions of such structures, which have subsequently required additional investments typically to rebuild, realign, or build upwards, underlines the point that once such a structure is introduced to a particular coast, then that coast will need to be carefully managed by its human occupants thereafter (Kraus 1988).

Much research has been carried out into recent land-use changes along Asia-Pacific coastal zones although there is a general lack of synthesis and discussion of

Figure 6.5. Artificial shoreline protection structures vary immensely in variety and appropriateness throughout the region. Once an artificial structure is built it affects coastal processes along a coastline, which creates a responsibility for future management of that coast. [photos: Patrick Nunn]. (A) Seawall (funded by Japanese Government aid) along the west coast of Malé Atoll, Republic of the Maldives. This type of structure, comprising a line of tetrapods to absorb wave energy backed by a solid structure to protect the unconsolidated land behind, was designed with knowledge of the coastal processes operating here. (B) Seawall composed of loose coral rubble scooped up from the adjacent lagoon floor to fill an embayment along the lagoon coast of northern Aitutaki Island, southern Cook Islands. This type of protection will likely have to be renewed on a regular basis

common problems (and solutions) within the region. Clearance of mangrove swamps for various reasons (for aquaculture development, paddy rice cultivation, firewood collection) have rendered associated shorelines generally more vulnerable to wave erosion, particularly to storm surges, and have had negative impacts on commercial fisheries involving species that require mangroves as nurseries. More than 200,000 hectares of mangroves were cleared from the coasts of both the Philippines and Thailand between 1961 and 1993 (GESAMP 1993). The Chakaria Sundarbans of eastern Bangladesh – an important buffer for storm surges – have been almost completely removed for aquaculture (ESCAP 1995).

Coastal reclamation has been a theme for research and management only by richer countries with limited land areas in critical locations in the Asia-Pacific region. Hong Kong (China), Japan, and Singapore all provide excellent examples of such coastal reclamation. In Hong Kong, the apparent falling sea level during the 20th century encouraged the territory's planners to reclaim land, especially at the foot of the steep sides of Victoria Harbour (Figure 6.6). Compaction and settlement (of fill) of reclaimed areas in Hong Kong may exacerbate the effects of future sea-level rise (Yim 1995).

Another major theme in the Asia-Pacific region has been the effects of future (accelerated) sea-level rise on coastal communities (see also Chapter 5). Sea level by the end of the 21st century may lie as much as 90 cm higher than its 1990 level (IPCC WGI 2001). This will pose new and often immense challenges for coastal-zone managers at every level, and requires cooperation, financial assistance, appropriate technology transfer, and the development of innovative solutions. Much emphasis has been placed on the effects of projected 21st century sea-level rise on

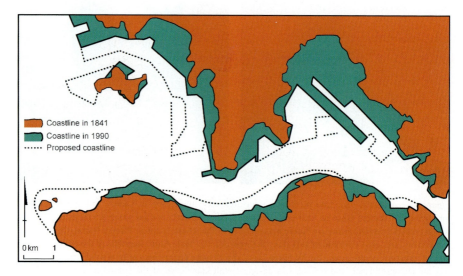

Figure 6.6. Map of Victoria Harbour, Hong Kong, showing changes in the position of the coastline. The change between in AD 1841 and AD 1990 was due to sea-level fall and reclamation. Further reclamation is proposed for port and airport development (after Yim 1995)

large cities and their support networks. Examples include the work on Shanghai and agricultural production in the Changjiang delta area associated with sea-level rise (summarized by Chen and Zong 1999) and the effects of reduced sediment supply to the Changjiang shoreline arising from completion of the massive Three-Gorges Dam project (Chen and Zong 1998).

While they occur only infrequently, tsunami(s) are major hazards in many coastal areas of the world. Capable of quite comprehensive destruction of people and their surroundings, tsunami have been studied at length. Most work has focused on either tsunami early-warning systems or hard solutions to tsunami impact along particularly vulnerable coasts. Early warning of tsunami clearly benefit large numbers of people in threatened areas, although most such systems cannot give adequate warning of locally generated (near-source) tsunami, only those which originate at great distances offshore. Tsunami-protection structures are often unsightly, restricting the appeal and usage of the coast, and expensive, commonly only affordable by the richer countries in the Asia-Pacific region. Anthropological studies of disaster (such as tsunami) impacts in less developed countries suggest that, contrary to popular belief, their societies have generally been better able to absorb the effects of disaster than those in more developed countries (Oliver-Smith 1996). Tsunami provide an excellent illustration of this point (see Case study 6.4) for, while invariably causing immense loss of life and often obliterating the means of livelihood for coastal dwellers, rarely succeed in fundamentally altering the lifestyles of coastal people or permanently displacing them from their coastal sites.

CASE STUDY 6.4   RESPONSE TO CATASTROPHIC COASTAL CHANGE: THE 1998 AITAPE TSUNAMI, PAPUA NEW GUINEA

The north coast of the island of New Guinea is especially prone to locally generated tsunami, and one affects part of this coast every 15–70 years. In most areas the recurrence time is many hundreds of years, preventing communal memory of tsunami (Davies 2002). On 17 July 1998, a locally generated tsunami swept across the Aitape coast killing around 2,000 people and destroying villages, infrastructure, and food resources (Figure 6.7). The survivors relocated to villages at the foot of hills 3–6 km inland, but they have since moved back to new villages in safer locations near the coast and, "We can imagine that with the passage of time . . . permanent settlements will be re-established on the coast" (Davies 2002, p 36).

This response to tsunami impact is similar to those of earlier such events. For example, the 1907 earthquake and subsidence that formed nearby Sissano Lagoon led the Arop people to move inland immediately after the event; however, a few months later they returned to their former coastal location (Welsch 1998). It is clear that along many Asia-Pacific coasts that have experienced tsunami, there is a communal memory within some long-established traditional coastal communities

CASE STUDY 6.4 (Continued)

that enables them to recognize the warning signs of tsunami and thereby minimize the impacts of these events by taking appropriate early action but this memory is insufficient to dissuade such communities from re-establishing themselves in tsunami-vulnerable areas.

Figure 6.7. Two views of devastation caused by the 1998 Aitape Tsunami (photos: courtesy of James Goff). Top – Hundreds of people used to live on this spit at Arop. Everything except a few palm trees was washed away in the tsunami. Bottom – View of the sand barrier from the lagoon that shows the low-lying surface of the spit. The bucket in the tree indicates the height attained by the tsunami wave

The known record of tsunami impact in the Asia-Pacific region was dwarfed by the 26th December 2004 Indian Ocean Tsunami that killed around 280,000 people, most in the region under study. This tsunami, which was generated by submarine landslides following an earthquake off the northwest part of Sumatra, led to apparently unprecedented amounts of inundation and destruction in Indonesia, India, Sri Lanka, and neighbouring countries.

6.4.2   Key Regional Areas of Research

Those who study the interactions between the earliest people in the Asia-Pacific region and its changing coasts have tended to focus on the continental margin, particularly areas like Japan where lifestyle changes were imposed on coastal people as islands were gradually created by flooding during postglacial sea level rise. Archaeologists studying the Palaeolithic in Japan focus on the cultural sequence and the first settlement of the Japanese islands. The existence or non-existence of landbridges and the first appearance of watercraft and over-water travel, are also aspects of the discussions dealing with the first settlement of Japan. Archaeologists seem to assume that all Palaeolithic coastal sites are now underwater, and probably also under deep accumulations of sediment, making consideration of how Palaeolithic humans might have responded to coastal changes an impossible question to approach. But this assumption ignores the fact that some of Japan's Pleistocene coast is now above the present sea surface because of uplift.

A good example of how Neolithic people in the Asia-Pacific region responded to coastal changes is provided by the study of the Incipient Jomon culture of Japan, discussed in Case study 6.5. The Incipient Jomon (approximately 13–9.5 ka BP or 16–11.2 ka cal yr BP) is conspicuous in Japanese prehistory because of the first appearance of pottery. The Incipient and Early Jomon evolved during a period of rapid coastal change, dominated by postglacial sea-level rise but including the Younger Dryas regression (see above). Much of the rest of the Jomon culture also evolved – and shows signs of having adapted to – rising postglacial sea level and associated land inundation, although its later parts (during the middle to late Holocene) show signs of having adapted to the falling sea level and the associated extension of coastal lands, particularly around river mouths.

CASE STUDY 6.5   THE JOMON CULTURE OF JAPAN

More than 50 years ago, studies of shellmounds in the Tokyo Bay area (Esaka 1943, 1954) showed that over time the shoreline, located by shellmounds, moved increasingly inland until about the end of Early Jomon about 5000 BP (Kobayashi et al. 2003), and then closer and closer to today's coast after that.

Studies of the earliest pottery in Japan suggest that its main use was related to fish processing (Taniguchi 2002). This oldest pottery is found in sites from central Hokkaido in the north to Kyushu in the southwest, with dates of ca. 13.0 ka BP (ca. 16.0 ka cal yr BP) (Keally et al. 2003). The Maeda Kochi site in

CASE STUDY 6.5 (Continued)

western Tokyo yielded thousands of salmon teeth – direct evidence of fishing even though such sites were often located far from the coast (Keally and Miyazaki 1986).

The quantities of potsherds in the earliest sites (ca. 16–11 ka cal yr BP) are extremely low (Keally et al. 2003, Taniguchi 2004), but quantities increase rapidly around the beginning of the Holocene, about the same time the first shellmounds appear. Nothing in the literature indicates the use of marine molluscs before the earliest shellmounds, suggesting a connection between use of marine molluscs and the increase in pottery in the sites.

Shellmounds of the Kanto Plain

Shellmounds are the oldest certain evidence of significant marine adaptations in Japan. The oldest of the shellmounds are found in the eastern Kanto Plain, where they date from ca. 9.5 ka BP (ca. 11.0 ka cal yr BP) (Keally et al. 2003). The Natsushima Shellmound (Sugihara and Serizawa 1957) is on the western side of Tokyo Bay and is still a coastal site. The equally old Nishinojo Shellmound is on the Tone River about 30 km from the present Pacific coast and about 50 km up the present river (Esaka 1967). The Hanawadai Shellmound is a few centuries younger and also on the Tone River about 50 km from the present coast and 70–75 km up the present river. Geological reconstruction of the coastline when these sites were occupied is poor, but they were probably near the heads of estaries that were encroaching further inland as sea level rose.

Shellmounds are not only oldest in the Kanto Plain region, but they are also by far the most numerous there (Oikawa et al. 1980). This situation is probably a direct result of the conditions in that region, particularly in the eastern part of the plain, which seem to have been the most favourable in all of Japan for humans to exploit marine resources.

The eastern Kanto Plain is a relatively flat dissected plain where, even in the early stages of the postglacial transgression, rich coastal environments would have existed at reasonably convenient distances from high, dry bluffs, where humans could live comfortably. Today, the Ara river floodplain, as far as 50 km upstream from Tokyo Bay, is only 10 m above sea level. At that point, the floodplain is 4–5 km wide from bluff to bluff, but most other river floodplains inundated by the transgression are considerably narrower. The other rivers and streams in the eastern Kanto Plain have only slightly steeper gradients such that very little sea-level rise would cause a sea transgression quite far inland. The sea would flood narrow valleys on the dissected plain creating a complex coastline with multitudes of small bays, estuaries, and coves. However, even as the postglacial transgression flooded lowlands further inland, the earliest bluffs with shellmounds would have remained convenient places to live, just as other bluffs farther inland also became convenient places to exploit the rich marine resources in adjacent tidal flats nearby.

Settlement patterns in the Murata Drainage Basin

There are very few studies of Jomon settlement patterns that are useful to understanding how the Jomon peoples responded to coastal change. Most of these studies, done by archaeologists, use too few sites to produce meaningful models.

One example is of Jomon sites in the small Murata river valley in Chiba Prefecture, just east of Tokyo (Koike 1986). The Murata river drains into the northeastern corner of Tokyo Bay and dissects a level upland, with a main valley about 10 km long and numerous short branch valleys, creating many peninsula-like headlands 20–40 m above narrow, marshy lowlands. There are 29 known Jomon sites in this small drainage area; 18 of these sites can be classified as shellmounds. During the Nojima to Kayama phases of the late Earliest Jomon Period (ca. 6.5–7.5 ka cal yr BP) there was only one shellmound in the drainage basin and four other small sites. The shellmound was on a bluff facing the bay. This timing precedes, or coincides with, the Holocene transgression maximum. The sea had evidently invaded the drainage basin creating a complex coastline consisting of an estuary and many coves with numerous tidal flats – targeted for shellfish – along much of the shore.

During the early part of the Middle Jomon period (late Holocene), the sea regression began and the coastline, now in the lowlands, became less convoluted. Human activity in the drainage increased considerably in the Kasori E phase in the later half of the Middle Jomon Period (ca. 4.0–4.5 ka cal yr BP). The Murata river drainage saw major human activity during the Late Jomon Period (ca. 3.0–4.0 ka cal yr BP). During this period, the regression advanced and the lowlands, or estuaries, became brackish swamps, and inter-tidal flats with sandy bottoms formed over a considerable area. In the Shomyoji phase at the beginning of the period, there were eight sites, none of them shellmounds. In the following Horinouch phase there were 19 shellmounds and only one other site. In the next Kasori B phase, there were 17 shellmounds and four other sites. The Latest Jomon Period (ca. 2.5–3.0 ka cal yr BP) saw only one site, and this occupation occurred in the first, Angyo IIIa phase of the period. This site was at the head of a branch valley; shellmounds are not reported.

Probably many factors influenced the changes Koike reports in the settlement patterns in this small drainage basin on Tokyo Bay. First, the scarcity of sites in the Early Jomon Period and the beginning of the Middle Jomon Period, and the abundance of sites in the Late Jomon Period, contradicts the population trends reported by Koyama (1978). But this period when sites were scarce does seem to coincide rather closely with the transgression maximum and immediate post-maximum, when the lowlands were flooded and likely not very productive in the resources the humans sought. Nevertheless, the increase in sites, especially shellmounds, in the late Middle Jomon and Late Jomon periods seems quite out of line with the known considerable decrease in local population in Late Jomon, even though the lowlands and sandy tidal flats that developed at this time certainly made the area more productive than at earlier times. The overall impression of Koike's (1986) report is

CASE STUDY 6.5 (Continued)

that the humans were responding to the coastal changes that occurred in the Murata river drainage area.

The Niitano Shellmound

The study of the Niitano Shellmound (Akazawa 1980) is a good example of what can be learned about human responses to coastal change when the details from one site are examined carefully. The Niitano Shellmound site is located on a bluff above the valley of the Isumi River in the highlands of the southern Boso Peninsula in Chiba Prefecture, about 10 km upstream from the present Pacific coast. The site was occupied in the Early and Middle Jomon periods. Akazawa treated these as two independent occupations for his analysis. Pottery types suggest about 1,000 years separation in the two occupations, the early one at about 6000 BP and the later one at about 5000 BP (uncalibrated radiocarbon ages). Based on the habitats of the molluscan and fish species found in the shell deposits of the two occupations, Akazawa concluded that (1) the Early Jomon occupation occurred about the time of the maximum Holocene transgression, at which time the site was probably on a bay or estuary, and (2) the Middle Jomon occupation probably coincided with the beginning of the post-maximum minor regression, leaving the site at the head of a bay or estuary near the mouth of the river.

The two most common molluscan species in the Niitano shell deposits were *Cyclina sinensis* Gmelin (brackish water) and *Corbicula japonica* Prime (riverine). In the Early Jomon shell deposits, these species made up 46% and 42% of the shells respectively, but, in the Middle Jomon deposits, *Corbicula japonica* Prime made up 99% of the shells. The fish species showed no change in time, perhaps because, despite changed habitats, most fish species continued to come close to the site at least seasonally.

Akazawa (1980) concluded that, when the environment/habitat below the site changed from Early to Middle Jomon, the people (female) changed the primary species of molluscs they collected in order to continue collecting very near the site, but, on the other hand, the people (male) continued to take the same fish species, although they probably had to go farther to catch them. Nevertheless, Akazawa found that all important mollusc and fish species used for food could be obtained within less that 10 km of the site. In other words, when the coast near the site changed, the humans simply changed the species of molluscs they took, in order to continue to live at the same location and in much the same way as before.

Conclusions

Studies of Jomon adaptations to the sea are common. These studies tend to concentrate on detailed analyses of fishing gear and macrofaunal remains from shellmounds. There are only a few studies of coastal settlement patterns, watercraft, and changes in seasonal food collecting activities. Studies of stable isotopes in human skeletons are becoming more common. The interpretations drawn from these studies of Jomon use of marine food resources are essential to understanding Jomon responses to coastal change.

Stable isotope studies reveal an unanticipated degree of variation in Jomon diets. For example, (1) the Jomon inhabitants of Hokkaido depended very heavily on sea mammals and some fish, with little use of plants, land animals, or molluscs; (2) the Honshu coastal Jomon peoples emphasized C3 plants (nuts) and land mammals with moderate quantities of fish, but only a few C4 plants (millets) and molluscs; (3) the Honshu inland Jomon peoples depended on C3 plants with relatively small quantities of C4 plants, meat, molluscs, and a few fish; and (4) the western Japan Jomon peoples used fish somewhat more than their eastern cousins (Chisholm et al. 1992, Minagawa 2005).

In considering the responses of people to coastal change since AD 1900, most research in the Asia-Pacific region has focused on its most densely populated parts. This generalization must, of course, be tempered by understanding that there is a difference in emphasis depending on who does the research.

- Most national government departments charged with managing a nation's coasts will focus on its most valuable areas. While these usually equate to the most densely populated areas, there are examples of less densely populated areas being targeted, particularly those valued for agriculture. In this context, it should be noted that there is immense variation among governments in the importance, indicated by the investment of resources, given to understanding and managing coastlines.
- Most regional organizations carry out research based on priorities set by their sponsors, and particular areas may be favoured over others.
- Finally, there are academic studies (such as this) which endeavour to give appropriate weighting to particular parts of the region depending on their relative importance, something most uncontroversially measured by population density.

This subsection examines three areas where attention has focused on understanding and planning human responses to coastal change – China, Southeast Asian lowlands, and urban coasts of islands in Southeast Asia and Japan. The latter is accompanied by Case study 6.6 dealing with the Manila Bay Coastal Strategy.

China

Comparatively speaking, a significant amount of research has been carried out on China's coast, mainly because it is so large, so densely populated, and facing so huge a challenge in terms of its sustained human occupancy. Some of the largest cities along the China coast – Guangzhou, Qingdao, Shanghai, Tianjin – were built close to the mouths of large rivers on deltaic sediments. Natural subsidence of these areas has been exacerbated both by the weight of buildings and infrastructure and by the extraction of groundwater to supply burgeoning human demands (Table 6.4). Upstream river diversion and dam construction also causes environmental problems

Table 6.4. Subsidence caused by groundwater extraction in parts of Southeast and East Asia

Location	Amount (m)	Dates of measurement (years AD)	Rate (mm a^{-1})
SOUTHEAST ASIA			
Bangkok, Thailand	0.7	1978–1988	70
Ramkhamhaeng, Bangkok Thailand	0.853	1978–1988	85.3
Manila, Philippines		1902–1962	2
Manila, Philippines		1963–1980	20
Davao, Manila, Philippines		1948–1967	2.1
Davao, Manila, Philippines		1968–1980	3.2
EAST ASIA			
Tokyo, Japan	4.0	1892–1972	500
Osaka, Japan	>2.8	1935–1972	76
Shanghai, China	1.64	1921–1992	23
	0.65	1921–1948	24
	0.28	1949–1956	40
	0.44	1957–1961	110
	0.21	1962–1965	69
	0.07	1966–1992	2.5
Niigata, Japan	>1.5	–	–
Tanggu Harbour, China	1.1	1975–1991	68
Tianjin, China	0.83	1975–1991	52

Data from sources in Goudie (1995) except for Shanghai (data from Wang, 1998), Bangkok (UNEP, 2001) and Manila (Siringan et al., 2000).

by limiting sediment discharge and subsequently causing erosion at the river mouths where these cities are located.

In Shanghai, which with a total population of 16.74 million (2000 census) is one of the largest cities in the Asia-Pacific region, land subsidence as high as 287 mm a^{-1} was recorded between 1957–1961, but has since been significantly checked by artificial recharge of aquifers. Yet relative sea-level rise (sea-level rise plus subsidence) of 6.5–11.5 mm a^{-1} continues to affect Shanghai and the surrounding area, particularly through saline contamination of groundwater and the effects of floods that can extend 170 km up the Changjiang River (Wang 1998). Although diked to heights of 3 m above mean sea level, the lower areas of Shanghai City are still flooded regularly. People have responded to this flooding by bricking up basement and street-level windows. Two elaborate tidal gates have been constructed to prevent flooding when storms surges travel up the Changjiang and Huangpu rivers (Han et al. 1995).

In Tianjin, located on the old Huanghe delta, subsidence has also been reduced to comparatively low rates, but there is little in the long term that can be done to reduce the relative sea-level rise which continues at rates of 4.5–5.5 mm a^{-1} and poses huge challenges for the future of Tianjin itself, for the development of the Shengli oil field, and for agriculture over a vast area (Han 1994). In the short term though, there have been numerous responses to threats of inundation from the ocean and from rivers (Yang et al. 2002). The seaward margin of this coastal plain is diked with a crest

2–3 m above mean sea level, enough to protect the delta plain from regular sea flooding, but not enough to prevent the passage of storm surges (Han et al. 1995).

Southeast Asian lowlands

A second regional focus of research has been the lowland areas at the heads of the large bays of Southeast Asia (Saito 2001), particularly the Chao Phraya (Bangkok) lowlands at the head of the Gulf of Thailand; the Ganges-Meghna-Brahmaputra lowlands, mostly in Bangladesh, at the head of the Bay of Bengal; and, with slightly less emphasis, the Red River lowlands (including the Vietnamese cities Hanoi and Haiphong) as well as the Irrawaddy delta, at the edge of which lies Burma's capital city of Yangon (Rangoon) (Tanabe et al. 2003, Thanh et al. 2002).

Urban coasts of islands in Southeast Asia and Japan

A third regional focus has been on islands in Southeast Asia (including the Philippines) and Japan where long-established coastal cities, in particular, are threatened by rising sea level and a host of other challenges associated with rapidly increasing populations. Examples of such areas include the Singapore metropolitan area, Jakarta (Indonesia), Manila (Philippines), Taipei (Taiwan), and Osaka and Tokyo (Japan).

A fine example of the type of initiative intended to ameliorate the relations between large numbers of people and a deteriorating ecosystem, on which they all depend, is the Manila Bay Coastal Strategy, discussed below in Case study 6.6. As an exercise in practical research, the Manila Bay Coastal Strategy exemplifies many of the factors that are driving similar practical research elsewhere. As an exercise in the implementation of research findings, it is a model for research elsewhere.

CASE STUDY 6.6   THE MANILA BAY COASTAL STRATEGY AND ITS DEVELOPMENT

Situated in the southwest portion of Luzon island, the largest island in the Philippines, Manila Bay (120°28' to 121°15' E; between 14°16' to 15°N) serves as the premier harbour centre of the country. Twenty-six catchments empty over 17,000 km^2 of watershed through seven major rivers scattered along the Bay's 190 km coastline.

Politically and administratively, the Bay region area is divided into thirty-three (33) coastal local government units (LGUs) and one hundred seventy-five (175) inland local government units. About 23 million people (or 30% of the country's total population) live in the surrounding Bay catchments.

Economically, the Bay region accounts for 55% of the Gross Domestic Product (GDP); the area is also responsible for a majority of economic contributions from industries (63%) and services (60%) and about 28% from agriculture. As such, developmental activities have brought overwhelming pressures on its resources and severely tested its natural resilience. These pressures include use conflicts, overextraction and overexploitation of its resources, pollution, habitat degradation, flooding, trans-boundary environmental issues, and garbage disposal. Presently,

> **CASE STUDY 6.6** (Continued)
>
> the institutional mechanisms and policy environment are inadequate to reduce these pressures. To emphasize the point, a class suit has been filed in a local court against the seeming negligence of several government and quasi-government offices in addressing the problems in the Bay.
>
> Accepting the challenge of managing Manila Bay as a "Pollution Hotspot" site, the Government of the Philippines initiated the development of a unifying framework – "The Manila Bay Coastal Strategy". The framework is based on sharing responsibilities, accepting that these are beyond the capacity of any single government agency, community, group, or individual.
>
> The process of developing the Coastal Strategy involved stakeholders in a long and comprehensive series of activities for consensus-building, review, analysis, consolidation, integration, revisions, validation, and, finally, endorsement/approval. The process stretched from August 2000 to October 2001 and culminated in a Manila Bay Declaration, which was presented to and accepted by the President.
>
> The four key components of the Strategy are PROTECT, MITIGATE, DIRECT, and COMMUNICATE. These embody the social, economic, and environmental aspects of Manila Bay in relation to its people's values, external and internal threats to its wellbeing, together with the actions required to improve this. The Coastal Strategy addresses the challenge by tasking every level of society with the responsibility of managing the Bay. This approach includes changes in attitude and behaviour by individuals and associations.
>
> Further, the strategy is based on several distinct features.
> - **Partnership** – The Strategy was developed and is implemented by stakeholders through partnerships;
> - **International conventions and agreements provide guidance** for actions within the framework;
> - **the National plans and programs**, which have consolidated sectoral initiatives, provide the direction and the forum for inter-agency and multi-stakeholder information sharing and decision making; and
> - **Capacity building of the stakeholders** is inherent in the strategy.
>
> Many difficulties and hard lessons were experienced in developing the Strategy. Despite the limited resources, the ever-sensitive political environment, and the complexity of issues in eco-governance over the Bay, its stakeholders remain resolute that the Strategy will be successfully implemented.

6.4.3 Challenges in Responding to Coastal Change

The major challenges (research gaps) in many parts of the region lie in rural and/or peripheral areas, particularly in those countries with lower GDPs and higher insularity indices that lack sufficient resources and personnel to effectively manage all parts of their coastline. The problem is often exacerbated in countries with highest insularity indices (see Table 5.1), a situation that complicates understanding of how coastal dwellers are interacting with coastal resources and ensuring sustainable resource use through the implementation of appropriate measures. An obvious gap is also the ineffectiveness of much problem-sharing and solution-formulating between the countries of the Asia-Pacific region and beyond.

Pacific Islands

In most rural parts of Southeast Asia, coastal problems can not be effectively addressed separately from issues of growing coastal populations and requirements for increased national food production and GDP. While most Pacific Island nations can be considered rural by some measures, such as population density (cf. Table 5.1), each consists of a center and a periphery. The remoteness of the periphery from the center is often greater than might be expected on the basis of distance, owing to the archipelagic character of most Pacific Island nations. It is common, for example, to find infrastructure and telecommunications of a high standard on (part of) the main island, while these are often rudimentary or absent on most other islands in an archipelagic Pacific nation.

Remoteness also has implications for the ways Pacific Island inhabitants interact with the coastal zone. In most rural (peripheral) parts of Pacific Island nations, people largely follow a subsistence lifestyle, which generally consist of a high dependence on marine foods and terrestrial crops. Key crops include cassava, taro, and yams that are grown wherever possible in lowland areas. The locations of most rural settlements reflect these lifestyle requirements, and preferred settlement locations are traditionally on coastal plains as close as possible to the ocean.

Sea-level rise during the 20th century has impacted both the location of settlements and the lifestyle of island residents. The changes caused by sea-level rise have been exacerbated by the removal of shoreline vegetation (particularly mangroves) and the degradation of offshore coral reefs (Nunn 2000b). In many rural communities, this habitat degradation has resulted from environmentally destructive activities, such as uncontrolled logging, fish poisoning, or reef mining, that provide short-term cash windfalls to local communities. However, the costs of these activities are evident in the increased vulnerability of coastal villages to sea-level rise and its consequences, notably erosion. For instance, many coastal villages occupy coastal plains that rise no more than two metres above high-tide level and are being eroded in many places by the effects of sea-level rise. Another problem associated with sea-level rise is the salinization of lowland groundwater, which is causing problems for many Pacific Island farmers, both subsistence and commercial. On atoll islands, the effects of sea-level rise are being expressed as shoreline erosion and by the increasing range and power of storm surges.

Most rural communities in the Pacific Islands have attempted to solve problems associated with recent sea-level rise by building artificial structures, notably seawalls (Figure 6.8A). In the vast majority of cases, these seawalls, while completed and "opened" with great fanfare, have deleterious effects on nearshore areas (and their ecosystems) and commonly collapse within about 18 months due to undermining by the sea. A cheaper and more effective long-term solution is planting mangroves along

Figure 6.8. Adaptation to erosion along an island shoreline at Yadua Village, Viti Levu Island, Fiji [photos: Patrick Nunn]. (A) The earliest response was to build a seawall using largely rock from the fringing coral reef. The seawall was undermined by wave erosion, collapsed, and rebuilt repeatedly until it became clear that this approach was not an effective long-term solution. This view shows part of the degraded seawall and the land behind it. Storm waves often impact this coast and erode the coastal flat on which the village lies. (B) The villagers are now replanting a mangrove forest along the worst affected part of the shoreline. This option is sustainable, effective and enhances the nearshore ecology. The main challenge with this strategy is that it may take 25 years for the mangrove fringe to reach maturity. Mangroves are grown in a nursery (pictured in middle ground) and then planted at regular intervals when they are about 80 cm tall

affected shorelines. A 30 m broad mangrove forest will absorb wave energy, effectively reverse shoreline erosion, and considerably enhance marine and littoral biodiversity with significant benefits for people occupying such coastal areas (Figure 6.8B).

Bangladesh

So much has been written about the environmental vulnerability of Bangladesh (e.g. Broadus et al. 1986, Mahtab 1989, Huq et al. 1995) that it may seem absurd to include it in a section on research gaps. Yet almost all that has been written outlines the manifest and extraordinary causes of its vulnerability to coastal change on both short and long time spans. Hardly anything has been written on how Bangladesh and its people are going to cope with the future. In this regard, the case of Bangladesh is typical of much that has been written on coastal vulnerability in the developing countries of the Asia-Pacific region. Much is known, future problems are predictable, but there are no workable blueprints for the future of these coasts. The same is true across the region, from Indonesia to Kiribati. The problems are twofold, they are physical and political, illustrated in the following paragraphs.

The modern coastline of Bangladesh is low, and many mangrove areas (the Sundarbans) have been cleared for aquaculture. Behind the coast, in the low delta of the Ganges-Meghna-Brahmaputra, live millions of subsistence farmers who are clearly unable to take any action (apart from moving to already crowded inland upland areas in the north of the country) to protect themselves from the short-term effects of storm surges and the longer-term effects of sea-level rise. They look to the government for help, but the government, understanding the magnitude of the problem in the context of its budget, can do little; "Clearly, Bangladesh is incapable of mitigating the effects of sea-level rise without external assistance" (Huq et al. 1995 p. 50).

One proposed action, suggested by Huq et al. (1995), is to build an artificial shoreline along the Bangladesh coast. This endeavour would involve a total embankment (seawall) length of 8685 km, including 4837 km of existing embankments. Total costs for the project in 1995 were around US$1000 million, an unrealistic figure for this nation (see Table 5.1).

The issue of Bangladesh and its future begs some troublesome questions. In the context of this book, the most pertinent enquiry is, "How can humans respond in the future to undesired coastal changes (sea-level rise, inundation, increased reach of storm surges) that seem impossible to prevent?"

Burma and North Korea

For two nations in the Asia-Pacific region – Burma and North Korea – there is comparatively little information available on the state of the coastal zone and the likelihood of a sustainable future for its inhabitants. Yet, given the other difficulties faced by these nations and their broadcast priorities, it is plausible to suppose that the coasts of these nations are currently being managed in unsustainable ways and that the future of their inhabitants looks bleak. Part of the problem may well be that the adaptation options open for these people are fewer than for their counterparts in other countries of the region. For example, there appears to be insufficient money available (from Table 5.1) to implement many costly solutions. Additionally, in the case of North Korea, restrictions on the movement of people within the country (and beyond) is likely to reduce both the number of adaptation options for coastal communities and the degree to which knowledge of shared problems can be exchanged (and collectively addressed) between affected communities.

6.5   Future Research Priorities

The lives and livelihoods of coastal populations in the Asia-Pacific region are vulnerable to a variety of natural hazards. Some of the threats faced are directly associated with the processes widely identified as aspects of global change, such as climate change, and associated sea-level rise and coral bleaching and death, sediment consolidation, shoreline erosion, increasing pollution, demographic changes, and the socio-economic effects of globalization. Others, such as increased flooding and storm damage, landslides, giant waves including tsunami, depletion of natural resources, and shortages of clean water are simply exacerbated by these processes. In the past, such threats have required the strengthening of existing coastal management systems, as well as the development of new research on the processes and sensitivity of coastal ecosystems, new coastal management partnerships, and a new understanding of the origins of current threats to the coastal environment – all developments that will have to continue for the foreseeable future.

REFERENCES

Ahn, S.M. 1993. Origin and differentiation of domesticated rice in Asia – A review of archaeological and botanical evidence. Unpublished PhD thesis, Institute of Archaeology, University College, London.
Akazawa, T. 1980. Fishing adaptation of prehistoric hunter-gatherers at the Niitano site, Japan. *Journal of Archaeological Science*, 7, 325–344.
Arthurton, R.S. 1998. Marine-related physical natural hazards affecting coastal megacities of the Asia-Pacific region – awareness and mitigation. *Ocean and Coastal Management*, 40, 65–85.
Bahn, P.G. and Flenley, J. 1992 *Easter Island, Earth Island*. London: Thames and Hudson.
Birdsell, J.B. 1977. The recalibration of a paradigm for the first peopling of Greater Australia. In: Allen, J., Golson, J. and Jones, R. (eds). *Sunda and Sahul*. London: Academic Press, 113–167.
Blanchon, P. and Shaw, J. 1995. Reef drowning during the last deglaciation: evidence for catastrophic sea level rise and ice-sheet collapse. *Geology*, 23, 4–8.
Boyd, W.E. and Lam, D.D. 2004. Holocene elevated sea levels on the north coast of Vietnam. *Australian Geographical Studies*, 42, 77–88.
Boyd, W.E., Higham, C.F.W. and Thosarat, R. 1996. The Holocene palaeogeography of the southeast margin of the Bangkok Plain, Thailand, and its archaeological implications. *Asian Perspectives*, 35, 193–207.
Broadus, J., Milliman, J., Edwards, S., Aubrey, D. and Gable, F. 1986. Rising sea level and damming of rivers: Possible effects in Egypt and Bangladesh. In: Titus, J.G. (ed.). *Effects of Changes in Stratospheric Ozone and Global Climate*, Volume 4, *Sea Level Rise*. Washington DC: UNEP and US Environmental Protection Agency, 165–189.

Chen, X. and Zong, Y. 1998. Coastal erosion along the Changjiang deltaic shoreline, China: History and prospective. *Estuarine, Coastal and Shelf Science*, 46, 733–642.

Chen, X. and Zong, Y. 1999. Major impacts of sea level rise on agriculture in the Yangtze delta area around Shanghai. *Applied Geography*, 19, 69–84.

Chisholm, B., Koike, H. and Nakai, N. 1992. Carbon isotopic determination of paleodiet in Japan: Marine versus terrestrial sources. In: Aikens C.M. and Rhee S.N. (eds), *Pacific Northeast Asia in Prehistory: Hunter-Fisher-Gatherers, Farmers, and Sociopolitical Elites*. Pullman, Washington: Washington State University Press, pp. 53–57.

Cho, K. 2003. Vulnerability of Korean coast to the sea level rise due to 21st global warming, Proc. International Symposium on Diagnosis, Treatment and Regeneration for Sustainable Urban Systems, Ibaraki University, pp.139–146.

Christie, P. and White, A.T. 1997. Trends in development of coastal area management in tropical countries: From central to community orientation. *Coastal Management*, 25, 155–181.

Davies, H. 2002. Tsunamis and the coastal communities of Papua New Guinea. In: Torrence, R. and Grattan, J. (eds), *Natural Disasters and Cultural Change*. Routledge: New York, 28–42.

deMenocal, P.B. 2001. Cultural responses to climate change during the late Holocene. *Science*, 292, 667–673.

Dickinson, W.R. 2001. Paleoshoreline record of relative Holocene sea levels on Pacific Islands. *Earth-Science Reviews*, 5, 191–234.

Dickinson, W.R. 2003. Impact of mid-Holocene hydro-isostatic highstand in regional sea level on habitability of islands in Pacific Oceania. *Journal of Coastal Research*, 19, 489–502.

Dobson, M. and Kawamura, Y. 1998. Origin of the Japanese land mammal fauna: Allocation of extant species to historically-based categories. *Daiyonki Kenkyu (The Quaternary Research)*, 37, 385–395 (in English with Japanese abstract).

Esaka, T. 1943. Minami Kanto Shinsekki jidai kaizuka yori kantaru Chusekisei ni okeru kaishin kaitai (Holocene sea transgressions and regressions as seen in the Neolithic shellmounds of South Kanto). *Kodai Bunka*, 14(4) (in Japanese).

Esaka, T. 1954. Kagansen no shintai kara mita Nihon no Shinsekki jidai (The Japanese Neolithic as seen from the changing coastline). *Kagaku Asahi*, 14(3) (in Japanese).

Esaka, T. 1967. Seikatsu butai (Context of livelihood). In: Yoshimasa Kamaki, (ed.), *Nihon no Kokogaku*, II, *Jomon Jidai (Japanese Archaeology*, II, *The Jomon Period)*. Tokyo: Kawade Shobo, pp. 399–415 (in Japanese).

ESCAP 1995. *State of the Environment in the Asia-Pacific, 1995*. Bangkok.

ESD-CAS [Earth Science Division, Chinese Academy of Sciences (eds.)] 1994, Impact of sea level rise on the deltaic regions of China and its mitigation. Beijing, China: Science Press, 355 p (in Chinese).

Field, J.S. 2004. Environmental and climatic considerations: A hypothesis for conflict and the emergence of social complexity in Fijian prehistory. *Journal of Anthropological Archaeology*, 23, 79–99.

Flierl, G.R. and Robinson, A.R. 1972. Deadly surges in the Bay of Bengal, dynamics and storm tide table. *Nature*, 239, 213–215.

Gellert, P.K. 1998. The Limits of capacity: The political economy and ecology of the Indonesian timber industry, 1967–1985. Unpublished PhD dissertation, University of Wisconsin, Madison.

GESAMP 1993. Joint group of experts on the scientific aspects of marine protection, impact of oil and related chemicals and wastes on the marine environment. Reports and Studies No. 50. IMO. London.

Green, R.C. 1997. Linguistic, biological and cultural origins of the initial inhabitants of Remote Oceania. *New Zealand Journal of Archaeology*, 17, 5–27.

Han, M. 1994. Impact of sea-level rise on the North China coastal plain and the cost-benefit analysis of the prevention measures concerned. In: ESD-CAS (eds). Impact of sea-level rise on the deltaic regions of China and its mitigation. Beijing: Science Press, 339–355.

Han, M., Hou, J. and Wu, L. 1995. Potential impacts of sea level rise on China's coastal environment and cities: A national assessment. *Journal of Coastal Research*, Special Issue, 14, 79–95.

Hance, D.S. 1995. The role of state in the technical and general management of the oceans. *Ocean and Coastal Management*, 29, 5–14.

Harunari, H., Imamura, M., Fujio, S., Sakamoto, M. and Kobayashi, K. 2004. Yayoi jidai no jitsunendai (The true date of the Yayoi period). *Nihon Kokogaku Kyokai dai-70-kai Sokai* (The 70th General Meeting of the Japanese Archaeological Association), resumes, 73–76 (in Japanese).

Haug, G.H., Günther, D., Peterson, L.C., Sigman, D.M., Hughen, K.A. and Aeschlimann, B. 2003. Climate and the collapse of Maya civilization. *Science*, 299, 1731–1735.

Hong, S.Y. and Chang, Y.T. (eds.) 1997. Special issue. Marine and coastal policies in East Asia. *International Journal of Marine and Coastal Law*, 12, 156 p.

Huang, J.C.K. 1997. Climate change and integrated coastal management: A challenge for small island nations. *Ocean and Coastal Management*, 37, 95–107.

Huq, S., Ali., S.I. and Rahman, A.A. 1995. Sea level rise and Bangladesh: A preliminary analysis. *Journal of Coastal Research, Special Issue*, 14, 44–53.

Indonesia 1995. BAPEDAL-The Environmental Impact Management Agency. Government of Indonesia.

IPCC WGI 2001. Climate Change 2001: The Scientific Basis. Cambridge: Cambridge University Press, 881p.

Kaluwin, C. and Smith, A. 1997. Coastal vulnerability and integrated coastal zone management in the Pacific Island region. *Journal of Coastal Research*, Special Issue, 24, 95–106.

Kawamura, Y. 1998. Daiyonki ni okeru Nihon retto e no honyurui no ido (Immigration of mammals into the Japanese islands during the Quaternary). *Daiyonki Kenkyu* (The Quaternary Research), 37, 251–257. (in Japanese with English abstract).

Kawamura, Y. 2001. Honyurui (Mammals). *Gekkan Kokogaku* (*Archaeology Quarterly*), 74, 36–40. (in Japanese).

Keally, C.T. and Miyazaki, H. 1986. A terminal Pleistocene salmon-fishing and lithic worksite at Maeda Kochi, Tokyo, Japan. *Current Research in the Pleistocene*, 3, 96–97.

Keally, C.T., Taniguchi, Y. and Kuzmin, Y. 2003. Understanding the beginnings of pottery technology in Japan and neighboring East Asia. *The Review of Archaeology*, 24(2), 3–14.

Kikuchi, T. 2001. Boso Hanto no chikei kara yomu Chu-Koki Koshinsei no kaisuijun to tekutonikusu (Some notes on Middle to Late Pleistocene topographic history inferred from the landforms of Boso Peninsula, Kanto District, Japan). *Daiyonki Kenkyu* (*The Quaternary Research*), 40, 267–274 (in Japanese with English abstract).

Kobayashi, K, Nishimoto, T., Imamura, M. and Sakamoto, M. 2003. AMS ^{14}C nendai ni yoru Jomon doki keishiki no henka no jikanhaba (The time span of Jomon pottery types according to AMS ^{14}C dating). *Nihon Kokogaku Kyokai dai-69-kai Sokai* (The 69th General Meeting of the Japanese Archaeological Association), resumes, pp. 29–32. (in Japanese).

Koike, H. 1986. Prehistoric hunting pressure and paleobiomass: An environmental reconstruction and archaeozoological analysis of a Jomon shellmound area. In: Akazawa, T. and Aikens, C.M. (eds), *Prehistoric Hunter-Gatherers in Japan: New Research Methods*. Bulletin no. 27. Tokyo: The University Museum, The University of Tokyo, pp. 27–53.

Koyama, S. 1978. Jomon subsistence and population. *Senri Ethnological Studies*, 2, 1–65.

Kumar, R., Nunn, P.D., Field, J.E. and de Biran, A. 2006. Human responses to climate change around AD 1300: a case study of the Sigatoka Valley, Viti Levu Island, Fiji. *Quaternary International*.

Kraus, N.C. 1988. The effects of seawalls on the beach: An extended literature review. *Journal of Coastal Research*, Special Issue, 4, 1–28.

Locker, S.D., Hine, A.C., Tedesco, L.P. and Shinn, E.A. 1996. Magnitude and timing of episodic sea level rise during the last deglaciation: *Geology*, 24, 827–830.

Lucking, L.J. and Parmentier, R.J. 1990. Terraces and traditions of Uluang: ethnographic and archaeological perspectives on a prehistoric Belauan site. *Micronesica*, Supplement 2, 125–136.

Mahtab, F.U. 1989. Effect of climate change anc sea level rise on Bangladesh. London: Commonwealth Secretariat, Expert Group on Climate Change and Sea level Rise, 230 p.

Mannion, A.M. 1999. Domestication and the origins of agriculture. *Progress in Physical Geography*, 23, 37–56.

Matsuda, F., Fujiwara, O., Sakai, T., Araya, T., Tamura, T. and Kamataki, T. 2001. Chiba-ken Kujukurihama heiya no Kanshinto no hattatsu katei (Progradation of the Holocene beach-shoreface system in the Kujukuri strand plain, Pacific coast of the Boso Peninsula, central Japan). *Daiyonki Kenkyu* (*The Quaternary Research*), 40, 223–233 (in Japanese with English abstract).

Midun, Z. and Lee, S.-C. 1995. Implications of a greenhouse-induced sea level rise: A national assessment for Malaysia. *Journal of Coastal Research*, Special Issue, 14, 96–115.
Miles, G., Fuavao, V. and Smith, A. 1995. Implementing Agenda 21: Oceans, coasts and the Barbados outcomes in the Pacific region. *Ocean and Coastal Management*, 29, 125–138.
Miller, L. and Douglas, B.C. 2004. Mass and volume contributions to twentieth-century global sea level rise. *Nature*, 428, 406–409.
Minagawa, M. 2005. Kodaijin wa nani o tabeteita ka: Hoppo senshijin no shokusei bunseki kara (What did the ancient peoples eat?: An analysis of the subsistence of prehistoric northern peoples). *Nihon Bunkazai Kagakukai Kaiho (Bulletin of the Japanese Society for Scientific Studies on Cultural Property)*, no. 49, pp. 7–16 (in Japanese).
Mitchell, J.K. 1981. Natural hazard management in the coastal zone: An overview. In: Valencia, M.K. (ed.), Proceedings of the workshop on Coastal Area Development and Management in Asia and the Pacific. Manila, Philippines, 3–12 December, 1979. Honolulu: Hawaii University Press, pp. 7–24.
Nakano, S., Tadokoro, S., Uno, K. and Fujiki, M. 2002. Long-term variation in the mean sea level and effects of global warming, Proc. Coastal Engineering, JSCE, 49, pp.1351–1355 (in Japanese).
Naruse, Y. 1981. Climate and sea level changes during the Quaternary. *Recent Progress of Quaternary Research in Japan*. Reprinted from: *Recent Progress of Natural Sciences in Japan*, vol. 6. Tokyo: Science Council of Japan, pp. 191–202 (in English).
Nerem, R.S. 1999. Measuring very low frequency sea level variations using satellite altimeter data, *Global and Planetary Change*, 20, 157–171.
Niigata Kosakyu Gurupu 1974. Niigata sakyu to jinrui iseki – Niigata sakyu no keiseishi I (Niigata sand dunes and archaeological relics – the geohistory of the formation of Niigata sand dunes, part I). *Daiyonki Kenkyu (The Quaternary Research)*, 13, 57–69 (in Japanese with English abstract).
Nunn, P.D. 1988. Recent environmental changes along south-west Pacific coasts and the prehistory of Oceania: Developments of the work of the late John Gibbons. *Journal of Pacific Studies*, 14, 42–58.
Nunn, P.D. 1990. Recent environmental changes on Pacific islands. *The Geographical Journal*, 156, 125–140.
Nunn, P.D. 1995. Holocene sea level changes in the South and West Pacific. *Journal of Coastal Research, Special Issue*, 17, 311–319.
Nunn, P.D. 1999. *Environmental Change in the Pacific Basin: Chronologies, Causes, Consequences*. London: Wiley, 357 pp.
Nunn, P.D. 2000a. Environmental catastrophe in the Pacific Islands about AD 1300. *Geoarchaeology*, 15, 715–740.
Nunn, P.D. 2000b. Coastal changes over the past two hundred years around Ovalau and Moturiki Islands, Fiji: Implications for coastal-zone management. *Australian Geographer*, 31, 21–39.
Nunn, P.D. 2000c. Illuminating sea level fall around AD 1220–1510 (730–440 cal yr BP) in the Pacific Islands: implications for environmental change and cultural transformation. *New Zealand Geographer*, 56, 46–54.
Nunn, P.D. 2003. Nature-society interactions in the Pacific Islands. *Geografiska Annaler*, 85 B, 219–229.
Nunn, P.D. 2004a. Understanding and adapting to sea-level change. In: Harris, F. (ed.), *Global Environmental Issues*. Chichester: Wiley, 45–64.
Nunn, P.D. 2004b. Through a mist on the ocean: Human understanding of island environments. *Tijdschrift voor Economische und Sociale Geografie*, 95, 311–325.
Nunn, P.D. 2005. Reconstructing tropical paleoshorelines using archaeological data: examples from the Fiji Archipelago, southwest Pacific. *Journal of Coastal Research*, Special Issue, 42, 15–25.
Nunn, P.D. and Peltier, W.R. 2001. Far-field test of the ICE-4G (VM2) model of global isostatic response to deglaciation: empirical and theoretical Holocene sea level reconstructions for the Fiji Islands, Southwest Pacific. *Quaternary Research*, 55, 203–214.
Oikawa, A., Miyamoto, S. and Koyama, S. 1980. Kaizuka detabesu – sono sakusei to oyo (A Jomon shellmound database: Its compilation and application). Kokuritsu *Minzokugaku Hakubutsukan Kenkyu Hokoku (National Museum of Ethnology Research Reports)*, 5(2), 439–470 (in Japanese with English abstract).

Oliver-Smith, A. 1996. Anthropological research on hazards and disasters. *Annual Review of Anthropology*, 25, 303–328.
Ongkosongo, O.S.R. 1979. Human activities and their environmental impacts on the coasts of Indonesia. In: Valencia, M.K. (ed.), Proceedings of the workshop on Coastal Area Development and Management in Asia and the Pacific. Manila, Philippines, 3–12 December, 1979. Honolulu: Hawaii University Press, pp. 67–74.
Pavlides, C. and Gosden, C. 1994. 35,000-year-old sites in the rainforests of west New Britain, Papua New Guinea. *Antiquity*, 68, 604–610.
Paw, J.N. and Chua, T.-E. 1991. Climate changes and sea level rise: implications on coastal area utilization and management in South-east Asia. *Ocean and Shoreline Management*, 15, 205–232.
Pringle, H. 1998. The slow birth of agriculture. *Science*, 282, 1446–1450.
Roy, P. and Connell, J. 1991. Climatic change and the future of atoll states. *Journal of Coastal Research*, 7, 1057–1075.
Ruangchotivit, T. 1979. A coastal land development project. In: Valencia, M.K. (ed.), Proceedings of the workshop on Coastal Area Development and Management in Asia and the Pacific. Manila, Philippines, 3–12 December, 1979. Honolulu: Hawaii University Press, pp. 102–106.
Saito, Y. 2001. Deltas in Southeast and East Asia: Their evolution and current problems. Proceedings of APN/SURVAS/LOICZ Joint Conference on Coastal Impacts of Climate Change and Adaptation in the Asia-Pacific Region, APN, Kobe, Japan, 185–191.
Sato, H. 2001. Nihon Retto no Zenki-Chuki Kyusekki Jidai o kangaeru – Fujimura-shi hikanyo shiryo kara no mitoshi (Considering the Japanese Early and Middle Palaeolithic from materials not associated with Fujimura). *Dai-15-kai Tohoku Nihon no Kyusekki Bunka o Kataru Kai – Yokoshu* (Resumes from the 15th Meeting for the Discussion of the Palaeolithic Culture in Northeastern Japan), held in Akita City, December 22–23, pp. 127–142 (in Japanese).
Sato, Y. and N. Mimura, 1997. Environmental problems and current management issues in the coastal zones of south and southeast Asian developing countries. *Journal of Global Environmental Engineering*, 3, 163–181.
Siringan, F.P., Maeda, Y., Rodolfo, K.P. and Omura, A. 2000. Short-term and long-term changes of sea level in the Philippine Islands. In: Global Change and Asia Pacific Coast, Proceedings of the APN/SURVAS/LOICZ Joint Conference on the Coastal Impacts of Climate Change and Adaptation in the Asia-Pacific Region, APN and Ibaraki University, pp 143–149.
Soeda, Y. and Akamatsu, M. 2001. Hokkaido tobu Saroma-ko shuhen-iki ni okeru 10 - 17 seiki no kaisuijun hendo (Sea level change from the 10th to 17th centuries around Lake Saroma, eastern Hokkaido). *Daiyonki Kenkyu (The Quaternary Research)*, 40, 423–430 (in Japanese with English abstract).
Solomon, S.M. and Forbes, D.L. 1999. Coastal hazards and associated management issues on South Pacific Islands. *Ocean and Coastal Management*, 42, 523–554.
Somboon, J.R.P. and Thiramongkol, N. 1992. Holocene highstand shoreline of the Chao Phraya delta, Thailand. *Journal of Southeastern Asian Earth Sciences*, 7, 53–60.
Spalding, M., Blasco, F. and Field, C. 1997. *World Mangrove Atlas*. Okinawa, Japan: International Society for Mangrove Ecosystems.
Statistics Bureau of the Ministry of Internal Affairs and Communications (eds) 2004. *Japan Statistical Yearbook*. 54th Edition (2005). Tokyo: Japan Statistical Association (in English and Japanese).
Sugihara, S. and Serizawa, C. 1957. *Kanagawa-ken Natsushima ni okeru Jomon Bunka Shoto no Kaizuka* (The Shell Mounds of the Earliest Jomon Culture at Natsushima, Kanagawa Pref., Japan). Meiji Daigaku Bungakubu Kenkyu Hokoku, Kokogaku Dai-2-satsu (Reports on the Research by the Faculty of Literature, Meiji University, Archaeology No. 2). Tokyo: Meiji University (in Japanese with long English summary).
Tanabe, S., Saito, Y., Sato, Y., Suzuki, Y., Sinsakul, S., Tiyapairach, N. and Chaimanee, N. 2003. Stratigraphy and Holocene evolution of the mud-dominated Chao Phraya delta, Thailand. *Quaternary Science Reviews*, 22, 789–807.
Taniguchi, Y. 2002. Nihon oyobi Kyokuto ni okeru doki shutsugen no nendai (Dating of the oldest pottery culture in Japan and the Far East). *Kokugakuin Daigaku Kokogaku Shiryokan Kiyo* (Memoir of the Museum of Archaeology, Kokugakuin University), 18, 45–67 (in Japanese).

Taniguchi, Y. 2004. Nihon Retto shoki doki-gun no kyaribureshon 14C nendai to doki shutsudo-ryo no nedai-teki suii (Radiocarbon calibration for the early potteries in Japan pre-dating ca. 10,000 BP and the temporal changes in quantities of pottery use). *Kokogaku Janaru (The Archaeological Journal)*, 519, 4–10 (in Japanese).

Thanh, T.D., Huy, D.V., Nguyen, V.L., Ta, T.K.O., Tateishi, M. and Saito, Y. 2002. The impact of human activities on Vietnamese rivers and coasts. In: Hong, G.H., Kremer, H.H., Pacyna, J., Chen, C.-T.A., Behrendt, H., Salomons, W. and Crossland, J.I.M. (eds), East Asia Basins: LOICZ Global Assessment and Synthesis of River Catchment-Coastal Sea Interaction and Human Dimensions. LOICZ Reports & Studies No. 26 (LOICZ IPO, Texel, The Netherlands), 179–184.

Uda, T., Ito, H. and Ohtani, Y. 1992, Mean sea level change along the Japanese coasts since 1955, Proc. Coastal Engineering, JSCE, 39, pp.1021–1025 (in Japanese).

Umitsu, M. 1991. Holocene sea level changes and coastal evolution in Japan. *Daiyonki Kenkyu (The Quaternary Research)*, 30, 187–196 (in English).

UNEP 1997. Asia-Pacific Environment Outlook. Bangkok: UNEP.

UNEP 2001. Bangkok: State of the Environment. Bangkok: UNEP.

Valencia, M.J. (ed.) 1981. Proceedings of the workshop on Coastal Area Development and Management in Asia and the Pacific. Manila, Philippines, 3–12 December, 1979. Honolulu: Hawaii University Press.

Wang, Y. 1998. Sea-level changes, human impacts and coastal responses in China. *Journal of Coastal Research*, 14, 31–36.

Welsch, R.L. 1998. *An American Anthropologist in Melanesia: A.B. Lewis and the Joseph N. Field South Pacific Expedition 1909-1913.* 2 volumes. Honolulu: University of Hawaii Press.

WRI 2001. Reefs at Risk in Southeast Asia (Chapter 8: Monitoring and improving the knowledge base Management of coastal resources). Available online only at http://www.wri.org/reefsatrisk/management_01.html viewed 31/01/04.

WRI/UNEP/UNDP/World Bank 1996. *World Resources 1996–97*. New York and London: Oxford University Press.

Wyrtki, K. 1990, Sea level rise: the facts and the future. *Pacific Science*, 44, 1–16.

Yamaguchi, M., Sugai, T., Fujiwara, O., Ohmori, H., Kamataki, T. and Sugiyama, Y. 2003. Nobi Heiya boringu koa kaiseki ni motozuku Kanshinto no taiseki katei (Depositional process of the Holocene Nobi Plain, Central Japan, reconstructed from drilling core analysis). *Daiyonki Kenkyu (The Quaternary Research)*, 42, 335–346 (in Japanese with English abstract).

Yan, W. 1993. Origins of agriculture and animal husbandry in China. In: Aikens, C.M. and Song, N.R. (eds), *Pacific Northeast Asia in Prehistory*. Pullman: Washington State University Press, 113–123.

Yang, Z., Saito, Y., Liu, B., Zhang, J. and Wang, H. 2002. Decadal and millennial time-scale changes of water and sediment discharge of the Huanghe (Yellow River) caused by human activities. In: Hong, G.H., Kremer, H.H., Pacyna, J., Chen, C.-T.A., Behrendt, H., Salomons, W. and Crossland, J.I.M. (eds), East Asia Basins: LOICZ Global Assessment and Synthesis of River Catchment-Coastal Sea Interaction and Human Dimensions. LOICZ Reports & Studies No. 26 (LOICZ IPO, Texel, The Netherlands), 118–121.

Yim, W. W.-S. 1995. Implications of sea level rise for Victoria Harbour, Hong Kong. *Journal of Coastal Research*, Special Issue, 14, 167–189.

Yim, W.W.S. 1996. Vulnerability and adaptation of Hong Kong to Hazards under climatic change conditions. *Water, Air, and Soil Pollution*, 92, 181–190.

Yim, W.W.-S. and Huang, G. 2002. Middle Holocene higher sea level indicators from the South China coast. *Marine Geology*, 182, 225–230.

Young, E. 2004. Last of the great migrations. *New Scientist*, 24th April, 38–41.

Zamora, P.M. 1979. The coastal zone management program of the Philippines. In: Valencia, M.K. (ed.), Proceedings of the workshop on Coastal Area Development and Management in Asia and the Pacific. Manila, Philippines, 3–12 December, 1979. Honolulu: Hawaii University Press.

CHAPTER 7

HOT SPOTS OF POPULATION GROWTH AND URBANISATION IN THE ASIA-PACIFIC COASTAL REGION

POH POH WONG[1], LEE BOON-THONG[2] AND MAGGI W.H. LEUNG[3]
[1]*Department of Geography, National University of Singapore, Singapore*
[2]*Department of Geography, University of Malaya, Malaysia*
[3]*Department of Geography and Resource Management, The Chinese University of Hong Kong, China*

7.1 INTRODUCTION

In many countries, human settlements are increasingly concentrated along the coastline. The average population density in the coastal areas is about 80 persons/km^2, which is twice the world's average population density (UN System-Wide Earthwatch 2005). If the current trend continues, it is speculated that 75 percent of humanity will reside in the coastal areas within three decades (Hinrichsen 1995).

Coastal zones have always attracted human development, and population centres have established there. In this century, the urbanisation of coastlines has grown dramatically, which follows worldwide trends toward urbanisation. Coastal urbanisation is driven by explosive population growth and migration to areas within 100 km of the coastline (Kullenberg 2001). On a global scale, this shift is reflected in the urbanisation and development of coastal cities that have populations well over one million. In the 21st century, there are 27 coastal cities with populations above 1 million, 12 in the range of 1–10 million, 13 in the range of 10–20 million, and 2 above 20 million (Kullenberg 2001).

Along coastal zones around the world, population growth and the human activities associated with it deserve more attention from planners and policy-makers because they can impose serious challenges on coastal and marine ecosystems as well as socio-economic systems. The Asia-Pacific coastal region is home to many of the largest urban agglomerations including Tokyo, Mumbai, Shanghai, Dhaka, Jakarta, Osaka, Beijing, Karachi, Manila, and Seoul. While the growing population in the coastal zone can be considered as a capital for further economic growth, such a concentration of human settlements and activities have also created social and environmental stress in many urban and rural communities. Indeed, with 69 percent of its coastal ecosystems at risk, the Asian coast is the second most threatened region in the world following Europe, where 86 percent of the coasts are estimated to be at risk (UN System-Wide Earthwatch 2005).

This chapter focuses on the hot spots of urbanisation and population growth in the Asia-Pacific coastal zone. It first discusses population patterns, growth, and related issues. Next it examines the urbanisation processes and the growth of the megacities and mega-urban areas. The implications of rapid urbanisation and megacities are then discussed. The chapter ends with an overview of the research on population, urbanisation, and megacities.

7.1.1 Definitions

There is no common definition for what constitutes a coastal region in the world, which causes estimates of the coastal population to vary. Most estimates are based on an area within 60 to 200 kilometers of the shoreline. Three billion people, or about half of the world's population, live within 200 kilometres of a coastline today. The figure is predicted to double by 2025 (Creel 2003).

Definitions of the Asia-Pacific region vary considerably (Smith 2001), and ESCAP's definition probably includes the largest geographical area. In this paper, the Asia-Pacific coastal countries start from China and end at Pakistan. All Southeast Asian countries and Pacific islands are included in the region; although, Hawaii, Australia, and New Zealand are not. The Asia-Pacific region is highly heterogeneous with regard to demographic, economic, socio-cultural, and political conditions. For instance, while mortality and fertility have declined in many parts of the region, they remain high in others. Although a rapid economic boom in many of the Asia-Pacific economies has successfully raised the general income level, the region remains home to a billion people living in poverty. While more and more people are able to read, write, and pursue tertiary education, the illiteracy rate is still considerable.

7.2 POPULATION

The Asia-Pacific region is home to over 3.7 billion people, which is roughly 60 percent of the world's population. The present world population of 6.1 billion is projected to surpass 9.3 billion by 2050. Of the ten most populous countries, six are in Asia: China, India, Indonesia, Pakistan, Bangladesh, and Japan. In the past two decades, Asia has shown a rapid decline in average population growth rate, which at 1.3 percent today is the lowest among low-income countries and compares closely with the world average of 1.2 percent. Since population growth is generally perceived as a serious threat to sustained economic growth, many Asian countries developed population strategies and adopted public family planning programmes decades ago. Several countries succeeded in reducing total fertility levels significantly. The examples include China, North Korea, Iran, South Korea, Thailand, Vietnam, Indonesia, Malaysia, India, Sri Lanka, and Bangladesh (UNFPA 2002).

While population growth has been slowing down in many of the Asia-Pacific countries, both population size and density in the coastal population have steadily risen in the past few decades. In many cases, the increase is less due to natural population growth and is mainly the result of migration from non-coastal areas. This trend will be further explored in the sections on urbanisation, megacities, and urban areas.

7.2.1 Demographic Transition

The demographic transition has been more rapid and dramatic in Asia than in any other part of the world. The population growth rates in the Asia-Pacific region reached a peak about 35 years ago and have declined ever since (Jones 2003). However, countries are at different stages of the demographic transition and enormous differences still exist (Table 7.1). Two key demographic variables affect population growth rate—fertility and mortality. New medical and public-health technology has led to a rapid decline in infant mortality. As elsewhere in the world, economic development, social modernisation, and access to new technology caused a drop in the death rate before they affected the birth rate (East-West Center 2002).

Several countries in the Asia-Pacific coastal region are considered to have completed their demographic transition, i.e. low levels of both mortality (life expectancy at birth above 70 years) and fertility (totality fertility rate below 2.1 children per woman). These include Japan, with the highest life expectancy in the world at 82 years, Hong Kong (81 years), and Singapore (79 years). At the other extreme, life expectancies are particularly low (in the 50s) in Cambodia, Burma, and East Timor. In the largest two countries in the region, life expectancy is 71 in China and 64 in India (Population Reference Bureau 2004).

Fertility decline generally began in the 1960s, although, it started much later in some countries. Fertility at the replacement level is defined at 2.1 children per woman. China, Hong Kong, Taiwan, Korea, Thailand, Vietnam, Singapore, and Sri Lanka have fertility of less than or equal to the replacement level. In Southeast Asia, fertility ranges from 1.4 children per woman in Singapore to 4.6 in Cambodia. In South Asia, Sri Lanka has the lowest fertility rate at 1.9, and Pakistan has the highest rate at 4.9. For Asia's largest countries, fertility averages 1.8 children per woman for China and 2.9 in India. Pakistan, Cambodia, and a number of Pacific islands have fertility of 4 or more (Population Reference Bureau 2004).

Most future population growth is expected in South Asia, although China contributes a substantial part of the region's growth by virtue of its large population. Six countries are projected to contribute to more than 80 percent of Asia's population growth from 2000 to 2050: India, Pakistan, China, Bangladesh, Indonesia, and the Philippines. Japan is projected to have a population decline (ESCAP 2004).

Table 7.1. Main features of the population in Asia-Pacific coastal countries

Region or country	Population, mid-2004 (millions)	Total fertility rate (per woman)	Life expectancy at birth (years) total	Percentage aged, 65+	Projected population 2025 (millions)
EAST ASIA					
China	1,313.3	1.8	71	7	1,445.1
DPR Korea	22.8	2.0	63	7	24.7
Hong Kong	6.9	1.0	81	12	0.8
Japan	127.8	1.3	82	19	121.1
Macau	0.4	1.1	77	8	0.5
R Korea	48.2	1.2	77	8	50.6
Taiwan	22.6	1.2	76	9	24.4
SOUTHEAST ASIA					
Brunei	0.4	2.4	76	3	0.5
Cambodia	14.5	4.6	57	3	21.9
East Timor	0.8	3.7	49	3	1.2
Indonesia	222.6	2.3	68	5	270.1
Malaysia	25.5	3.0	73	4	33.5
Burma	50.1	2.8	57	5	59.8
Philippines	81.4	3.1	70	4	108.6
Singapore	4.3	1.4	79	8	4.9
Thailand	63.8	1.7	71	7	70.1
Vietnam	81.5	1.9	72	5	102.9
SOUTH ASIA					
Bangladesh	137.5	3.3	60	3	192.2
India	1,081.2	2.9	62	5	1,369.3
Pakistan	157.3	4.9	61	4	249.8
Sri Lanka	19.2	1.9	72	7	21.5
PACIFIC					
Am Samoa	0.06	4.0	—	3	0.09
Cook Islands	0.02	3.1	—	7	0.18
Fiji	0.85	2.8	67	4	0.96
Fr Polynesia	0.25	2.4	72	5	0.32
Guam	0.17	2.8	78	6	0.22
Kiribati	0.09	4.3	63	4	0.15
Marshall Is	0.06	5.7	69	2	0.08
Micronesia	0.11	4.1	67	3	0.15
Nauru	0.01	3.4	61	2	0.02
New Caledonia	0.24	2.4	73	6	0.32
N Mariana Is	0.08	1.6	—	2	0.15
Palau	0.02	2.5	70	5	0.03
Papua New Guinea	5.8	3.9	57	2	8.4
Samoa	0.18	4.6	73	5	0.22
Solomon Is	0.49	4.3	61	3	0.78
Tonga	0.11	3.6	71	6	0.12
Tuvalu	0.01	3.8	—	6	0.01
Vanuatu	0.22	4.0	67	3	0.33

Sources: ESCAP 2004; Population Reference Bureau 2004.

7.2.2 Migration

In the Asia-Pacific region, the last three decades have seen a substantial increase in the scale, diversity, and complexity of population movements between geographic regions. There has been a significant redistribution in the population from rural to urban areas, which has contributed much to the growth of the urban population. In addition, there has been an upsurge of international movements within the region and to other parts of the world. The migration experience and spatial distribution have been divergent among the countries in the Asia-Pacific region (Anh 2003).

Internal migration

Migration is one of the major forces that accounts for the population growth in the coastal zones around the world. Due to the inequitable distribution of resources, services, and opportunities in different part of many countries, coastal cities in the Asia-Pacific region have witnessed immense population inflow. Each day in China, about 1,000 people leave their homes to seek a new life in large coastal cities. Similar situations have also been noted in Vietnam and the Philippines (Creel 2003).

Ever since the 1980s, migration movements in the Asia-Pacific region have been female dominated. The feminisation of the labour force has been brought about by the deregulation of labour markets that prompted employers to hire "unskilled" female workers. Because coastal urban areas are likely to have a higher concentration of industries, larger service sectors, and higher concentrations of foreign investment, they become the logical destinations of many female migrants. The export-led economic development model adopted by many governments in the Asia-Pacific region has given rise to the demand for large numbers of female workers in the coastal zone through the establishment of export-processing zones (EPZs). According to AFL-CIO (2005), 90 percent of the 27 million workers in the EPZs are women, most of them between 16 and 25 years old. Women are considered more desirable than men as workers because they are perceived to be willing to accept lower wages and flexible work contracts, less likely to engage in disputes with employers and managers, and more suited to the repetitive and often detailed tasks required in light manufacturing.

The Chinese case is particularly interesting as far as internal migration is concerned. Huge numbers of rural Chinese leave their villages for cities in the coastal provinces, mainly because of dire conditions and lack of job opportunities in the rural areas. The influx of migrants, a so-called "floating population" estimated at 30 million according to the Population Reference Bureau, largely accounts for the fast population growth in certain coastal provinces. Guangdong, a prosperous province on the south-eastern coast, recorded 37 percent growth in the last 10 years, which is the fastest growth rate in the country.

Chinese rural-urban migrants tend to be young (between 20 and 35 years old), and move from the central and western regions to the coastal provinces for low-skilled work, such as construction work for male migrants and factory jobs for females (Goodkind and West 2002). The migrants in the coastal areas earn US$50 to US$70 a month (far exceeding the average income from farming) and work more than the maximum 49-hour work week (*Migration News* 2005). Unfair treatments against migrant workers are common due to their illegal status. The migrants are attracted not so much by high wages, but by high gross incomes and entrepreneurial activity, a situation that is consistent with much of the evidence found in the literature for other developing nations (Chen and Coulson 2002).They appear relatively indifferent to the quality of urban life, suggesting that it is primarily economic goals that motivate them to leave their family homes for the cities (Chen and Coulson 2002).

China regulates population mobility within the country through the "household registration system" (*hukou* system), which was introduced in the 1950s. According to the system, every citizen is listed in a specific location, usually his or her place of birth. The registration guarantees one's rights to housing, education, and government services. Since 1994, it is possible for rural residents to apply for city work permits. Nevertheless, most rural migrants cannot afford the permits, which cost from US$6,000 to US$12,000. Due to their lack of proper registration papers, many migrant workers live in poor conditions, are often cheated out of their pay, and are harassed by the police. Many migrants have trouble collecting their wages because most employers keep two months wages and refuse to pay their workers if they do not stay as long as the employer wishes. The *Beijing Review* reported in December 2003 that 72 percent of migrant workers are owed a total of $12 billion in back wages (*Migration News* 2005).

However in recent years, several other factors are working against continued increases in the floating population. First, a number of cities are experiencing sizeable lay-offs of workers from state-owned enterprises and are reacting by limiting migrant access to jobs. Secondly, technological change and the progressive development of the economy may reduce the demand for unskilled labour, especially in the coastal cities. Economic restructuring since the late 1970s has brought about the loss of large numbers of jobs in the state sector. By the beginning of 2004, about 30 million workers had been laid off from state-owned enterprises since 1998 (*China Daily* 8.1.2004, Roberts 2002). According to *China Daily* (8.1.2004), as many as three million workers will further be laid off each year until 2006 when the restructuring of state-owned firms is expected to finish.

In spite of such harsh and unfair treatments, the rural exodus is likely to continue after China's accession to the World Trade Organisation (WTO), which will accelerate China's economic restructuring process. Bound by the WTO membership, China will become more attractive for foreign investors in the industrial sector, will further reduce agricultural subsides, and will reduce the number and size of state-owned enterprises. Economic growth is predicted to accelerate in the 15 Free Trade Zones and five Special Economic Zones (Shenzhen, Zhuhai and Shantou in Guangdong, Xiamen in Fujian, and Hainan provinces), which are all located in the coastal provinces. China's future population growth is also expected to concentrate along the coast.

Vietnam has a similar version of the Chinese model of "hukou" aimed at controlling population mobility, especially spontaneous outflows of rural residents. The economic reforms (*Doi Moi*), officially launched in 1986, consist of structural changes that represent a shift from a centralized planning system to a more market-oriented economy. It provides increased economic opportunities and greatly promotes the out-migration of rural labour (Dang 1999). Today, household registration procedures are not applied the way they were in the past. The result is a hybrid system, and a key element is the issuance of temporary residence to stay in the cities. However, the government has not been able to reverse or profoundly modify the direction of population migration (Anh 2003).

Intra-regional migration

According to estimates by the United Nations, there are about 19 million international migrants in the Asia-Pacific region, which at 2.3 percent of the total population is relatively low (UN Habitat 2004). Intra-regional migration movements within the region can be considered a key factor in our globalising world. A large number of long-term migrants, refugees, and migrant workers—male and female, documented and irregular, unskilled and highly skilled—move around and live in many countries, especially in urban areas in the region. The "export" of nationals can almost be considered a development strategy among some low-income countries. The Philippines is perhaps the most well-known case. Remittances from Filipino domestic workers and industrial workers in Southeast Asia, the Gulf States, and Japan contribute to a large proportion of the Philippines' foreign exchange earnings. Also, a large number of migrant women from Indonesia, Thailand, and increasingly, China leave their economically stricken homes to work, often as domestic helpers, in other parts of the region, in particular Japan, Hong Kong, Taiwan, Korea, Singapore, and Malaysia.

Many migrants do not succeed in escaping from poverty. In many Asian cities, the living conditions of migrants are poor, characterised by overcrowded and unhygienic housing. Squatter settlement is common. For instance, in Bangkok there is a very high concentration of Burmese migrants in the slum of Klong Toey. In Karachi there are about 1.5 to 2.5 million undocumented migrants, mainly Bengalese, Afghan, and Burmese, living mostly in the unplanned areas (*katchi abadis*) of Baldia and Orangi. Due to their illegal status, poor migrants can be considered the most disenfranchised social group among the urban poor.

7.2.3 Ageing Population

The population's shifting age structure is an important feature of the demographic dynamics in the Asia-Pacific region. While disparity in fertility and life expectancy exists across the region, there is generally a drop in fertility and an increase in longevity. As a result, the proportion of the population under age 15 years has been falling, and the proportion of population in the age group 65 years and over has been rising. The two population groups are expected to make up relatively equal proportions of the population by 2050 (ESCAP 2003).

One of the implications of the decline in fertility and mortality is population ageing. In the ESCAP region as a whole, the proportion of the population aged 65 years and over has reached 6 percent. Japan is the most aged country, with 19 percent of its population over 65 years old (Table 7.1). As a result of improved health conditions, the number of older persons is increasing annually at the rate of 2.8 percent, which is more than twice as high as the growth rate of the total population. Not only are the higher-income societies such as Japan, Hong Kong, and Singapore experiencing an increasingly grey population, China has also registered a similar situation.

The ageing trend in China is, however, geographically varied. While the issue was noticed in the coastal power-house of Shanghai by the late 1970s, similar demographic shifts are not likely to be experienced in less-developed provinces, such as Qinghai, until 2010, according to the Chinese Institute of Population Research (*China Daily* 5.11.2003). Growth in ageing populations implies a shrinking labour force and a growing number of older, dependent persons. This demographic profile could result in declines in productivity, public revenues, and rates of savings and investments as well as a higher demand for public expenditures on social security and health care.

The ageing trend in the Asia-Pacific coastal zone warrants attention. Policies promoting intergenerational equity could be formulated to assist the elderly to remain healthy and prolong their productive, independent years. A number of places in the region, like Singapore and Hong Kong, have also attempted to formulate new social and population policies that encourage childbirth and/or attract skilled migrants.

7.3   URBANISATION

The ability to examine trends in urbanisation in the Asia-Pacific region is challenged by the availability of statistics (Cohen 2004). While the lion's share of increase in urban populations over the next 15 years is expected to be in towns and cities with fewer than 1 million inhabitants, there is not currently a database for cities below 750,000. Also, there is no standard criterion for defining city boundaries. For example, Shanghai's city boundaries include a large population outside the city proper, such as land that is basically used for agricultural purposes. Other cities, such as Bangkok, Jakarta, Manila, and Taipei may be twice as large as the official, tightly drawn city boundaries and may omit important peri-urban areas just outside the city.

Generally, economic growth and urbanisation are positively correlated; however, patterns vary within the Asia-Pacific region due to the diversity of political, economic, and social conditions. Trends in the region also vary from those found in other parts of the world (Tables 7.2 and 7.3; Smith 2001). Within the Asia-Pacific region, China's demographic and urbanisation trajectories stand out from those of the other countries (Pannell 2003) and are thus given specific attention in section 7.3.2.

Hot Spots of Population Growth and Urbanisation

Table 7.2. Urban and rural population and economic growth rates for the Asia-Pacific region

Region	Mid-year population (millions)				Economic growth rate (percent)		
	1950	1975	2000	2030	1950–70	1975–2000	2000–30
East Asia & Pacific							
Urban	103	258	703	1,358	3.7	4	2.2
Rural	639	1,008	1,113	870	1.8	0.4	−0.8
South Asia							
Urban	72	164	372	849	3.3	3.3	2.7
Rural	392	645	982	1,176	2	1.7	0.6
World	2,520	4,066	6,057	8,270	1.9	1.6	1

Source: Summarised from Cohen 2004.

Table 7.3. Urbanisation of selected Asia-Pacific coastal countries

Region or country	Percentage urban			
	1960	1980	2000	2005*
EAST ASIA				
China	16.0	19.6	35.8	40.5
Japan	43.1	59.6	65.2	65.7
R Korea	27.7	56.9	79.6	80.8
SOUTHEAST ASIA				
Brunei	43.4	59.9	73.9	77.6
East Timor	10.3	8.5	7.5	7.8
Indonesia	14.6	22.1	42.0	47.9
Malaysia	26.6	42.0	61.8	65.1
Philippines	30.3	37.5	58.5	62.6
Thailand	19.7	26.8	31.1	32.5
Vietnam	14.7	19.4	24.3	26.7
SOUTH ASIA				
Bangladesh	5.1	14.9	23.2	25.0
India	18.0	23.1	27.7	28.7
Pakistan	22.1	28.1	33.1	34.8
Sri Lanka	17.9	21.6	21.1	21.0
MELANESIA	8.1	18.1	19.3	19.6
MICRONESIA	37.9	57.3	67.3	70.2
POLYNESIA	28.5	38.4	43.1	44.0

Source: UN 2004b.
* Projection.

7.3.1 Urban Transition

Urbanisation has increased rapidly on a global scale. In 1900, less than 14 percent of the population was urban. By 2005 half of world's population is expected to live in cities, and the Asia-Pacific region is a major contributor to the global urban transition

(Douglass 2000). Economic development has transformed the region and many of its cities at a speed and scale never before witnessed, with China accounting for a large share of the growth. There is also a strong, positive correlation between urbanisation and economic development. Urbanisation is a necessary condition for economic development because agglomeration in urban areas creates economies of scale which promotes economic development (Yap 2003).

Unlike the urbanisation experience of Europe and the United States in the first half of 20th century, urbanisation in the Asia-Pacific region has several features. The urban population growth rate has been rapid and has exceeded the rate of growth occurring in either the total population or the rural population. Cities have increased in absolute number and have reached unprecedented sizes. The highest urban growth has been in rapidly industrialising cities, located mainly in Southeast Asia. In China, economic development has transformed the coastal region, and cities have urbanised at an unparalleled rate and scale. Since the Chinese government embarked on a pattern of gradual economic reform in 1978, coastal cities, such as Shenzhen, Guangzhou, and Xiamen, have grown at record rates and have been transformed physically and economically into modern cities.

Urban population growth is the result of three processes: natural urban population growth, rural-to-urban migration, and reclassification of rural areas as urban areas. Often, it is difficult to separate the effects of the last two processes (Yap 2003). Two distinctive features can be observed in the urbanisation in the Asia-Pacific region: the degree of urban primacy and the increasing size and complexity of urban entities (Smith 2001). Both features overlap, but they are not the same. Urban primacy refers to the large proportion of the national population living in a single city, which in many cases is the capital city. A principal example of primacy is Bangkok, which is about 50 times larger than the second most populous city in the country. Urban primacy is not necessarily equated with the overall size of a settlement, but principally refers to its dominance as a population centre within a national context.

City-centred regional growth, a phenomenon characterised by a complex of cities, towns, and urban-oriented rural populations, has become more pronounced, and it is now identifiable as a new scale of development. Known by various terms, this region-based, rather than city-based, urbanisation has evolved differently in the Asia-Pacific region and represents something distinctive from urbanisation in other parts of the world (Webster 1995).

7.3.2 China's Urbanisation

China's urbanisation merits separate treatment because it has generated a lion's share of urban growth in the Asia-Pacific region, and it has followed a different demographic and urbanisation trajectory (Pannell 2003). Telling the story of China's urbanisation is somewhat hampered by data reliability, as a result of changes in the definitions used for counting urban populations. China has persistently undercounted its urban population, partly a Maoist legacy associated with the policy of controlled urbanisation and socialist industrialisation (Pannell 2003). But there is no doubt that the speed and scale of rural-to-urban transformation has been breathtaking. "In urban terms, China is going where no other country has gone before in terms of the number, size, and sheer human scale of its experience" (Pannell 2003: 480). Relative to recent changes in the spatial patterns of capitalist cities, the scope, speed, and degree of China's urban space transformation away from the socialist pattern has been profound. The "change has been more revolutionary than evolutionary" (Ma 2004: 241), and there is an impressive corpus of literature on China's urban transformation since the initiation of economic reforms in 1978.

Prior to the 1978 economic reforms, the system of cities in China was dominated by large and extra-large cities for optimum industrialisation. For reasons of national defense, most of the new cities were developed in the central and western interior rather than the eastern coast (Lin 2002). With market reforms and relaxation of state control over local development since the late 1970s, a large number of small cities and towns have flourished on the basis of bottom-up, rural transformative development. The influence of global market forces has helped re-consolidate the dominance of the eastern coast in China's urban development (Lin 2002).

Several factors have driven this rapid urbanisation and growth of cities and towns, especially in the coastal areas: (1) continuing, although slowing, population growth; (2) migration of rural people, as regulations on rural and urban household registration change; (3) rapid structural shifts in employment activities and the decline of farm employment; (4) foreign trade and foreign investment, especially in the coastal areas; (5) restructuring of state-owned enterprises and growth of private enterprises and activities; and (6) allocation of domestic funds in fixed assets for urban infrastructure, also concentrated in the coastal areas (Pannell 2002).

There is no doubt that urbanisation is related to modernisation, economic growth, structural changes, and inflows of foreign direct investments (FDI). "The causal link runs from economic growth to urbanisation and not *vice versa*" (Zhang 2002: 2001). Zhang and Song (2003) confirm that rural-urban migration has been a dominant mechanism for Chinese urban population growth, but that the causal link runs from economic growth to migration and not *vice-versa*.

Urbanisation is accompanied by spatial shifts with a bias to the coastal areas. Parallel with growing urbanisation is an economic shift in which large rural populations move to employment in commercialised agriculture and activities in urban areas (Pannell 2003). This rural-to-urban movement has involved a predominant west-to-east migration stream, as many migrants from impoverished interior areas have flocked to the dynamic eastern coastal provinces in search of better livelihoods. This change has occurred as the rules and regulations regarding movement of the rural population have been applied with growing flexibility (Pannell 2003). Further relaxation of the rights of rural migrants to work in cities in January 2003 has encouraged a move to smaller cities and towns and has helped to avoid the emergence of huge slum areas that are found in other Asian cities, e.g. Bombay, Jakarta, and Manila. These changes have also quickened the pace of urbanisation, especially in the coastal areas linked to the global and national economy (Pannell 2003).

The deliberate policy of rapid growth for China's coastal region started in 1978 with an open policy to attract foreign investment and technology that was initially limited to four Special Economic Zones (SEZs): Shenzhen, Zhuhai, Xiamen, and Shantou. Shenzhen benefited from proximity to Hong Kong and Xiamen to Taiwan. Gradually, other special zones were established (Cohen 2004). This experience was followed by an open policy for 14 coastal cities in 1984. As a result, the urban growth in the coastal regions of China is highest in the country. In 2000, Shanghai, Beijing, and Tianjin had the highest levels of urban growth, at more than 70 percent. The 2000 census data showed dynamic economic progress associated with migration and population growth was most closely associated with the eastern belt and coastal cities (Fan 2002). Coastal regions with the strongest urban growth were identified using time-series and cross-section analyses, based on data from 1978 to 2000. Provinces with rapid urban growth and an urban level exceeding 40 percent are Fujian, Guangdong, Jiangsu, Zhejiang, and Hainan. In 2000 the average level of urbanisation in the coastal areas was 44.6 percent compared to 30.92 percent in the interior provinces (Zhang 2002).

Chinese urbanisation has wrought a marked landscape transformation to the coastal regions. Part of this process has involved the in-situ urbanisation of rural areas, in contrast to the normal urbanisation process dominated by rural-urban migration (Zhu 2000). For example, Fujian Province is driven by development of townships, village enterprises, and inflow of foreign investment that have been facilitated by relevant government policies since 1978. The conversion of land to nonagricultural use has been another important process that has been more intense in the coastal provinces (Ho and Lin 2004). The important factors in the conversion of land to nonagricultural use are rural-urban migration, rapid economic growth, and increased investment in roads.

Urban growth in China has proceeded in step with the transition of its socialist economy and with its rapid economic growth. The 2000 census data indicates that China's urban population of 456 million, which is 36 percent of the total population, is increasing much more rapidly than the overall population (Pannell 2002).

7.4 MEGACITIES AND URBAN REGIONS

The past half century has been characterised by an increase in megacities through an urbanisation process that has led to half of the Asia-Pacific population living in cities with more than 10 million inhabitants. In 1950, the world had two megacities. This number increased to four in 1975 and to twenty in 2003, with half of these located in the Asia-Pacific coastal region (UN 2004b). With 35 million inhabitants in 2003, Tokyo is by far the most populous urban agglomeration, partially due to a new definition of metropolitan areas in the *2003 Revision* (UN 2004b) that comprises a greater number of cities and towns than previously associated with Tokyo.

7.4.1 Globalisation

Urban growth is more dependent on the global economy than before. The forces propelling much of the expansion of cities and urban networks now operate on an international level. Increasingly, the form and dominant activities within a given city are shaped by its linkages with globalising circuits of capital. Globalisation propels the urbanisation process by creating form and physical content in cities that are operating in an environment of intensified competition for global investment. "Urbanisation and globalisation are interdependent and mutually reinforcing" (Douglass 2000: 2315).

Globalisation is defined as the progressive integration of the world's economies, and it has accelerated in the past three decades. It is driven by an astonishing rate of technological change, particularly in areas of transportation and telecommunications. It has reshaped the organisation, management, and production of firms and industries (Cohen 2004). In economic terms, globalisation is defined as the accumulation through integration of three major circuits of capital—production, commodity trade, and finance—at the global scale (Douglass 2000). Its new focus is to create specific industrial requirements for reliable electric and water supplies, a diverse labour supply, transport structure, and higher order functions, such as hospitals, universities, parks, and centres for hosting international conference and spectacular world events (Douglass 2000).

Globalisation has clearly had a very uneven impact on various parts of the world, and Asia has benefited the most from globalisation (Cohen 2004). Up to the 1997/1998 financial crisis, it brought a broad set of globally driven arrangements that enhanced similar patterns of urban development in the Asia-Pacific region. The financial crisis itself was a watershed event in urban and regional development processes (Marcotullio 2003).

Within the Asia-Pacific region, the processes of urbanisation are primarily driven by global rather than national forces, and megacity development is linked to globalisation. Evidence about the impact of globalisation on urbanisation is varied. In the 1980s and the first half of the 1990s, the average annual GDP growth was more than 10 percent for China and 5–10 percent for all other Southeast Asian countries, except the Philippines, South Korea, and Hong Kong (Islam 1997). Much of the region's GDP growth has been attributed to FDI. The result is a structural change in employment patterns from agriculture to manufacturing and services, e.g. agricultural employment went down to 12 percent in ASEAN countries. Secondary employment increased from 12 to 22 percent in Malaysia (McGee and Robinson 1995). In fact, the manufacturing sector was the largest generator of employment opportunities in the 1980s and 1990s. Capital cities became focal points in the global economy not just for manufacturing, but also for finance and other specialised services.

Within the context of globalisation, researchers have posited that the Asia-Pacific region serves as a functional city system with distinctive features and global connections. Networks of cities are linked on the basis of economic or socio-political functions (Lo and Yeung 1996). These networks are often hierarchical and may be organised as a large number of smaller networks. Within the system it is possible to identify different types of centres, including: (1) command-and-control centres, e.g. Tokyo, Osaka, Seoul, and Taipei; (2) industrial development centres, e.g. Bangkok, Jakarta, and Shanghai; (3) international gateways, e.g. Hong Kong and Singapore (which, having developed their regional influence, have extended their metropolitan regions across international boundaries (Lo and Marcotullio 2000, Marcotullio 2003)).

In the future, growth in urban services (or tertiarisation) will be crucial to the development of Asia-Pacific city-regions. Such service industries extend from banking and finance, corporate control, and advanced technology to specialised producer services, creative services, education, culture and heritage, gateway roles, and tourism and conventions (Hutton 2004).

7.4.2 Megacities

A megacity is technically defined as one that had a projected population of over 8 million inhabitants in the year 2000, but the term is conventionally applied to cities with over 10 million people. Based on population data from the UN (2004a), cities in the Asia-Pacific with over 10 million inhabitants in 1975, 2003, and projected to 2015 are identified in Table 7.4. The growth evident in Table 7.4 has been propelled by rapid globalisation and the unprecedented growth that has occurred in the Asia-Pacific region over the last two to three decades, as discussed in the previous section. In fact, Asian countries have a greater proportions of their urban population located in megacities than any other region in the world.

Megacities and emerging megacities in the region have diverse characteristics. They range from very advanced city states (e.g. Hong Kong, Singapore) to very-low income cities that are starting on the development path (e.g. Hanoi, Yangon) (Figure 7.1). Some countries like Japan, South Korea, and Taiwan, have seen the growth of large urban corridors that are linked by fast rail and road transport (Tokyo-Osaka, Seoul-Pusan, and Taipei-Kaohsiung). Others have developed large urban agglomerations around their capital regions, as in Kuala Lumpur, Bangkok, and Jakarta. In some of these cases, a substantial share of the country's urban population is located in an extended metropolitan region (e.g. Bangkok 50%, Jakarta 20%, Manila 15%), rather than being

Table 7.4. Urban conglomeration in the Asia-Pacific coastal region

1975 (millions)		2003 (millions)		2015 (millions)	
Tokyo	26.6	Tokyo	35	Tokyo	36.2
Shanghai	11.4	Mumbai	17.4	Mumbai	22.6
		Calcutta	13.8	Dhaka	17.9
		Shanghai	12.8	Jakarta	17.5
		Jakarta	12.3	Calcutta	16.8
		Dhaka	11.6	Karachi	16.2
		Osaka-Kobe	11.2	Shanghai	12.7
		Karachi	11.1	Metro Manila	12.6
		Beijing	10.8	Osaka-Kobe	11.4
		Metro Manila	10.4	Beijing	11.1

Source: UN 2004b.

Hot Spots of Population Growth and Urbanisation

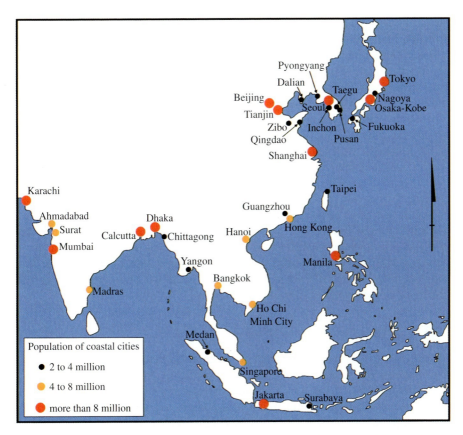

Figure 7.1. Coastal urban agglomerations with populations of two million or more in the Asia-Pacific region (Based on data from UN 2004a)

concentrated in the city proper or being more evenly distributed across several cities. (McGee and Yeung 1994, McGee and Robinson 1995).

Why have Asian-Pacific countries tolerated the development of huge cities? Basically, the economic growth paradigm has dominated the development of megacities in the Asia-Pacific region. Megacities make substantial contributions to the national output. For example, Osaka and Tokyo contribute 36 percent of Japan's GDP, Bangkok contributes 30 percent to Thailand's GDP, and Manila contributes 24 percent to the Philippines' GDP (Asian Development Bank 1997). Surproductivity (defined as the difference between the city productivity and the productivity of the rest of the country, divided by the productivity of the rest of the country) in megacities is two to four times that of national GDP per capita, e.g. in Shanghai, Manila, and Bangkok. Economic productivity is a far more important consideration than other things in the process of urbanisation. However, this explanation is not meant to imply that megacities are deliberately planned. The Asia-Pacific countries are more likely to be unable to control the growth of their cities.

Ensuring that megacities remain attractive for global transactions is a factor in developing implicit or explicit policies in transportation and housing. This dynamic sets up another cycle of investments. Globalisation brings new energy and resources to the urban system in the form of building construction, industrial production, traffic generation, tourism activities, and transnational mobility of labour and capital. A consequence of this momentum is the superimposition of global functions upon existing urban structures that have existed for a long time (Lee 2000a). The opulence of new global infrastructure, such as intelligent buildings, deluxe apartments, high-rise condominiums, and ultra-modern office and commercial complexes, is found coexisting with traditional structures and societies. Limited space means that the replacement of old buildings or new construction takes place wherever space allows. This situation creates a distinct dichotomy in the urban landscape, and is illustrated by cases where traditional squatters/ villages (*kampungs*) are sandwiched between luxury apartments and modern commercial complexes. This tension is compounded by a five percent per annum growth rate of squatters in some Asian cities.

7.4.3   Mega Urban Regions (MURs)

In the past two decades, researchers have examined new concepts and the interplay between globalisation and spatial transformation, particularly related to very large urban areas. This research has identified a new urbanisation phenomenon among megacities: an agglomeration of a number of cities, which are different sizes and retain a physical identity, that cumulatively constitute a population of 20 or even 30 million. McGee (1991) used the term *desakota* zone (derived from two Indonesian words for village, *desa*, and town or city, *kota*) for this city-centred regional growth, characterised by a complex of cities, towns, and urban-oriented rural populations. The concept of *desakota*, presumably outside the large cities, has been used interchangeably with many other concepts such as "extended metropolis", "extended metropolitan region", "metropolitan interlocking region", "core area" and "mega-urban region" (MUR), which all contain large cities. MUR is used in this chapter.

Globalisation plays a key role in the emergence of MURs in the Asia-Pacific region (Douglass 2000). The region has an increasing share of the world's urban population and hosts among the largest MURs. Tokyo is the world's largest MUR with 40 million people, or almost one-third of Japan's population. It is closely followed by Shanghai (37 million) and several other cities of more than 25 million in coastal China (Beijing, Hong Kong-Guangzhou, Shenyang-Dalian, and Quingdao-Jinan). In other countries, the principal MURs also account for substantial shares of national populations. Seoul has almost half of the population of South Korea. Taipei, Manila, Kuala Lumpur, and Bangkok, contain 37, 23, 21, and 20 percent, respectively (Douglass 2000). The MURs continue to grow faster than the national population growth rates, a trend that continues to increase the proportion of the national population that is within their boundaries. In Southeast Asia, the MURs of Jakarta, Bangkok, and Kuala Lumpur doubled their size in less than fifteen years (Douglass 2000).

MURs are one level in a hierarchical series of changes that take place on a spatial scale as part of economic development. They form core urban centres that are linked to each other through technological, economic, and social networks to create a series of nodes through which development occurs. It can be useful to think of MURs as existing along a continuum that ranges from "world cities" at the most sophisticated end of the spectrum to "regional cities" at the less developed. For example, Tokyo is at the highest level, most Asia-Pacific MURs are operating at the regional level, and Singapore is in the middle of the spectrum.

Importantly, MURs are becoming the foci for another layer of spatial development—formation of transborder regions. Examples of this trend include the growth triangles of the Sijori (Singapore, Johore (Malaysia), and Indonesia), the Pearl River Delta (China, Hong Kong), and the Yellow Sea Rim (China, Mongolia, South Korea, North Korea, and Russia) (Douglass 2000). These growth triangles are areas of economic integration that take place through accelerated urban transition.

A study of the Pearl River Delta, a rapidly growing MUR in a transitional socialist economy, reveals a distinct development pattern that does not conform to the prevailing city-based and economically deterministic interpretation of urbanisation processes (Lin 2001). Contrary to other findings, industrialisation has not concentrated a growing population into large cities, even after economic reforms. Urban and rural economies are not separated, but are intensely intersected. Global capitalism has not overridden local conditions, but has instead succumbed to them.

The spatial transformation in the Pearl River Delta is characterised by the relative decline of Guangzhou City as a dominant metropolitan centre along with the accelerated growth of the counties and cities located in the areas outside and between proximate metropolitan centres. At the same time two newly established Special Economic Zones (Shenzhen and Zhuhai) lead the whole region in the growth of industrial and agricultural production. Economic activities are concentrated in the triangular zone bordered by Hong Kong–Guangzhou–Macao. These changes have been the result of the interaction of various internal and external forces, cheap labour, improved transportation, and extensive pre-existing personal connections.

With the rapid expansion of urban areas, especially in China, East Asian MURs are introducing urban issues on geographic scales that are unprecedented in world history. By 2007, for the first time in human history, more than half the people in the world are predicted to be urban dwellers (UN 2004). It is noteworthy that the vast majority of the largest cities in the world are located along the coast. The rapid growth of some urban areas in the Asia-Pacific region has been called "hyper-urbanisation". Bangkok, for example, grew from 67 km^2 during the late 1950s to 426 km^2 by the mid-1990s. According to a UN-Habitat (2004), two-thirds of the entire ASEAN urban population will live in only five MURs by 2020: the Bangkok-centred MUR (30 million), the Kuala Lumpur–Klang MUR (6 million), the Singapore Triangle (10 million), the Java MUR (100 million), and the Manila MUR (30 million). More breathtaking growth is expected in the coastal zone in East Asia. The Tokyo-Osaka-Kyoto-Kobe-Nagoya MUR is expected to reach the size of 60 million, the Hong Kong-Shenzhen-Guangdong MUR will grow to 120 million people, and the Greater Shanghai MUR is estimated to reach 83 million.

7.4.4 World Cities

The term "world city" expresses the notion that a city has become relevant within the context of the larger structure of the world-system (Smith 2004). The concept was coined as early as 1915 by Geddes, but Friedmann's (1986) work on the "World City Hypothesis" is generally considered the starting point for modern thinking about world cities (Douglass 2000). World cities are increasingly viewed as "places with enlightened mode of governance ... where technological and economic advancement sustain global and local development, thereby enriching socio-economic, human, cultural and environmental capital" (Ng and Hills 2003: 161). As in the case of urbanisation, there are problems of measurement, and it is a difficult task to measure and map world cities. Regardless of the assessment method used, it is clear that Tokyo is a world city and that Seoul and Taipei are now next on the list from the Asia-Pacific region (Smith 2004).

Although Tokyo is Asia's first world city, and in the ranks of New York and London, its position in any hierarchy of world importance is not a static phenomenon. It changes its strategic urban policy to adapt to economic conditions. During the 1970s and 1980s, the Tokyo metropolitan area recorded a faster growth than any other metropolitan area in Japan for the obvious reason that it plays a leading role in Japanese politics, administration, finance, and business, in addition to being an international centre for finance, business, research and development, and media (Osada 2003). The concept of "world city" was used to legitimise dramatic development projects in the city twenty years ago. However, when the economic bubble burst in the early 1990s followed by a recession, the urban policy in Tokyo entered a phase of inertia. Tokyo is now entering a third phase in which a new competitive attitude is emerging regarding its world city role, and this resurgence is leading to changes in strategic urban policy (Saito and Thornley 2003).

Both Tokyo and Seoul depart from the world-city model which assumes convergence in economic base, spatial organisation, and social structure. The differences are due to late industrialisation and the relationship between industrial policy and the organisation of its finance (Hill and Kim 2000). There is also intense competition between megacities and emerging megacities in the Asia-Pacific region to become world cities. Within Asia, some comparison has been made between Tokyo, Hong Kong, Singapore, Taipei, and Shanghai (Ng and Hills 2003).

Among Chinese cities, Hong Kong, Shanghai, and Beijing have the possibility of becoming world cities (Shi and Hamnett 2002). Some additional insights come from a study of Beijing, Shanghai, and Guangzhou as world cities (Lin 2004).

Chinese documentation of world cities has been overwhelmingly concerned with what qualifies a city to be a "world city", and hence what should be done to promote a Chinese city to the world city rank. For the Chinese cities, the dramatic urban transformation after the 1980s and the country's entry to World Trade Organisation have implications for urban problems (e.g. land property speculation, chaotic land use, social tension, uncontrolled urban sprawl, congestion, and environmental degradation) and new challenges (management of the continuing growth of population, maintenance of urban liveability and environmental sustainability, accommodation of migrants and marginalised groups, and promotion of public participation in the planning processes). Innovative formulation of policies and planning strategies are needed to address these issues (Lin 2004).

7.5 ISSUES

Rapid urbanisation poses a serious threat to the immediate and adjacent environment, often causes high levels of water and air pollution, and produces land degradation. Rapid growth puts an even heavier burden on social, health, housing, and sanitation services in urban areas, especially those in low-income countries. The "ecological footprint" is a good indicator to illustrate the high ecological impact of cities. It estimates the area of biologically productive land and water needed to supply the resources and assimilate the wastes generated by that population, using the prevailing technology (Wackernagel and Rees 1996). For example, the ecological footprint of Hong Kong, according to various studies, is 148 to 267 times the area of the city itself (Hong Kong Planning Department 2002). Such "ecological deficits" are recorded for all urban areas around the world. As urban systems are not self-contained, their problems and solutions are part of and impinge on other areas beyond their geographical boundaries. This extension of influence further complicates the already demanding planning and management processes.

Asian cities face a wide range of environmental crises created by rapid, uncontrolled economic and physical growth (Heineke 1997, Yeung 2001, Daniere and Takahasi 2002). The key issues are not only in the realm of urban planning and management, but also in state-civil society relations, environmental sustainability, and a host of issues related to economic growth, social justice, and equity. These issues can be grouped under five categories: (1) governance, referring to more participatory systems of governance that will fundamentally change the politics of economic growth and urban development; (2) liveable cities and environmental management; (3) sustainable economies; (4) social justice and urban poverty, particularly the social and political questions regarding the growing presence of foreign workers; and (5) uneven spatial development and rural neglect (Douglass 2000).

It can be argued that the MUR is but a temporary pressure valve for urban problems. Worse, city problems are actually exacerbated, and not ameliorated, by and through the MUR. The outer rings of the metropolitan urban regions become the foci of attention, as they are more dynamic, to the neglect and expense of the central urban cores. The increasing visibility of development in the extended peripheral areas, in conjunction with increasing information technology (IT) development, suggests that centrality and accessibility have become less important. The dynamics of growth soon exhaust the availability of land within the extended metropolitan regions. Urban sprawl causes more traffic and environmental problems. The phenomena of globalised growth patterns create new challenges within the central urban cores that demand attention if Asia-Pacific megacities are to remain sustainable (Lee 2000b).

7.5.1 Liveable and Environmentally Sound Cities

The issues of liveability and environmental degradation have been long neglected in Asian-Pacific cities in favour of economic growth models. Environmental and social issues become subordinate to economic efficiency. For example, in the midst of bustling Tokyo, there are some 10,000 pavement sleepers living in "cardboard cities". In China, as many as 100 million people are leaving their impoverished villages in search of the "Chinese Dream" in the big cities (Roberts 2000). This floating population of migrants, in cities such as Shanghai, Beijing, Tianjin, Dongguan, and Guangzhou, live in squalid dormitories, deprived of health care, education, and pensions. They often work in factories and sweatshops that are exploitative of their labour.

Another aspect of liveability is the loss of heritage buildings. The use of urban space in Asian-Pacific cities has changed over the last three decades. Increasing modernisation, verticalisation, and gentrification have led to pressures in the inner city areas, especially where historic buildings have—until now—withstood the ravages of modern demands. The loss of heritage buildings is robbing Asian cities of their Asian character (Lee 2000a). Almost all Asia-Pacific cities have some older housing stock that has existed for centuries, such as those in Shanghai. These heritage buildings are the links with the past and may be architectural gems or pieces of history that enrich the urban experience and give character to Asian cities. Heritage buildings are now endangered under the onslaught of globalisation. Hanoi's Ancient Quarters is in danger of being modernised. In Georgetown, Penang, some 10,000 pre-war buildings, representing the very heart and cultural soul of the city, are in danger of being replaced.

7.5.2 Urban Poor

There has been an unprecedented decline in poverty in the Asia-Pacific region. Recent progress in poverty reduction in the region has been described as "one of the largest decreases in mass poverty in human history" (UN Habitat 2004). During the early 1970s, more than half the population was poor, average life expectancy was 48 years, and only 40 percent of the adult population was literate. Today, about one-fourth of the population is poor, life expectancy has increased to 65 years, and about 70 percent of adults are literate (UN Habitat 2004).

Despite this remarkable progress in reducing poverty in Asia over the last 25 years, the region still accounts for two-thirds of the world's poor, or as many as 800 million people, of whom 240–260 million reside in urban areas. The majority of the urban poor are concentrated in Bangladesh, India, Indonesia, Pakistan, and China. Six coastal megacities in the region—Beijing, Mumbai, Calcutta, Jakarta, Shanghai, and Tianjin—have the largest concentrations of urban poor. Nearly one-third of these poor people lack access to safe water and more than two-thirds lack access to adequate sanitation. According to United Nations statistics, one billion people are slum dwellers today. The number is expected to double by 2030. It is estimated that about half of the world's urban slum dwellers live in Asia and the Pacific. In 2001, 38 percent of the region's 3.7 billion people were living in urban areas, with 43 percent of its 1.2 billion urban residents residing in slums. It is estimated that only 38.3 percent of these informal settlements in the Asia-Pacific region have piped water connection and only 7.4 percent are served with proper sewerage disposal. Reflecting the size of their population, India and China together accounted for 65 percent of the overall Asian urban slum population (UN-Habitat 2003, 2004).

Particular attention should be paid to poor women in urban areas. Most of the migrant women in lower-income countries end up living in the urban slums (UN-Habitat 2003). In many of the rapidly urbanising cities, poor women tend to suffer the most. In addition to the general hardship of an unhygienic and unsafe environment, urban-poor women also have to bear the double burden of working for money and raising children under the most unfavourable circumstances. Women are more likely to suffer from poverty, malnutrition, and diseases. They are vulnerable to crime and violence and, because women are often less educated than men, they have fewer opportunities to escape from the cycle of poverty.

7.5.3 Environmental Issues

Megacities cause environmental problems on different geographical scales. On a local scale the problems are related to air and water pollution and to solid waste production. Regionally additional problems are water supply, protection of groundwater resources, saline intrusion of fresh water supplies, and land subsidence. Further, more than 3 billion people rely on coastal and marine habitats for food, building sites, transportation, recreation, and waste disposal. On a global scale the urban flow of energy, water, and materials contributes to climate change and sea level rise, depletion of the ozone layer, and loss of biodiversity.

Clearly, coastal cities face problems of how to manage rapidly increasing population growth and a variety of environmental risks. The concerns for coastal populations include access to fresh water, water quality, waste and sewage disposal, and health care (Kullenberg 2001). The problems common to megacities, particularly in the developing countries, include inadequate land for development, underdeveloped infrastructure, water shortages, poor sanitation, air pollution, and traffic congestion (Kullenberg 2001).

The coastal megacities in Asia face severe problems as a result of utilisation and misuse of resources in their immediate areas and hinterlands. They are experiencing additional problems related to their coastal location, including: coastal erosion, the potential impact of sea water intrusion to freshwater supply, the loss of natural habitats, the depletion of fishery resources as a food supply, public health problems related to sea food contamination, land subsidence due to construction and water extraction, the deterioration of the marine environment as an area for recreation and tourism, natural disasters and extreme weather, harmful algal blooms, global warming and sea-level rise, conflicting uses of coastal areas, and so on (IOC 2000). Jiang, Kirkman and Hua (2001) provide more detail on existing land-based pollution problems, including its causes, pollutant pathways to the marine and coastal environment, and recommended solutions.

Globalisation and rapid development have also accompanied some of the worst environmental problems of cities. It is possible to identify the typologies of urban development and environmental conditions within cities (Marcotullio 2003). First, environmental issues are classified as "brown" (water supply, sanitation, and infectious diseases), "gray" (air and water pollution, other negative aspects of industrial processes) or "green" (consumption-related problems, ecosystem health, ozone depletion, and greenhouse gas (GHG) emissions). Using these categories, the coastal cities in the Asia-Pacific region can be grouped into "hotspots" in terms of the seriousness of environmental issues: (1) low income cities outside the regional city system struggling with basic "brown" issues (Hanoi, Vientiane, Yangon) or a mix of "brown" and "gray" environmental issues (Phnom Penh and Ho Chi Min City); (2) rapidly industrialising cities (Bangkok, Jakarta, Shanghai) with uncontrolled population growth face a mix of environmental problems, e.g. air pollution, water pollution, serious lack of sewer service, and solid waste; (3) high income cities (Tokyo, Seoul, Taipei) with associated environmental issues such as air pollution and water pollution; and (4) non-hotspots (Singapore, Hong Kong) are exceptions with their well-managed environments. The problems of megacities were highlighted in the first Coastal Zone Asia-Pacific Conference, held in Bangkok, May 2002 (Chuenpagdee and Pauly 2004).

7.5.6 Examples: Bangkok, Jakarta, Mumbai, and Shanghai

The following megacities are examples to show the variety of urban environmental issues faced in the Asia-Pacific region. They illustrate not only the sustainability of large cities, but also issues related to population growth, urbanisation, environment, globalisation, and other aspects.

"Bangkok is almost invariably cited as an example of all that is wrong with unfettered urban growth" (Smith 2001: 443). It is one of the most primate cities in the world, containing 70 percent of the country's urban population. Bangkok is about 50 times larger than the second largest city in Thailand. The lack of a coordinated central administration gives rise to many problems. Its basic problems include congestion and shortfalls in providing basic needs. It is infamous for its traffic congestion and environmental pollution is severe. Only two percent of the industrial and domestic wastes are treated, and the rest are discharged by major waterways to the sea. The poorest households experience the worst environmental problems (Smith 2001).

During the 30 years preceding the Asian financial crisis of 1997/1998, urban development in the Jakarta Metropolitan Region (JMR) was stimulated by both foreign and domestic investment. Rapid economic growth induced migration to the city, and in 1999 the JMR had an estimated population of nearly 20 million (Li 2003). The economy of the JMR was hurt badly by the economic crisis, which resulted in increased unemployment and poverty. Urban poverty is the most pronounced issue in JMR development and planning (Firman 1999). In short, the Asian economic crisis has now adversely changed JMR from a "Global City" into a "City of Crisis", similar to the major cities in Latin America and Africa during the 1980s (Firman 1999). Jakarta is an example of a third world metropolis with problems of overpopulation and inadequate housing, employment, transportation, and environmental quality (Cybriwsky and Ford 2001). The impact of the economic crisis on urban development also shows that globalisation has both positive and negative impacts on urban development (Firman 1998).

The Mumbai Metropolitan Region (MMR), covering 4,400 km^2, is the largest coastal city in India and the second largest coastal city in world. The population of the MMR is about 16 million, 68 percent of which is contained in greater Mumbai—an area of only 468 km^2 (Murthy, Rao and Inamdar 2001). The high population density and the constant influx of people are responsible for its overall environmental degradation. The main problems in the MMR coastal region are a lack of land-use zoning, inadequate residential and industrial water supply and waste disposal, degraded air quality as a result of auto pollution, soil pollution, noise pollution, coastal marine pollution, and the depletion of important coastal habitats (wetlands and mangroves). Nearly 60 percent of greater Mumbai's population lives in slums and squatter settlements, which generate around 7,500 tons of solid waste daily. This partially treated waste is dumped at four sites along the creeks. The MMR coastal region is unsuitable for recreation and tourism as a result of the fouling smell and unaesthetic conditions. Facilities are available to treat 390 million litres per day (MLD) of wastewater out of 1,200 MLD generated (http://oceansatlas.org). (http://www.oceansatlas.org/servlet/CDSServlet?status=ND0xOTYyOCY2PWVuJjMzPSomMzc9a29z).

Shanghai is a coastal city and the largest megacity in China, with a population of some 14 million. It has achieved a GDP average annual growth rate of over ten percent from 1992–2000. It is rapidly developing into an international city and a high-ranking international shipping, banking, and trade centre (Li 2003). However, it is saddled with growing environmental pollution, including land-based pollution that eventually ends up in the coastal areas. For historical reasons, factories and residential quarters are interlocked in some areas of Shanghai, where pollution affects the life of residents. The city discharges 2.5 million tons of industrial sewage and 3 million tons of human sewage into the Yangtze River every day. It has a capacity to treat about half a million tons of sewage daily. As a result, red tides have occurred in 2000 and 2001 (http://oceansatlas.org). (http://www.oceansatlas.org/servlet/CDSServlet?status=ND0xOTY0MiY2PWVuJjMzPSomMzc9a29z).

7.6 RESEARCH

Research on population, urbanisation, coastal cities, and the coastal areas has been conducted sectorally with a strong separation between various disciplines. In recent years, there are more integrated or inter-disciplinary approaches to the issues of coastal cities, especially coastal megacities.

7.6.1 Population and Urbanisation

Some generalisations can be made on the topics and the agencies involved in existing research on population, urbanisation, and coastal megacities as hot spots. Of the various research topics on urbanisation in the Asia-Pacific region, China's urbanisation has attracted a lot of research (Ma 2002) as its trajectory of urban development is different from the experiences of other countries in the region.

International and regional organisations provide important databases on population and urbanisation for research by other specialised centres. For example, the UN Statistical Division and the Population Reference Bureau provide population and urbanisation data. Within the Asia-Pacific region there are a number of ongoing research programmes and centres conducting research on population and urbanisation, and some of these are described below.

The UN Human Settlements Programme (UN-Habitat), which started in 1978, has been a full-fledged programme in the UN system since 2002. Its goals on urban development are clearly stated in the Millennium Declaration (UN's development agenda for the next 15–20 years). One major goal is to improve the life of slum dwellers. Overall, UN-Habitat focuses on monitoring global trends and conditions and assessing progress in implementing the Habitat Agenda at the international, regional, national, and local levels. Its regional office for the Asia-Pacific region is in Fukuoka, Japan (http://unhabitat.org) (http://www.unhabitat.org/about/about_overview.asp).

UN-Habitat runs a number of programmes related to cities and urbanisation. Of these, the Urban Management Programme (UMP) represents a major effort by UN-Habitat and the United Nations Development Programme (UNDP), together with external support agencies, to strengthen the contribution that cities and towns in developing countries make towards economic growth, social development, and the alleviation of poverty. Now in Phase 4 (2002–2006), the programme is working to institutionalise UMP participatory processes, consolidate experiences, and deepen knowledge and understanding of urban management. One regional network in the Asia-Pacific region, the Urban Resource Centre in Asia Pacific (URCAP) is based in Bangkok, Thailand, with a sub-regional support office in the AIILSG (All India

Institute of Local Self-Government) in Mumbai, India (http://www.unhabitat.org). (http://www.unhabitat.org/programmes/ump/).

Specific to the Asia-Pacific region is the ESCAP Statistical Division, which maintains population-related bibliographic information on a computerised database. ESCAP's Population and Social Integration Section (PSIS), among many other objectives, conducts research and facilitates the exchange of information on population and social integration and the setting of regional goals (http://www.unescap.org). (http://www.unescap.org/esid/psis/population/popin/asia.asp).

The ESCAP Statistical Division publishes the *Asia-Pacific Population Journal* (quarterly) and the *Asia Population Studies Series* that cover topics on population and development, population and environment, migration and urbanisation, mortality and ageing, among others (http://www.unescap.org) (http://www.unescap.org/esid/psis/population/popseries/bytopics.asp).

The East-West Center in Hawaii lists population and health as one of its four study areas. It conducts basic and applied research on population, offers professional education and training, and facilitates the exchange of information. Research collaborations have been with institutions in the United States, the Asia-Pacific region, Canada, and Europe. Further details of research can be found on their webpage (http://www.eastwestcenter.org). (http://www.eastwestcenter.org/res-ph.asp). Research is discussed in *Asia-Pacific Population Research Reports*, published since 1995, and a four-page policy brief, *Asia-Pacific Population & Policy*, is issued quarterly.

The Australian Population Research Network is hosted by the Australian Centre for Population Research (ACPR) at the Research School of Social Sciences, Australian National University in Canberra. It seeks to build on its leadership in population research to encourage greater inter-disciplinary work and more collaboration with international researchers. The participants in the Network include researchers in New Zealand, United States, United Kingdom, Singapore, and Japan (Australian Population Research Network n.d.).

In Tokyo, Japan, the Nihon University Population Research Institute of the Nihon University College of Economics is the only university research institute in the country to specialise in international research on population, mainly in the Asia-Pacific region. It publishes a "Research Paper Series" (http://www.eco.nihon-u.ac.jp). (http://www.eco.nihon-u.ac.jp/english/research_institute/research_institute.html).

The Hong Kong Institute of Asia-Pacific Studies, Chinese University of Hong Kong carries out research that includes urbanisation, regional development, and Asia-Pacific economies and publishes the *Asian Economic Journal* (http://www.cuhk.edu.hk) (http://www.cuhk.edu.hk/hkiaps/homepage.htm).

7.6.2 Megacities

In recent years, international and specialised organisations have regarded coastal cities in the Asia-Pacific region as a research "hot spot". The continued growth of MURs poses new challenges to urban and regional planning in the Asia-Pacific region, and increasing recognition is being given to the regional dimensions and the governance systems of MURs (Smith 2001). More importantly and fundamentally is the prominent issue that faces all the coastal megacities: how to balance rapid economic development and the preservation of coastal environments through sustainable development, as called for by Agenda 21, Chapter 17. This is where the concept of Integrated Coastal Area Management (ICAM) becomes pertinent, and where scientists and managers working together may be able to provide some answers.

In the 2003–2005 UNEP Strategy for the Asia-Pacific region, the key issues related to the coastal cities have been identified for the subregions. The key issues for Southeast Asia are urban air pollution, urbanisation and waste disposal/management, and degradation of the coastal and marine environment. Urbanisation and poverty are the issues for South Asia, and degradation of the marine environment is a priority focus for the South Pacific (UNEP 2003).

Work carried out by the Asia-Pacific Forum for Environment and Development (APFED), established at the 10th Environment Congress for Asia and the Pacific in October 2001, aims to propose a model of equitable and sustainable development for the Asia-Pacific region. It held its first substantial meeting in January 2002 and identified five major issues that require priority attention, namely freshwater resources, renewable energy, trade, finance, and urbanisation. It encourages countries in the region to take an integrated approach to manage urbanisation and to strengthen local initiatives in the management of urban environments. In relation to the above, it aims to collect and analyse best policy practices (http://www.iges.or.jp) (http://www.iges.or.jp/apfed-message/apfed.html).

Air Pollution in the Megacities of Asia (APMA) is a project sponsored by the Ministry of Korea and Swedish International Development Cooperation Agency. It was initiated in 2000 and provides the foundation for an online information network (http://www.asiairnet.org/). It completed a workshop on air pollution of major and megacities of Asia (Haq, Han and Kim 2002). Its key activity is to establish a benchmark for the status and management of urban air quality in major megacities of Asia, including the following coastal megacities: Bangkok, Beijing, Calcutta, Guangzhou, Hong Kong, Manila, Mumbai, Osaka, Seoul, Shanghai, Singapore, Taipei, and Tokyo.

The Institute for Global Environmental Strategies (IGES) was established as a research institute by the Japanese Government in 1998 to conduct strategic policy research supporting sustainable development in the Asia-Pacific region (http://www.iges.or.jp/en/outline/index.html). Its Urban Environmental Management Project proposes practical strategies for urban environmental management with the aim of making cities in Asia more sustainable; addressing specific issues, such as air pollution, solid waste, and wastewater management; and investigating cross-cutting issues, such as regulatory and institutional arrangements and financial mechanisms (http://www.iges.or.jp/en/about_project/index.html). The project develops useful ideas and models that serve as guidelines of urban environmental management policy in the Asian region. In particular, these models are relevant to local initiatives, which are increasingly recognised as important for achieving sustainable development. Research has been conducted in phases lasting three years: Phase 1 (1998–2001), Phase 2 (2001–2004), and (on-going) Phase 3 (2004–2007) (http://www.iges.or.jp) (http:// www.iges.or.jp/en/ue/index.html).

The IGES also supports the Kitakyushu Initiative, which is a mechanism to assist in the priority implementation of a regional programme (2001–2005) with specific focus on environmental quality and human health in urban areas. (http://www.iges.or.jp/kitakyushu/main_outline_background.htm) Further details of pilot activities of this initiative in the Asia-Pacific region are available on its website (http://www.iges.or.jp). (http://www.iges.or.jp/kitakyushu/main_pilot_projects.htm).

The GEF (Global Environmental Facility) is supporting a project called the Transboundary Diagnostic Analysis for the South China Sea, which aims to reverse environmental degradation in the South China Sea and the Gulf of Thailand. The project has produced a report that identifies 35 pollution hot spots and 26 sensitive and high-risk spots around the South China Sea. Many of these areas are associated with coastal megacities (Talaue-McManus 2000).

PEMSEA or Partnerships in Environmental Management for the Seas of East Asia is implementing the Bohai Sea Project as part of its Sustainable Development Strategy. The objective of the project is to develop collaborative actions among adjacent provinces and municipalities in order to reduce waste discharges. Beijing, Tianjin and Shenyang are within the coastal area of the Bohai Sea, which is threatened by pollution (http://www.pemsea.org). (http://www.pemsea.org/abt%20pemsea/sites/demo_hotspt_bohai.htm).

The United Nations University in Tokyo carries out research on sustainable management of the coastal hydrosphere in the Asia-Pacific region. These studies cover several areas relevant to megacities: population growth; urbanisation; and industrialisation and large-scale development, e.g. reclamation (Adeel and King 2002).

Several conferences have focused specifically on megacities:

1. The Mayors' Asia-Pacific Environmental Summit, 31 January to 3 February, 1999, held in Honolulu, Hawaii, was aimed to promote sustainable cities and share best practices in pursuing sustainable development at the local level. It was an opportunity for businesses to exhibit much needed technologies, products, and services (Hausker and Khan 1999).
2. The International Workshop on Coastal Mega-cities, 27–30 September 1999, was held in Hangzhou, China (http://www.icsea.or.id). (http://www.icsea.or.id/sea-span/0399/MA0510LL.htm) in co-operation with the IOC (Intergovernmental Oceanographic Commission) and the Chinese government's State Oceanic Administration. The workshop focused on the rapid development of coastal megacities and the related environmental problems (IOC 2000).
3. The Asia Research Institute, National University of Singapore, held a conference on Growth Dynamics of Mega-Urban Regions in Asia in June 2004, and the papers are to be published in a book.

The Hangzhou meeting produced the Hangzhou Declaration, which is the first declaration addressing the issues of coastal megacities (Li 2003). The declaration can serve as a firm basis for further research and recommends the following:
- Recognise integrated coastal area management, including sustainable urban and coastal systems, as an appropriate tool to affectively address management and planning of coastal areas;
- Increase education, training, and public awareness to better understand interactions between the coastal and urban environments;
- Form a network of coastal megacities for a global form to exchange ideas on various issues; and
- Carry out case studies/assessments in order to identify indicators on interrelationships between population pressures, environmental conditions, and economic development (http://icm.noaa.gov) (http://icm.noaa.gov/globalinfo/hangzhou.html).

The research on megacities is facilitated by some new methodologies. In addition to the traditional analyses of data from primary sources (e.g. population census) or secondary sources (UN sources), some new methodologies have emerged on the study of urban growth of MURs. Heikkila, Shen and Yang (2003) proposed a method using the mathematics of fuzzy sets to measure and characterise periurbanising (*desakota*) systems of China, Southeast Asia, and other areas experiencing rapid urbanisation. Other mathematical tools include a methodology to generate a World City Index, using Shanghai as case study (Cai and Sit 2003) and using an empirical test to relate primacy with economic development for countries in Asia and the Americas (Moomaw and Alwosabi 2004).

Remote sensing is a potential tool for collecting observations that are relevant to the study of urban growth and environmental conditions and are not available from other sources (Miller and Small 2003). One area of particular interest is the use of remote sensing to monitor urban sprawl. GIS-based spatial analysis and modeling can be used to study ecological impacts and reveal salient features of landscape dynamics in the *desakota* regions. Remote sensing also supports studies for actual and effective water resources management, e.g. water quality, water quantity, floods, droughts, accessibility, pollution disasters, sediment fluxes, etc. Risk assessment studies can also be made in relation to natural hazards and global change.

New concepts in urban planning are being tried out, such as mixing or merging urban and rural land for an ordered growth in order to reduce serious environmental problems (Yokohari et al. 2000). Other strategies include improving the rural-urban development partnership, developing income-earning opportunities in the rural areas to stem the tide of migration into larger cities, and increasing elements of urbanisation in rural life through micro enterprises, etc. (Epstein and Jezeph 2001). Integrated Coastal Management remains a significant overall framework (Murthy, Rao and Inamdar 2001).

In its new research agenda, the APN (Asia-Pacific Network for Global Change Research) has recognised the increasing population and urbanisation pressure at the coastal zone and the issues related to coastal cities, especially megacities (Harvey, Rice and Stevenson 2005). Arising from the earthquake and tsunami of

26 December 2004 off the west coast of Sumatra, research on coastal cities need to consider aspects such as coastal habitat restoration, coastal protection, and hazard mitigation and management.

7.7 CONCLUSION

The Asia-Pacific coastal countries display highly varied demographic and socio-economic levels and trends, but they face the challenges of an endangered coastal zone. Rapid urbanisation, development, and population growth in the coastal areas have resulted in massive destruction of the ecosystems, such as the loss of coastal wetlands, coral reefs, fisheries and land degradation. Coastal cities have consumed vast amounts of natural resources and produced worrisome volumes of wastes and pollution. The existence of extremely poor dwellers in coastal cities in low-income countries has also rung a bell about the non-sustainability of the current development model. More sustainable development of the coastal region requires more comprehensive, integrated, and forward-looking strategic planning and management. Rather than only focusing on short-term economic gains, serious consideration and priority should also be given to the environmental and social well-being of communities.

Continued urbanisation is likely to accelerate the growth of more megacities and see more growth in medium-sized cities. The Asia-Pacific megacities have been configured and will continue to be reconfigured by globalisation processes. Globalisation has not created a single path for urban development and has also produced some of the worst environmental conditions (Marcotullio 2003).

The major question to be asked is whether the megacities in the Asia-Pacific are sustainable cities? What must be done, and done urgently, to assure sustainability? Will the Asia-Pacific megacities be synonymous with gargantuan, squalid cities?

It is clear that what is required is commitment and good governance not just to implement reactive strategies and *ad hoc* policies, but also to give cognisance to issues of livelihood, environmental issues, and conservation of heritage buildings. It would be beneficial to share lessons and expertise with developed countries that have well-established research, experience, and knowledge in sustainability issues. Learning from developed countries, development planning must account for the costs of preventing the destruction of natural resources, and managing pollution, and protecting the underprivileged in urban areas (Sazanami 2000).

In developing solutions to the issues described in this chapter, the importance of multi-layered governance and integrated policy responses at all scales of governance should be recognised (Marcotullio 2003). It is also necessary to bring an integrated understanding to solving environmental problems, such as an ecosystem approach that considers both the complex interactions between the hinterlands and the marine environments (Marcotullio 2003). Integrated policies are needed for the environmental problems of cities. Also, the management of urban environments needs to recognise various scales: the physical environment within cities, the environmental systems of cities and their regional hinterlands, and the global environmental created by a network of megacities (Miller and Small 2003). Ready technologies and new techniques are available. Regional programmes require further coordinated research inputs from various groups of scientists.

REFERENCES

Adeel, Z. and King, C. (eds) 2002. *Conserving Our Coastal Environment*. United Nations University, Tokyo.

AFL-CIO Working Women's Department 2005. *Workers' Rights are Women's Rights*. (http://www.aflcio.org/yourjobeconomy/women/global/WorkersRights.cfm)

Anh, D.N. 2003. Internal migration policies in the ESCAP region. *Asia-Pacific Population Journal*, 18(3): 27–40.

Asian Development Bank 1997. The Asian Development Bank on Asia's megacities. *Population and Development Review*, 23: 451–459.

Australian Population Research Network n.d. (http://www.geosp.uq.edu.au/qcpr/aprn/overview.pdf).

Cai, J. and Sit, V.F.S. 2003. Measuring world city formation – the case of Shanghai. *Annals of Regional Science*, 37: 435–446.

Chen, A. and Coulson, E. 2002. Determinants of urban migration: evidence from Chinese cities. *Urban Studies*, 39: 2189–2198.

China Daily, 5.11.2003. Loosening family planning policy unconducive to ageing problem. (http://www.chinadaily.com.cn/en/doc/2003–11/05/content_278715.htm)

China Daily, 8.1.2004. Millions to be laid-off as SOEs streamline. (http://www.chinadaily.com.cn/en/doc/2004–01/08/content_296959.htm)

Cohen, B. 2004. Urban growth in developing countries: a review of current trends and a caution regarding existing forecasts. *World Development*, 32: 23–51.

Chuenpagdee, R. and Pauly, D. 2004. Improving the state of the coastal areas in the Asia-Pacific region. *Coastal Management*, 32: 3–15.

Creel, L. 2003. *Ripple effects: Population and coastal regions*. Population Reference Bureau policy brief. (http://www.prb.org/pdf/RippleEffects_Eng.pdf)

Cybriwsky, R. and Ford, L.R. 2001. Jakarta. *Cities*, 18: 199–210.

Dang, A. 1999. Market reforms and internal labour migration in Vietnam. *Asian and Pacific Migration Journal*, 8: 3181–409.

Daniere, A. and Takahashi, L.M. 2002. *Rethinking Environmental Management in the Pacific Rim*. Ashgate: Aldershot.

Douglass, M. 2000. Mega-urban regions and world city formation: globalization, the economic crisis and urban policy issues in Pacific Asia. *Urban Studies*, 37: 2315–2335.

East-West Center 2002. A 'snapshot' of populations in Asia. *Asia-Pacific Population and Policy*, April 2002, no. 59: 1–4. (http://www2.eastwestcenter.org/pop/misc/p&p-59.pdf)

Epstein, T.S. and Jezeph, D. 2001. Development – there is another way: a rural-urban partnership development paradigm. *World Development*, 29: 1443–1454.

ESCAP 2003. Highlights of the 2003 ESCAP Population Data Sheet, *Asia-Pacific POPIN Bulletin* (2003) Vol. 15. No. 1, Jan.–April. (http://www.unescap.org/esid/psis/population/popin/bulletin/2003/v15n01.htm#n)

ESCAP 2004. *Population Data Sheet*. (http://www.unescap.org/esid/psis/population/database/data_sheet/2004/index2.asp)

Fan, C.C. 2002. Population change and regional development in China: insights based on the 2000 census. *Eurasian Geography and Economics*, 43: 425–442.

Firman, T. 1998. The restructuring of Jakarta Metropolitan Area: A 'global city' in Asia. *Cities*, 15: 229–243.

Firman, T. 1999. From 'Global City' to 'City of Crisis': Jakarta Metropolitan Region under economic turmoil. *Habitat International*, 23: 447–466.
Friedmann, J. 1986. The world city hypothesis. *Development and Change*, 17: 69–83.
Goodkind, D. and West, L.A. 2002. China's floating population: definitions, data and recent findings. *Urban Studies*, 39: 2237–2250.
Haq, G., Han, W.J. and Kim, C. 2002. *Urban Air Pollution Management and Practice in Major and Mega Cities of Asia*. Korea Environment Institute, Seoul.
Harvey, N., Rice, M. and Stevenson, L. (eds) 2005. *APN Global Change Coastal Zone Management Synthesis Report*. Asia-Pacific Network for Global Change Research (APN), Kobe.
Hausker, K. and Khan, S. (eds) 1999. *Mayors' Asia Pacific Environmental Summit. Summit Report.* (http://www.csis.org/e4e/MayorReport.pdf)
Heikkila, E.J., Shen, T.Y. and Yang, K.Z. 2003. Fuzzy urban sets: theory and application to desakota regions in China. *Environment and Planning B-Planning and Design*, 30: 239–254.
Heinke, G.W. 1997. The challenge of urban growth and sustainable development for the Asian cities in the 21st century. *Environmental Monitoring and Assessment*, 44: 155–171.
Hill, R.C. and Kim, J.W. 2000. Global cities and development states: New York, Tokyo and Seoul. *Urban Studies*, 37: 2167–2195.
Hinrichsen, D. 1995. *Coasts in Crisis*. (http://www.aaas.org/international/ehn/fisheries/hinrichs.htm)
Ho, S.P.S. and Lin, G.C.S. 2004. Converting Land to Nonagricultural Use in China's Coastal Provinces: Evidence from Jiangsu. *Modern China*, 30: 81–112.
Hong Kong Planning Department 2002. *Ecological Footprint*. Working Paper 10. (http://www.info.gov.hk/hk2030/hk2030content/wpapers/wpaper_10/e_wpaper10.htm)
Hutton, T.A. 2004. Service industries, globalization, and urban restructuring within the Asia-Pacific: new development trajectories and planning responses. *Progress in Planning*, 61: 1–74
IOC 2000. *Intergovernmental Oceanographic Commission Workshop Report No. 166*. IOC-SOA International Workshop on Coastal Megacities: *Challenges of Growing Urbanisation of the World's Coastal Areas*, Hangzhou, People's Republic of China, UNESCO, Paris, 27–30 September 1999.
Islam, R. 1997. Economic development in Asia and the Pacific in the 21st century: issues and challenges. Paper presented at the *ILO Workshop on Employers' Organizations in Asia-Pacific in the Twenty-First Century*, Turin, Italy, 5–13 May 1997.
Jiang, Y.H., Kirkman, H. and Hua, A. 2001. Megacity development: managing impacts on marine environments. *Ocean and Coastal Management*, 44: 293–318.
Jones, G. 2003. Population and poverty in Asia and the Pacific. In *Fifth Asian and Pacific Conference Selected Papers*, Asian Population Studies Series No. 158, UN, pp. 31–51.
Kullenberg, G. 2001. Contributions of marine and coastal area research and observations towards sustainable development of large coastal cities. *Ocean and Coastal Management*, 44: 283–291.
Lee, B.T. 2000a. The borderless economy, rent decontrol and the urban morphology in Penang. *Malaysian Journal of Tropical Geography*, 33: 45–53.
Lee, B.T. 2000b. Core issues in Asian megacities: urban sustainability and the superinduced development of the central urban core. Paper presented at the International Symposium on *21st Century Asia: Economic Restructuring and Challenges of Mega-Cities*, Institute for Economic Research, Osaka, 26–27 Sept. 2000.
Li, H. 2003. Management of coastal mega-cities—a new challenge in the 21st century. *Marine Policy*, 27: 333–337.
Lin, G.C.S. 2001. Metropolitan development in a transitional socialist economy: spatial restructuring in the Pearl River Delta, China. *Urban Studies*, 38: 383–406.
Lin, G.C.S. 2002. The growth and structural change of Chinese cities: a contextual and geographic analysis. *Cities*, 19: 299–316.
Lin, G.C.S. 2004. A theme issue on planning for China's large cities in the era of globalization. *Progress in Planning*, 61: 137.
Lo, F.C. and Marcotullio, P.J. 2000. Globalisation and urban transformation in the Asia-Pacific region: a review. *Urban Studies*, 37: 77–111.

Lo, F.C. and Yeung, Y.M. (eds) 1996. *Emerging World Cities in Pacific Asia*. United Nations University Press, Tokyo.

Ma, L.J.C. 2002. Urban transformation in China, 1949–2000: a review and research agenda. *Environment and Planning A*, 34: 1545–1569.

Ma, L.J.C. 2004. Economic reforms, urban spatial restructuring, and planning in China. *Progress in Planning*, 61: 237–260.

Marcotullio, P.J. 2003. Globalization, urban form and environmental conditions in Asia-Pacific cities. *Urban Studies*, 40: 219–247.

McGee, T.G. 1991. The emergence of desakota regions in Asia: expanding a hypothesis. In N. Ginsburg, B. Koppel and T.G. McGee (eds), *The Extended Metropolis: Settlement Transition in Asia*. Honolulu: University of Hawaii, pp. 3–25.

McGee, T.G. and Robinson, I.M. 1995. *The Mega-Urban Regions of Southeast Asia*, Vancouver: UBC Press.

McGee, T.G. and Yeung, Y.M. 1994. Urban futures for Pacific Asia: towards the 21st Century. *Urban Voice*, 6: 1–4.

Migration News, 2005. (http://migration.ucdavis.edu/mn).

Miller, R.B. and Small, C. 2003. Cities from space: potential applications of remote sensing in urban environmental research and policy. *Environmental Science and Policy*, 6: 129–137.

Moomaw, R.L. and Alwosabi, M.A. 2004. An empirical analysis of competing explanations of urban primacy evidence from Asia and the Americas. *Annals of Regional Science*, 38: 149–171.

Murthy, R.C., Rao, Y.R. and Inamdar, A.B. 2001. Integrated coastal management of Mumbai metropolitan region. *Ocean and Coastal Management*, 44: 355–369.

Ng, M.K. and Hills, P. 2003. World cities or great cities? A comparative study of five Asian metropolises. *Cities*, 20: 151–165.

Osada, S. 2003. The Japanese urban system 1970–1990. *Progress in Planning*, 59: 125–231

Pannell, C.W. 2002. China's continuing urban transition. *Environment and Planning A*, 34: 1571–1589

Pannell, C.W. 2003. China's demographic and urban trends for the 21st century. *Eurasian Geography and Economics*, 44: 479–496.

Population Reference Bureau 2004. *World Population Data Sheet*. http://www.prb.org/pdf04/04WorldDataSheet_Eng.pdf

Roberts, D. 2000. The great migration: Chinese peasants are fleeing their villages to chase big-city dreams, *Business Week Online*, 11 December. (http://www.businessweek.com/archives/2000/ b3711018.arc.htm)

Roberts, D. 2002. The other specter: Labor unrest. *Business Week Online*. 25 November. (http://www.businessweek.com/magazine/content/02_47/b3809167.htm)

Saito, A. and Thornley, A. 2003. Shifts in Tokyo's world city status and the urban planning response. *Urban Studies*, 40: 665–685.

Sazanami, H. 2000. Challenges and future prospects of planning for a better living environment in the large cities in Asia. In Yeh, A.G.O. and Ng, M.K. (eds) *Planning for a Better Urban Living Environment in Asia*, Ashgate: Aldershot, pp. 5–22.

Shi, Y. and Hamnett, C. 2002. The potential and prospect for global cities in China: in the context of the world system. *Geoforum*, 33: 121–135.

Smith, D.A. 2004. Global cities in East Asia: empirical and conceptual analysis. *International Social Science Journal*, 56: 399–412.

Smith, D.W. 2001. Cities in Pacific Asia. In R. Paddison (ed.) *Handbook of Urban Studies*, London: Sage Publications, 419–450.

Talaue-McManus, L. 2000. *Transboundary Diagnostic Analysis for the South China Sea*. EAS/RCU Technical Report Series No. 14. UNEP, Bangkok, Thailand.

UN 2004a. Urban Agglomerations 2003. (http://www.un.org/esa/population/publications/wup2003/ 2003 UrbanAgglomeration2003_Web.xls)

UN 2004b. *World Urbanization Prospects: The 2003 Revision*. UN Department of Economic and Social Affairs/Population Division, New York.

UN-Habitat 2003. *The Challenge of Slums: Global Report on Settlements*. Nairobi: United Nations Human Settlements Programme.

UN-Habitat 2004 *State of the World's Cities 2004/5: Globalization and Urban Culture.* Sterling, VA.: Earthscan.
UN System-Wide Earthwatch 2005. *Oceans and Coastal Areas.* (http://earthwatch.unep.net/oceans/coastalthreats.php)
UNEP 2003. *Strategy for UNEP Asia and the Pacific 2003–2005.* Bangkok, Thailand. (www.roap.unep.org/document/ROAP%Strategy%202003–05%20Final.doc)
UNFPA (United Nations Population Fund) 2004. *State of World Population 2004. The Cairo Consensus at Ten: Population, Reproductive Health and the Global Effort to End Poverty.* New York: UNFPA.
Wackernagel, M. and Rees, W. 1996. *Our Ecological Footprint: Reducing Human Impact on the Earth.* Gabriola Island, BC: New Society Publishers.
Webster, D. 1995. The urban environment in Southeast Asia: challenges and opportunities. In Institute of Southeast Asian Studies (ed) *Southeast Asian Affairs*, pp. 89–107. Singapore: ISAS.
Yap, K.S. 2003. Urbanisation and internal migration. In *Fifth Asian and Pacific Conference Selected Papers*, Asian Population Studies Series No. 158, UN, pp. 137–155.
Yeung, Y.M. 2001. Coastal mega-cities in Asia: transformation, sustainability and management. *Ocean and Coastal Management*, 44: 319–333.
Yokohari, M., Takeuchi, K., Watanabe, T. and Yokota, S. 2000. Beyond greenbelts and zoning: A new planning concept for the environment of Asian mega-cities. *Landscape and Urban Planning*, 47: 159–171.
Zhang, K.H. 2002. What explains China's rising urbanization in the Reform Era. *Urban Studies*, 39, 2301–2315.
Zhang, K.H. and Song, S.F. 2003. Rural-urban migration and urbanization in China: evidence from time-series and cross-section analyses. *China Economic Review*, 14: 386–400.
Zhu, Y. 2000. In situ urbanization in rural China: case studies from Fujian Province. *Development and Change*, 31: 413–434.

CHAPTER 8

PRESSURES ON RURAL COASTS IN THE ASIA-PACIFIC REGION

LIANA TALAUE-McMANUS
Division of Marine Affairs, Rosenstiel School of Marine and Atmospheric Science, University of Miami, USA

8.1 INTRODUCTION

The rural coast of the Asia-Pacific region is home to about 1 billion inhabitants, accounting for about a third of the region's population and contributing one-sixth of the planet's population. Of this number, around 5.5 million live along the rural coasts of small island developing nations in the Indian and Pacific Oceans. The coastal rural population consists mostly of farmers, fishers, fish farmers, and small-scale tourism entrepreneurs, all of who depend on the soil and sea for their livelihoods and are most vulnerable to global environmental change and an increasingly globalising trade. The devastating tsunami that struck Indonesia, Thailand, and parts of South Asia in December 2004 dramatised the lack of institutional safety nets that are most crucial to enhancing the resilience of coastal societies to cope with global change in this region.

This chapter provides an overview of the natural resource-dependent rural economies of the region and how these interact with the coastal environment in creating tightly linked issues of population, poverty, and environmental health. It discusses the major livelihoods of farming, fishing, and tourism; attendant environmental costs; and the all-prevailing conditions of poverty. When one overlays global environmental change and international trade against this backdrop, it becomes obvious that creative ways to strengthen institutions and capacity should be sought to increase the potential for sustainable futures in this region, as well as others in the tropical developing world.

Throughout this chapter, the term Asia-Pacific region is used to include the developing and developed countries of East, Southeast, and South Asia, the Russian

Federation as well as those in the Pacific. This geographic domain was used in making quantitative estimates of population, consumption, and trade for the region. Where appropriate, data at country scales were compiled for the Asia-Pacific region for comparison with other developing and developed country groups. Where rural or agriculture-based data were available, these are presented together with regional averages for context. Given the focus on rural populations, economies of which are agriculturally based, data pertinent to the developing countries of the region are highlighted.

8.2 COASTAL RURAL POPULATION

The total population of the Asia-Pacific region constitutes about 57% of the global population (UN 2004, World Bank 2005). Because the region subsumes the most heavily populated countries of China and India, trends in the region's demographics profoundly influence global trends. Figure 8.1 shows a rapidly urbanising developing world across all regions, with highest growth rates evident in East and Southeast Asia for the period 1990 to 2003. Changes in rural populations, on the other hand, exhibit regional variability, with decreases in East-Southeast Asia and the Pacific (EAP), Europe, Central Asia, and in Latin America & Caribbean, and increases in South Asia, Middle East and North Africa, as well as

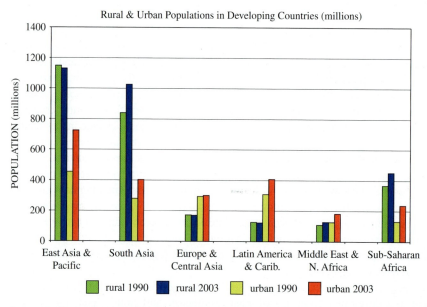

Figure 8.1. The rural and urban populations in developing countries (millions) (Data from World Development Indicators, World Bank 2005)

Table 8.1. Coastal Population in Asia-Pacific

Group	Coastal Urban[1]	Coastal Rural[2]	Total
Small Island Developing States (SIDS) (2003)	1,676,000	5,502,000	7,178,000
Non-SIDS Countries (2003)	373,199,000	1,011,703,000	1,384,902,000
Total	374,875,000	1,017,205,000	1,392,080,000
APN (2003)	(Urban) 1,396,092,000	(Rural) 2,211,491,000	(Total) 3,607,583,000
World (2003)	(Urban) 3,043,935,000	(Rural) 3,257,529,000	(Total) 6,301,464,000

[1] Country populations (UN 2004) were multiplied by percent population within 100 km of coast (Burke et al. 2000) to obtain coastal populations. Populations of coastal cities in each coastal state (Brinkhoff 2005) were added to estimate coastal urban population. Coastal cities included connected administrative units sensu Brinkhoff (2005).

[2] The difference between coastal total population less coastal urban population provide estimates for coastal rural population.

in Sub-Saharan Africa. Rural population in South Asia increased by 185 million from 1990 to 2003. In the EAP it decreased by 19 million over the same period. The UN (2004) projects that by the year 2030, the urban population worldwide will increase by 1.9 billion, while rural population will decrease by a meager 80 million.

Forty-six percent of the region's rural population, or slightly over 1 billion people, live within 100 km of the coast (Table 8.1). Six countries account for 80% of the coastal rural inhabitants. China contributes 250 million, India 206 million, Indonesia 180 million, Philippines and the Russian Federation 63 million each, and Vietnam 61 million.

8.3 ASIA-PACIFIC FISHERS

8.3.1 Fishing Capacity

FAO (2002) estimates that there were roughly 35 million fishers worldwide in 2000 and that 85% live in Asia. Assuming a family size of about 5 members, Delgado and co-authors (2003) believe that about 200 million people (fishers and their dependents) rely on fishing as primary income source. The Asia-Pacific fishing fleet consists of 1.1 million decked boats out of 1.3 million decked boats worldwide and 2.0 million out of 2.8 million undecked boats (FAO 2002). It is logical to assume that this fishing population accounts for a fifth of the total population living along rural coasts in the region.

8.3.2 Catch, Consumption, and Trade

How much do Asia-Pacific fishers catch, consume, and trade? Tables 8.2 to 8.6 provide the latest statistics, derived from Delgado et al. (2003). To highlight differences in catch, consumption, and trade, the data for food fish is disaggregated into 4 categories. High-value finfish include large pelagics and demersals, such tuna, cod, salmon and trout. Carp and small pelagics like herring and mackerel make up the low-value finfish group. Crustaceans and mollusks are the third and fourth categories, respectively. Capture and aquaculture production were aggregated for all categories.

Asia-Pacific fishers caught 6.4 million mt in 1997 or 24% of the world's high-value finfish production (Table 8.2). They consumed 4.9 million mt and exported 1.3 million mt. In contrast, the developed countries, hauled in 14.5 million mt in the same year or 57% of the global catch for this category. They consumed 4.2 million mt more than they caught by importing from the Asia-Pacific and other developing countries.

The picture dramatically changes for low-value finfish (Table 8.3). More specifically, the trend is the film negative of that for high-value food fish. Asia-Pacific fishers caught 71% of the global catch in 1997, and consumed 75%, the deficit made up by imports from the developed countries.

For crustaceans and mollusks, the trends are similar to high-value finfish. Fishers of the region caught 71% and 68% of global production for crustaceans and mollusks, respectively (Tables 8.4 and 8.5). They consumed less than they produced, and the surplus was exported to developed countries.

These patterns of consumption and trade raise immediate concerns about equity. Developing countries, including those in the Asia-Pacific, trade fishery products for foreign exchange and consume mostly cheap small pelagics including those imported

Table 8.2. High-value finfish production, consumption, and trade (Data from Delgado et al. 2003)

Region	Total Production (Capture + Aquaculture) (10^3 mt)			
	Actual			Projected
	1973	1985	1997	2020
Asia-Pacific (AP)	1974	3113	6351	9277
AP ÷ Developing World	*0.51*	*0.50*	*0.63*	*0.59*
AP ÷ Developed World	*0.12*	*0.17*	*0.42*	*0.56*
AP ÷ World	*0.09*	*0.13*	*0.24*	*0.29*
	Consumption (10^3 mt)			
Asia-Pacific (AP)	1479	2358	4935	8757
AP ÷ Developing World	*0.53*	*0.50*	*0.70*	*0.73*
AP ÷ Developed World	*0.08*	*0.12*	*0.26*	*0.45*
AP ÷ World	*0.07*	*0.10*	*0.19*	*0.28*
	Total Net Exports (10^3 mt)			
Asia-Pacific (AP)	428	695	1317	335
Other developing countries	536	682	2560	3006
United States	−1045	−565	−901	−1235
Japan	−245	−1037	−2073	−1903
European Union 15	−1140	−1231	−2521	−2081

Table 8.3. Low-value finfish production, consumption, and trade (Data from Delgado et al. 2003)

Region	Total Production (Capture + Aquaculture) (10^3 mt)			
	Actual			Projected
	1973	1985	1997	2020
Asia-Pacific (AP)	9384	13025	31609	46636
AP ÷ Developing World	0.66	0.63	0.80	0.81
AP ÷ Developed World	1.52	1.45	5.57	9.10
AP ÷ World	0.46	0.44	0.71	0.74
	Consumption (10^3 mt)			
Asia-Pacific (AP)	10242	14942	32583	46164
AP ÷ Developing World	0.66	0.63	0.79	0.78
AP ÷ Developed World	3.08	3.55	15.59	22.27
AP ÷ World	0.54	0.54	0.75	0.75
	Total Net Exports (10^3 mt)			
Asia-Pacific (AP)	−1163	−2022	−1192	162
Other developing countries	−1197	−3279	−1141	−2795
United States	146	367	422	690
Japan	966	3165	320	347
European Union 15	324	−377	129	144

Table 8.4. Crustacean production, consumption, and trade (Data from Delgado et al. 2003)

Region	Total Production (Capture + Aquaculture) (10^3 mt)			
	Actual			Projected
	1973	1985	1997	2020
Asia-Pacific (AP)	1110	1823	4986	7898
AP ÷ Developing World	0.75	0.77	0.86	0.87
AP ÷ Developed World	1.39	1.76	3.94	5.16
AP ÷ World	0.49	0.53	0.71	0.75
	Consumption (10^3 mt)			
Asia-Pacific (AP)	829	1314	4176	7490
AP ÷ Developing World	0.82	0.87	0.95	0.95
AP ÷ Developed World	0.65	0.69	1.58	2.76
AP ÷ World	0.36	0.39	0.59	0.71
	Total Net Exports (10^3 mt)			
Asia-Pacific (AP)	280	508	811	1218
Other developing countries	192	362	563	−30
United States	−187	−271	−432	−539
Japan	−159	−371	−712	−629
European Union 15	−151	−293	−518	−432

from developed nations. The latter, on the other hand, are net recipients with a preference for high-value, top-predator species. This preference results in a bias whereby the smaller populations of developed countries appropriate a significantly greater proportion of natural production capital (Table 8.6). The result is an inequitable situation in which the

Table 8.5. Mollusc production, consumption, and trade (Data from Delgado et al. 2003)

Region	Total Production (Capture + Aquaculture) (10^3 mt)			Projected
	Actual			
	1973	1985	1997	2020
Asia-Pacific (AP)	773	2192	9849	17768
AP ÷ Developing World	0.65	0.66	0.85	0.88
AP ÷ Developed World	0.27	0.58	2.50	3.96
AP ÷ World	0.19	0.31	0.68	0.72
	Consumption (10^3 mt)			
Asia-Pacific (AP)	771	2008	9267	17376
AP ÷ Developing World	0.71	0.76	0.88	0.90
AP ÷ Developed World	0.28	0.47	2.00	3.50
AP ÷ World	0.20	0.29	0.61	0.72
	Total Net Exports (10^3 mt)			
Asia-Pacific (AP)	1	184	582	390
Other developing countries	105	399	545	527
United States	−67	−256	−195	−444
Japan	−41	−252	−647	−478
European Union 15	−22	−239	−341	−74

Table 8.6. Per capita consumption of food fish products for the period 1996–1998 (Data from Delgado et al. 2003)

Region	Consumption (kg/person/year)				
	Low-value food fish	High-value food fish	Mollusks	Crustaceans	Total Food fish
Asia-Pacific					
China	15.8	1.5	6.8	2.4	26.5
Southeast Asia	15.9	3.8	1.4	1.9	23.0
India	3.6	0.9	0.0	0.2	4.7
Other South Asia	4.9	0.9	0.0	0.2	6.0
Latin America	5.3	1.7	0.7	0.2	7.8
West Asia & North Africa	5.1	0.8	0.1	0.1	6.2
Sub-Saharan Africa	5.9	0.6	0.1	0.0	6.7
United States	0.1	13.3	3.2	3.1	19.7
Japan	8.3	33.1	14.4	6.8	62.6
European Union 15	1.9	15.4	4.3	2.0	23.6
Eastern Europe & former Soviet Union	0.4	9.9	0.2	0.1	10.6
Other developed countries	1.2	9.9	2.2	1.4	14.7
Developing world	9.1	1.6	2.3	1.0	14.0
Developed world	1.6	14.5	3.6	2.0	21.7
World	7.5	4.4	2.6	1.2	15.7

poor are consuming five times as much low-quality fish food as their rich counterparts. Thus, a person living in a developing country consumes 9.1 kg of low-value food fish per year compared to his affluent brother who eats only 1.6 kg. The reverse is true for high-value finfish with developed world per capita consumption of 14.5 kg per person per year compared with 1.6 kg per person per year for developing countries.

Examining the resulting fish-based nutritional status in the Asia-Pacific region further highlights disparities in fish food consumption and protein supply (Table 8.7). Metrics for comparing fish-based nutritional status include daily per capita fish protein kcal, and its contribution to daily per capita caloric intake.

Protein should constitute about 8 to 14% of daily caloric intake to maintain relatively healthy metabolism and is obtained mostly from animal sources (livestock and fish) (FAO 2004). This requirement is met in all regions. Dramatic differences exist in terms of fish consumption. The developing states of East and Southeast Asia consumed 23 to 26 kg of food fish per person per year for the period 1990 to 2002, obtaining 30 to 34 fish protein kcal per person per day. The region derives 43–45% of its animal protein from fish over this period. In South Asia, a supply of 4-5 kg of fish per person per year translates to 5–7 fish protein kcal per person per day. Fish protein contributes only 13–14% of daily animal protein intake. In contrast, a Japanese on average consumes 68 kg of fish per year, or 100–109 fish protein kcal per day. Like in Southeast Asia, fish protein accounts for 45–48% of animal protein intake in Japan. However, its fish consumption by weight and by fish kcal is three-times those for Southeast Asia and almost fifteen-times those for South Asia. The disparities shown in these patterns of production, consumption, and trade suggest that the declining status, if not collapse, of global food fish stocks will leave rural dwellers of coastal Asia-Pacific, specially those in Southeast Asia, disproportionately vulnerable to the negative consequences of fish scarcity.

8.3.3 Aquaculture's Share

It is important to take stock of the increasing contribution of farmed products to total aquatic production in the rural Asia-Pacific. In the last two decades, the contribution of aquaculture to total aquatic production in the Asia-Pacific region has gone up from 15% in 1980 to 46% in 2002 (Table 8.8). Globally, aquaculture accounted for 30% of total aquatic production in 2002. Among developing countries aquaculture contributed 35% to total aquatic production, which is equal to 91% of global aquaculture production. The annual growth rate for aquaculture has not gone below 10% in developing countries for the last two decades. In 2002, China alone accounted for 79% of worldwide aquaculture production (FAO 2004).

Although growth rates for aquaculture are by far the fastest growing economic sector worldwide, i.e. more than that for tourism, there are clear limits to growth (FAO 2004). One limitation is the reliance on capture fisheries for feed and seed, especially for grow-out species such as peneaid prawns. Another concern is the environmental costs associated with maintaining water quality and reducing negative impacts to marine community structure, genetic diversity, and sediment geochemistry. At the scale of coastal rural communities, it is critical that aquaculture is

Table 8.7. Seafood contribution to nutrition in Asia-Pacific countries. Range of minimum dietary requirements are based on lowest and highest values for countries in a region

Region	Parameter	1990–1992	2000–2002
East & Southeast Asia (developing): *Minimum dietary requirements (2000–2002, kcal/caput/day) = 1720 to 1930*	Fish supply (kg/caput/year)	23	26
	Fish protein (kcal/caput/day)	30	34
	Fish protein as % of daily animal protein	45	43
	Total caloric intake (kcal/cap ut/day)	2519	2696
	Total protein kcal as % of daily total food kcal	10	10
South Asia (developing): *Minimum dietary requirements (2000–2002, kcal/caput/day) = 1760 to 1830*	Fish supply (kg/caput/year)	4	5
	Fish protein (kcal/caput/day)	5	7
	Fish protein as % of daily animal protein	13	14
	Total caloric intake	2328	2397
	Total protein kcal as % of daily total food kcal	10	10
Oceania (developing): *Minimum dietary requirements (2000–2002, kcal/caput/day) = 1770 to 1910*	Fish supply (kg/caput/year)	27	21
	Fish protein (kcal/caput/day)	29	26
	Fish protein as % of daily animal protein	17	16
	Total caloric intake (kcal/caput/day)	2452	2536
	total protein kcal as % of daily total food kcal	12	12
Japan: *Minimum dietary requirements (NA)*	Fish supply (kg/caput/year)	69	67
	Fish protein (kcal/caput/day)	109	100
	Fish protein as % of daily animal protein	48	45
	Total caloric intake (kcal/caput/day)	2813	2783
	total protein kcal as % of daily total food kcal	14	14
Oceania (developed): *Minimum dietary requirements (NA)*	Fish supply (kg/caput/year)	20	23
	Fish protein (kcal/caput/day)	18	22
	Fish protein as % of daily animal protein	6	8
	Total caloric intake (kcal/caput/day)	3814	3112
	total protein kcal as % of daily total food kcal	14	14
World: *Minimum dietary requirements (NA)*	Fish supply (kg/caput/year)	13	16
	Fish protein (kcal/caput/day)	16	19
	Fish protein as % of daily animal protein	15	16
	Total caloric intake (kcal/caput/day)	2704	2795
	total protein kcal as % of daily total food kcal	11	12

Source: FAOSTAT 2004.

appropriately evaluated for its costs and benefits from both the economic and ecological standpoints. The choice in favour of short-term profits derived from prawn ponds that are built by removing mangrove cover proved catastrophic when a tsunami struck Asia in December 2004.

Table 8.8. Share of aquaculture in aquatic production

Region/ Production Sector	Production (thousand MT)			Average Annual Growth Rate (%)	
	1980	1990	2002	1980–90	1990–2002
East & Southeast Asia:					
Aquaculture	2490	8668	31349	13.3	11.3
Capture Fisheries	13186	20392	33640	4.5	4.3
Total	15676	29060	64989		
South Asia:					
Aquaculture	464	1235	3016	10.3	7.7
Capture Fisheries	3136	4157	5951	2.9	3.0
Total	3600	5392	8967		
Oceania (developing):					
Aquaculture	No data	1	4	No data	12.6
Capture Fisheries	133	184	377	3.3	6.2
Total	133	185	381		
Asia-Pacific Region:					
Aquaculture (% of total)	2954 (15)	9904 (29)	34369 (46)	12.9	10.9
Capture Fisheries	16455	24733	39968	4.2	4.1
Total	19409	34637	74337		
Developing Region:					
Aquaculture (% of total)	3022 (9)	10213 (18)	36096 (35)	12.9	11.1
Capture Fisheries	29699	46537	65716	4.6	2.9
Total	32721	56750	101812		
Developed Region:					
Aquaculture (% of total)	1684 (4)	2867 (7)	3703 (12)	5.5	2.2
Capture Fisheries	37479	38229	27474	0.2	−2.7
Total	39164	41096	31177		
WORLD:					
Aquaculture (% of total)	4707 (6)	13080 (13)	39799 (30)	10.8	9.7
Capture Fisheries	67177	84765	93191	2.4	0.8
Total	71884	97845	132990		

Source: FAO 2004.

8.4 ASIA-PACIFIC FARMERS

Farmers make up 80% of rural coastal dwellers in the Asia-Pacific and number about 800 million. At least another 800 million of the region's agricultural population live inland, beyond 100 km from shore (FAO 2004). With increases in both East Asia and the Pacific and South Asia, the Asia-Pacific region grew by 111 million more farmers between 1990 and 2000 (World Bank 2005, Table 8.9). Worldwide, the farming population increased in numbers over this period. Percentage-wise, farmers made up 46% of the total global population in 1990 and decreased slightly to 42% in 2000. FAO (2004) projects that by year 2010, the global farming population will still be growing, but will account for only 38% of the total global population, estimated to

Table 8.9. GDP in developing economies by sector. Agriculture includes farming, fishing, hunting and forestry

Region	Total Population in Agriculture (10^3)		GDP per capita		% GDP from Agriculture		%GDP from Industry		%GDP from Services	
	1990	2000	1990	2003	1990	2003	1990	2003	1990	2003
East Asia & Pacific	1,091,142	1,122,962	417	1,096	25	14	40	49	35	36
South Asia	663,197	742,054	358	537	31	22	27	26	43	51
Europe & Central Asia	incomplete	83,953	2,379	2,970	16	8	43	31	41	61
Latin America & Caribbean	116,662	108,057	2,536	3,268	9	7	36	27	55	66
Middle East & North Africa	79,300	81,409	1,775	2,390	14	11	38	43	48	47
Sub-Saharan Africa	288,791	396,933	585	624	19	17	34	31	47	52
High income	133,993	99,752	19,941	30,183	3	2	33	27	65	71
WORLD	2,443,393	2,573,456	4,128	5,813	5	4	34	28	61	68

Source: WDI 2005 (World Bank 2005).

be 2.62 billion. Seventy-two percent (1.9 billion) of the global total are projected to be living in the Asia-Pacific region.

8.4.1 Farming Capital

The Asia-Pacific region's agricultural area of about 1 billion ha accounts for 20% of global farming area (FAO 2004). Throughout the world, arable land (that devoted to temporary crops and meadows) per person is decreasing. Because of a large agricultural population, the AP has the smallest per capita arable land at 0.10 ha per person. The same decreasing pattern is true for land allotted for cereal production, except in Latin America and in sub-Saharan Africa (Table 8.10). The region has the world's highest fertiliser consumption rate at 215 kg per ha of arable land for the period 2000–02; a consumption rate that is up by 38 kg/ha from a decade ago. South Asia consumes fertiliser at half the rate of East Asia. In terms of farm machinery, the AP region has the lowest number of tractors per thousand workers or per 100 km^2 of tilled acreage, just slightly above that of sub-Saharan Africa.

Table 8.10. Agricultural inputs

Region	Arable land (ha/caput)		Irrigated land (% of cropland)		Land for cereal production (10^3 km^2)		Fertiliser consumption (kg/ha)		Agricultural machinery			
									Tractors per thousand workers		Tractors per 100 km^2 arable land	
	1989–91	2000–02	1989–91	2000–02	1989–91	2001–03	1989–91	2000–02	1989–91	2000–02	1989–91	2000–02
By Developing Region:												
East Asia & Pacific	0.1	0.1	nd	nd	1418	1339	177	215	2	2	56	73
South Asia	0.2	0.1	34	41	1291	1253	74	105	4	5	62	92
Europe & Central Asia	0.6	0.6	10	11	1413	1163	99	34	108	111	194	184
Latin America & Caribbean	0.3	0.3	12	12	483	495	59	86	36	40	119	119
Middle East & N. Africa	0.2	0.2	34	38	278	243	70	86	22	25	107	136
Sub-Saharan Africa	0.3	0.2	4	4	638	843	15	13	2	1	21	15
By Income level: Low = < $825/person/year; Middle = $826 < X < $10065; High > $10066 (Note: Developing = Low + Middle income)												
Low income	0.2	0.2	24	27	2060	2287	55	67	4	4	50	66
Middle income	0.3	0.2	18	20	3460	3049	110	105	13	12	132	130
High income	0.4	0.4	11	12	1430	1316	125	120	642	918	433	436
World	0.3	0.2	18	20	6951	6651	100	99	22	20	191	190

Source: World Development Indicators 2005, World Bank 2005.

8.4.2 Agricultural Productivity

Tables 8.11 and 8.12 show trends for cereal and meat production. Globally, production growth rates in both crop and livestock production have dramatically slowed down over the period 1999 to 2002 (FAO 2004). This trend is true in both developed countries, which posted negative growth rates for cereal production over the last decade, and for the Asia-Pacific region. Over the last three decades in the Asia-Pacific region, peak growth rates were reached between 1980 and 1990, followed by declining rates from 1990 to 2002. This regional trend is largely

Table 8.11. Cereal production for period 1970–2002

Region	Production (Millions MT)				Average Annual Growth Rates (%)		
	1970	1980	1990	2002	1970–80	1980–90	1990–2002
East & Southeast Asia	284	389	548	634	3.2	3.5	1.2
South Asia	148	185	251	310	2.3	3.1	1.8
Oceania (developing)	0	0	0	0	0	0	0
Developing countries	587	766	1038	1237	2.7	3.1	1.5
Developed countries	606	784	913	849	2.6	1.5	−0.6
World	1193	1550	1952	2086	2.7	2.3	0.6

Source: FAO 2004

Table 8.12. Meat production for period 1970–2002

Region	Production (Millions MT)				Average Annual Growth Rates (%)		
	1970	1980	1990	2002	1970–80	1980–90	1990–2002
East & Southeast Asia	10.8	19.4	38.1	80.4	6.0	7.0	6.4
South Asia	2.9	3.7	5.8	8.4	2.7	4.5	3.1
Oceania (developing)	0.2	0.2	0.3	0.4	2.4	2.6	2.5
Developing countries	30.8	46.9	75.3	139.3	4.3	4.8	5.3
Developed countries	69.8	89.9	104.8	108.4	2.6	1.5	0.3
World	100.6	136.8	180.1	247.8	3.1	2.8	2.7

Source: FAO 2004.

attributed to diminishing production rates, especially in the cereal sector in China, which has accounted for 48% of the region's production for the period 1961 to 2002 (FAO 2004).

The trends in average annual production rates are different for total food production (Table 8.13) than those described above for cereal and meat production. Developing countries have shown an increase in average annual growth rate from 3% is the 1970s to 3.8% in the 1990s. In contrast, developed states showed a declining pattern with 2.0% average annual growth during the 1970s dropping to only 0.2% in the 1990s.

As a whole, agricultural production in developing countries has continued to grow at a sustained average annual rate of 3.6% over the last two decades (Table 8.14). Production rates in the developed world, however, have grown in the last 12 years at an average of 0.2%, which is 20% of their rate 20 years ago. Despite a slower growth rate in developed countries, much smaller agricultural populations have resulted in higher per capita production values for farmers in developed countries. The per capita production (in 1999–2001 international prices) for farmers in the developing countries, which is US$710, is less than 10% the developed country value of US$10,300 in 2002.

Table 8.13. Total food production for period 1970–2002

Region	Average annual growth rate (%)					
	Total Food Production			Per caput food production		
	1970–80	1980–90	1990–2002	1980	1990	2002
East & Southeast Asia	3.3	4.4	4.9	1.4	2.7	3.7
South Asia	2.7	3.8	2.9	0.4	1.5	1.0
Oceania (developing)	2.2	1.7	1.9	−0.1	−0.6	−0.4
Developing countries	3.0	3.6	3.8	0.7	1.5	2.1
Developed countries	2.0	1.0	0.2	1.2	0.3	−0.2
World	2.5	2.4	2.4	0.6	0.6	1.0

Source: FAO 2004.

Table 8.14. Total agricultural production for period 1970–2002

Region	Average annual growth rate (%)					
	Total Agricultural Production			Agricultural production per agricultural worker (1999–2001 International $)		
	1970–80	1980–90	1990–2002	1980	1990	2002
East & Southeast Asia	3.3	4.3	4.6	300	391	636
South Asia	2.6	3.8	2.8	345	457	510
Oceania (developing)	2.4	1.7	1.9	757	736	734
Developing countries	2.8	3.6	3.6	421	520	706
Developed countries	2.0	1.0	0.2	5684	7628	10312
World	2.4	2.4	2.3	788	883	1033

Source: FAO 2004.

8.5   TOURISM IN DEVELOPING COUNTRIES OF ASIA-PACIFIC

Like data for farming and fisheries, information in tourism databases has not readily segregated the rural from the urban elements of the industry. At the global scale, the tourism sector has been monitored as closely as the fishing and farming sectors, but the information and databases have yet to evolve into formats amenable for public access. The body responsible for providing a global forum on tourism is the World Tourism Organization, which became a specialised UN agency only in 2003; however it has its roots as the International Congress of Official Tourist Traffic Associations that was established in 1925 (www.world-tourism.org). Relevant data that are publicly available on its website are compiled and shown in Tables 8.15 and 8.16.

Table 8.15 shows the Asia-Pacific as receiving 13% of the world's total international tourist arrivals in 1990, a figure which grew to 19% by 2002. Of these regional visitors, 75.5 million travelled to coastal developing countries, which accounts for

Table 8.15. International tourist arrivals by region

Region	International tourist arrivals (millions)				
	1990	1995	2000	2001	2002
Asia-Pacific	57.7	85.6	115.3	121.1	131.3
Northeast Asia	28.0	44.1	62.5	65.6	73.6
Southeast Asia	21.5	29.2	37.0	40.2	42.2
Oceania	5.2	8.1	9.6	9.5	9.6
South Asia	3.2	4.2	6.1	5.8	5.9
Europe	280.6	322.3	392.7	390.8	399.8
Americas	93.0	108.8	128.0	120.2	114.9
WORLD	455.9	550.4	687.3	684.1	702.6

Source: WTO 2003.

Table 8.16. Major coastal developing country destinations in Asia-Pacific

Country	2002 GDP[1] (millions)	2002 International tourist arrivals[2] (10^3)	2002 Tourism GDP[3] (%)	Fisheries GDP[4] (%)
China	1 238 102	36 803	1.6	3.7 (2001)
India	600 637	2 370	0.5	1.8 (2000)[5]
Indonesia	152 910	5 033	2.4	4.1 (2001)
Malaysia	86 199	13 292	7.9	1.5 (2001)
Philippines	82 429	1 933	2.1	4.8 (2001)
Thailand	123 219	10 873	6.4	4.1 (2001)

[1] World Development Indicators data, World Bank 2005.
[2] World Tourism Organization 2003 Data at www.world-tourism.org
[3] Tourism 2003 Highlights, WTO 2003.
[4] Sugiyama et al. 2003.
[5] Bay of Bengal Programme at http://www.bobpigo.org/bobp_india.htm

about 58% of the total visitors to the region in 2002 (based on WTO 2003 data). The potential for tourism to contribute to a country's GDP is significant. Although data presented for fisheries and tourism in Table 8.16 is taken from different years, the available information suggests that tourism contributions to GDP are comparable to, if not greater than, those from fisheries.

Despite similar economic contributions, the tourism sector cannot necessarily be considered a substitute for fishing, and, in fact, the two sectors compete with each other for access to coastal areas. Fishers cannot necessarily become tourism entrepreneurs because they may lack appropriate capital and/or skills. They may be employed to run tour boats and do other tourism-related work, but the ability of the tourism sector to absorb displaced fishers may be limited. Another possibility is that tourism may provide employment for other members of fishing or farming families.

At the same time, tourism operations can and do displace fishers by appropriating coastal space for related coastal development, which can conflict with the latter's residential space and fishing grounds. Tourism is often developed with private capital, and it promises government a share in earned revenues that is greater than government earnings from fisheries. This situation creates an economic incentive for central and local governments to not fully take account of the conflicting interests between tourism and marginal populations, including fishers. Thus tourism and fisheries in a rural setting often share more points of conflict than consensus. A government that attempts to balance protecting the public good with encouraging private enterprise is required if goals for social equity and environmental quality are to be achieved.

8.6   ENVIRONMENTAL ISSUES AND PROBLEMS

In the Asia-Pacific region, the dual realities of high population growth and natural resource-dependant livelihoods illustrate the tight interconnections between economics and environment. Livelihood activities including farming, fishing, aquaculture, and tourism impose heavy environmental costs. When these environmental stressors exceed the natural buffering or carrying capacities of natural systems, the result is a negative feedback to the ecological systems on which economic production depends. Unless these costs are reduced, degraded natural systems can limit economic productivity and may require costly technological interventions, if they are even possible, to maintain current production status and levels of quality of life. Overfishing, eutrophication, habitat modification, and water-related issues dominate the list of major environmental problems that threaten the sustainability of future coastal scenarios.

8.6.1   Overfishing

In the Asia-Pacific region, evidence is becoming more unequivocal that overfishing is occurring in major fishing grounds of the region. For this section, latest available findings for the Gulf of Thailand and the rest of the South China Sea, the Bay of Bengal, and the West-Central Pacific will be discussed.

Gulf of Thailand and South China Sea

The Gulf of Thailand is a shallow subsystem of the South China Sea and overlies part of the Greater Sunda Shelf. In the 1960s, demersal trawling was introduced as a fishing method (Sugiyama et al. 2004). By 1983, fish biomass was reduced by 86% of the 1960 level (Silvestre et al. 2003; Figure 8.2). In addition to rapidly depleted biomass, ecosystem impacts of overfishing include the replacement of long-lived, high-trophic level species with small, short-lived, low-trophic level species (Christensen et al. 2003; Figure 8.3). While modeling the fisheries with scenarios of much reduced fishing effort by demersal gear, including bottom trawlers and push nets, indicates that there is the potential to recover from ecosystem overfishing in the Gulf of Thailand

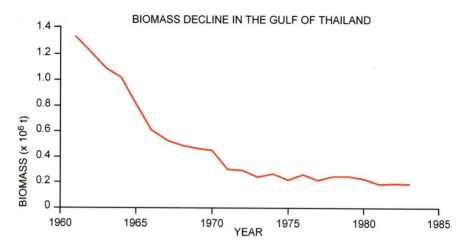

Figure 8.2. The fisheries in the Gulf of Thailand show biomass reduction over a 22-year period (Silvestre et al. 2003) (Reproduced with permission of the WorldFish Center and Elsevier)

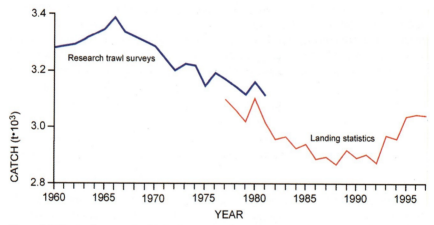

Figure 8.3. Decreasing mean trophic levels among catch of research trawl surveys and those landed. Both datasets excluded large pelagics. Landed catch included small pelagics and other organisms low in the food chain (Adapted by Christensen et al. 2003 from Pauly and Chuenpagdee 2003) (Reproduced with permission of the WorldFish Center and Elsevier)

(Sugiyama et al. 2004), the same may not be true for highly degraded coastal fishing grounds in the South China Sea. An exception may be the coastal waters of Brunei where exploitation rates have remained relatively low (Silvestre et al. 2003).

Western and Central Pacific and Pacific small-island nations

In the Western and Central Pacific, one gauge of the extent of fishing impacts is the status of tuna stocks, which are mostly exploited in offshore waters by the industrial fleets of 11 distant water fishing nations (Sugiyama et al. 2004). Japan, USA, Korea

RO and China PR catch the bulk of the tuna in the EEZs of small Pacific Island nations. In return, these developing nations negotiate for access fees that very often are below the economic value of the tuna catch of the foreign fleets. Such access fees are packaged as development aid monies, an arrangement that inappropriately treats access fees as a charitable donation, rather than an economic transaction between resource owners and distant water fishing nations (Petersen 2002).

Myers and Worm (2003) contend that the world ocean has lost more than 90% of its predatory fish biomass based on an analysis of catch per-unit-effort in major fishing grounds since the onset of industrial fisheries. In the case of tuna fisheries, both climate forcing and fishing need to be examined to identify the overriding drivers of net trends. Four tuna species (big-eye, yellow fin, skipjack and south pacific albacore) are commercially exploited within the climate-driven Western Pacific warm pool (Lehody et al. 1997). Big-eye seems overexploited with over a 35% reduction of its biomass by fishing (Figure 8.4, Hampton & Williams 2003, Hampton et al. 2005). Yellow fin tuna is fully exploited, but both south pacific albacore and skipjack may be quite easily sustained (Figure 8.5). These trends were reaffirmed by the latest stock assessments conducted in 2003 (Langley et al. 2005).

The coastal fisheries of small-island developing nations in the Pacific provide for domestic consumption as well as for export. Assessments of multi-gear and multi-species fisheries for artisanal and small-scale commercial fisheries are always difficult to make. Gillett (2002) provides the most thorough analysis of this sector to date. In

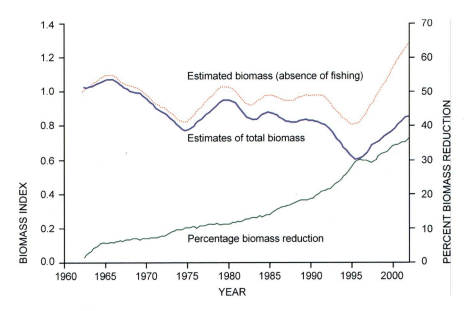

Figure 8.4. Uppermost line is the modeled biomass estimate for big-eye tuna without fishing. Second (darkest) line is model estimates of total biomass in the Western and Central Pacific Ocean. The third (grey) line represents the percent biomass reduction attributed to fishing, at over 35% in 2002 (Hampton & Williams 2003) (Reproduced with permission of the Secretariat of the Pacific Community)

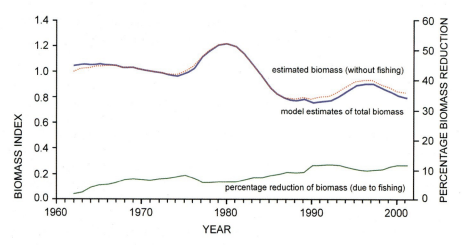

Figure 8.5. South Pacific albacore tuna. Trend in existing biomass closely tracks that of modeled estimate for unfished biomass. Over a 40-year period, fishing has reduced biomass by about 12% (grey line) (Hampton and Williams 2003) (Reproduced with permission of the Secretariat of the Pacific Community)

2002, coastal artisanal fisheries harvested 102,000 t or 71% of total coastal fisheries. Over more than 30 years, coastal fisheries catch increased from providing 10 kg of fish supply per capita per year in the 1960s to 18 kg per person per year in the 1990s. In terms of trade, major export items include sea cucumber, trochus shell, pearl shell, pearls, deep-water snappers (mainly in Tonga), giant clams (mainly in Fiji), live groupers, and aquarium fish. These species account for 29% of coastal fisheries production.

The increasing commodification of the artisanal catch is a major driver for increasing exploitation rates in coastal fisheries. The list of export items, therefore, provide an indication of organisms that are threatened by the lure of export markets. These include sea cucumbers, giant clams, pearl oysters, coconut crabs (terrestrial), and sea turtles, which have been overfished (Asia Pacific Fisheries Commission 2004).

Bay of Bengal

Very little systematic study has been done to assess both the inshore and offshore fishery resources of the Bay of Bengal. With the institutionalisation of the Bay of Bengal Programme through the establishment of the Bay of Bengal Programme Intergovernmental Organisation (BOBP-IGO) in Chennai in April 2003, a more systematic monitoring, assessment, and science-based management may be possible (Yadava 2003). The BOBP-IGO articulated an agenda, which includes a regional programme for fish stocks assessment in the Bay (Bay of Bengal News 2003). Fact sheets of countries that share the Bay of Bengal indicate heavy exploitation of coastal resources such as shrimps in the Sundarban-rich fishing grounds of Bangladesh; groupers, aquarium fish, beche-de-mer and giant clams from the reefs of the Maldives; and coastal resources from the inshore waters of India (Bay of Bengal News 2003). For these countries, the deep-sea resources offer greater scope for expansion. In the case of Sri Lanka, Maximum Sustainable Yields (MSYs) estimated for its coastal and deep-sea fisheries indicate near full exploitation of both.

8.6.2 Eutrophication

The heavy use of fertiliser in farming is the major driver behind coastal eutrophication in the Asia-Pacific region as it is worldwide (Tilman et al. 2001). The region has among the highest fertiliser consumption rates on a per hectare basis at 215 kg/ha/a in Southeast and East Asia and at 105 kg/ha/a in South Asia (Table 8.10; World Bank 2005). The mean rate for Southeast Asia exceeds the mean average use in high-income countries, which is 125 kg/ha/a. In a study of four rural sites in Southeast Asia, agriculture was the major source of dissolved inorganic nutrients (nitrogen and phosphorus) (Talaue-McManus et al. 2001). With excessive nutrient loading in estuarine and coastal waters, plant material responds by increasing biomass production. For phytoplankton species, population growth can reach bloom proportions, some of which have adverse health impacts on humans through toxins the cells produce. Phytoplankton blooms lead to large amounts of ungrazed biomass that reach the sediments as organic material. Decomposition of this material uses up oxygen. Under conditions of low physical energy, where oxygen cannot be easily replenished, or when conditions favour thermal water stratification, as in summertime, dissolved oxygen can plummet to physiologically limiting or hypoxic concentrations. These oxygen-limited conditions can lead to mass mortalities in benthic organisms and demersal fish.

In the Philippines, summer fish kills occurred in a number of embayments in the main islands of Luzon and Mindanao in 2002 (Figure 8.6). Among the most dramatic were a series of mass mortalities of pen-grown milkfish, Chanos chanos, that occurred in Bolinao, Pangasinan (Relox and Bajarias at http://fol.fs.a.u-tokyo.ac.jp/rtw/TOP/EXabst/019JuanRReloxJr.pdf). A bloom of *Prorocentrum minimum* occurred in the waters of Caquiputan Channel that were crowded with 1,170 fish pens and cages, twice the number allowed by a municipal ordinance (Pabico 2002). This number of pens crowded the channel and minimised the flow of water. Reduced water flow, exacerbated by summer temperatures, prevented physical replenishment of oxygen, which was rapidly being used up in decomposing both excess fish feeds and the *P. minimum* bloom. The hypoxic conditions eventually led to massive die-off of milkfish and to dire economic consequences for the fish farmers. In this case, the excessive use of feeds in fish farming and high density of fish pens and cages led to eutrophication and hypoxia. Subsequently, the habitat in the channel has been degraded to the extent that cultured fish survival and growth has been severely impaired.

Eutrophication in developing countries is poorly documented, as is the case in rural Asia-Pacific. Global maps of hypoxic events only include those recorded in industrialised countries with resources to monitor environmental quality (Burke et al. 2001). The economic, and to a lesser extent the ecological, impacts of nutrient-laden estuarine and coastal waters of developing nations are just beginning to be assessed. Of immediate concern is the lack of institutional capacity to mitigate both ecological and economic impacts of degraded environments. Without this capacity,

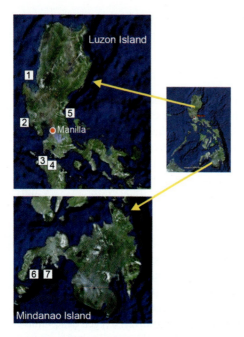

Figure 8.6. Fish kills occurred in the Philippines in 2002 in association with hypoxic events following algal blooms. 1. Bolinao, Pangasinan (Feb. 2002); 2. San Antonio, Zambales (June 22–27, 2002); 3. Nasugbu, Batangas (June 13–14, 2002); 4. Balayan Bay, Batangas (June 7–8, 2002); 5. Eastern Luzon (May and June, 2002); 6. Sibuguey Bay; 7. Dumanquilas Bay, Zamboanga del Sur (June 2002) (Modified from Relox & Bajarias at http://fol.fs.a.u-tokyo.ac.jp/rtw/TOP/EXabst/019JuanRReloxJr.pdf. Base maps from google earth at http://earth.google.com)

agricultural and aquaculture activities continue to be implemented in ways that are not sustainable, and the productivity of impacted coastlines is reduced.

8.6.3 Coastal Development and Habitat Modification

The best metric for determining the extent of coastal habitat modification in tropical and subtropical latitudes is the loss of mangrove forests. Because seagrasses and coral reefs are submerged ecosystems, it has proven harder to monitor net changes in cover and community structure. Advances in remote sensing of pigments associated with shallow-water macrobenthos may provide the synoptic means to monitor submerged ecosystems at operational scales in the future.

For countries in the region with multi-year data, rates of mangrove loss were highest for China and Pakistan at 4.9% and 2.3% per year, respectively. (Table 8.17; Valiela et al. 2001). Although the reported rates suggest uniform loss of area over the period covered by the early and recent estimates, it could well be that rate losses were non-uniform over these periods. In terms of area, India has the largest area lost to conversion at 112 km^2 per year, followed by Thailand and Pakistan at 64 and 58 km^2

Table 8.17. Estimates of existing mangrove cover and loss rates in selected countries in Asia-Pacific

Country	Early Estimate of Mangrove Cover (km^2, year reported)	Late Estimate of Mangrove Cover (km^2, year reported)	Loss rate (km^2 (% of early estimate) per year)
Bangladesh	6 400 (1980)	5 767 (1997)	37 (0.6)
China & Taiwan	670 (1980)	178 (1995)	33 (4.9)
India	6 820 (1963)	3 565 (1992)	112 (1.6)
Burma	5 171 (1965)	3 786 (1994)	48 (0.9)
Pakistan	2 495 (1983)	1 683 (1997)	58 (2.3)
Philippines	4 500 (1920)	1 325 (1990)	45 (1.0)
Thailand	3 724 (1961)	1 687 (1993)	64 (1.7)
Vietnam	4 000 (1945)	1 520 (1995)	50 (1.2)

Source: Valiela et al. 2001.

per year. Regionally, loss rates in the Americas was most severe with about 2,251 km^2 of mangroves lost per year (Table 8.18). Asia has lost 36% of its original cover and has continued to lose 630 km^2 per year over the last 24 years.

The anthropogenic drivers of mangrove loss have been enumerated by a number of reviews. Both deforestation and pond aquaculture are the most profligate of mangrove resources (Alongi 2002). Barbier and Cox (2003) formulated an economic model to identify determinants of mangrove appropriation. Shrimp aquaculture and agricultural expansion were significantly associated with mangrove loss across 89 countries examined. Among the factors associated with wider areas of conserved mangroves were high GDPs, political stability, and relatively high number of protected areas. In the case of countries in the Asia-Pacific, the production of shrimp to earn foreign exchange is the major economic activity behind mangrove conversion. Satellite images of before and after conversion are shown for the Sundarbans (Bangladesh) and the Gulf of Thailand in Figures 8.7 and 8.8 (UNEP 2005). With increasing wealth, the demand for cultured shrimp will remain high. However, given that shrimp ponds depend on wild seeds, the loss of mangroves will undermine the production of larvae, unless technologies are developed that can culture shrimps throughout their life cycle in artificial conditions. Apart from shrimp aquaculture, mangroves forests have given way to space for habitation and

Table 8.18. Regional comparison of loss rates of mangroves

Region	Present Area (km^2)	% Loss of Original Area	Annual rate of loss ($km^2\ yr^{-1}$)	% Loss yr^{-1}
Asia	77 169	36	628	1.52
Australia	10 287	14	231	1.99
Africa	36 259	32	274	1.25
Americas	43 161	38	2 251	3.62
WORLD	166 876	54	2 834	2.07

Source: Valiela et al. 2001.

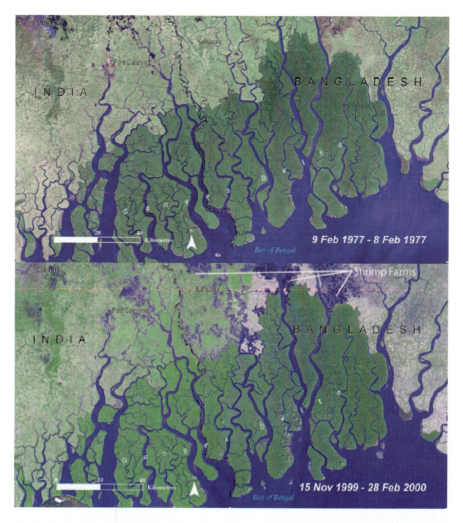

Figure 8.7. Cutting up the inner Sundarbans (mangrove forests) for shrimp ponds, Bangladesh (UNEP 2005 at www.na.unep.net/OnePlanetManyPeople/AtlasDownload/UNEP_Atlas/andwww.earthprint.com)

infrastructure, including hotels and resorts for tourism. As a consequence, the sediment binding and wave buffering capacities of forested coastlines are gone, which gives rise to increased sedimentation and heightened vulnerability to flooding and storm surges.

In the aftermath of the tsunami disaster in December 2004 that struck Indonesia, Thailand, Burma, Sri Lanka, the Maldives, and India, it became obvious that coastal ecosystems which buffer wave action (mangroves, seagrasses, and corals) have become too degraded to provide this function. The devastating

Pressures on Rural Coasts in the Asia-Pacific Region 219

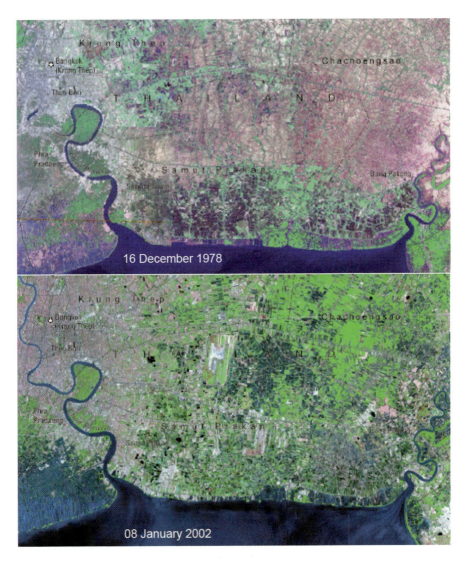

Figure 8.8. Coastal and inland aquaculture ponds replace mangrove vegetation along the coast fringing the Gulf of Thailand (UNEP 2005 at www.na.unep.net/OnePlanetManyPeople/AtlasDownload/UNEP_Atlas/and www.earthprint.com)

impacts to life and property of affected rural coastal communities were staggering, particularly given the absence of economic safety nets for these people. Yet despite the high costs of ecosystem degradation, the continuing pressures of increasing global consumption and increasing local populations may continue to drive the loss of these coastal forests.

8.6.4 Water Issues

Rural Asia-Pacific contends with a number of critical issues about water: its supply and quality relative to human and environmental water requirements. Salient points are discussed here and are based largely on an excellent analysis of global water use and food production by Rosegrant et al. (2002). Table 8.19 shows the amounts of total water withdrawal at 1995 levels and at projected rates in 2025 based on a business-as-usual scenario (BAU scenario). Globally and in Asia, water withdrawal will increase by 22% from 1995 to 2025. In evaluating water withdrawal, a criticality ratio (ratio of withdrawn and renewable water) of 40% or higher is used as an indicator of water scarcity stress. China and India are projected to reach 33% and 36% scarcity levels by 2025. Currently, dry regions of these 2 countries have already reached scarcity stress levels beyond 40%. To inform meaningful design of water governance, criticality ratios should be calculated at river basin and seasonal scales.

If one examines how humankind uses up the water it withdraws, agricultural consumption for irrigation and livestock accounted for 80% of total water consumption in 1995 and is projected to decrease to 72% in 2025, relative to other uses (BAU scenario; Table 8.20). In volume, however, agricultural consumption will increase 83 km^3 by 2025, 68% of which will be used for irrigating land. In Asia, consumption by volume shows a doubling for livestock, industries, and domestic use over the 30-year period. About 53% of the increase (147 km^3) will be accounted for by increasing domestic consumption, which includes water for drinking, sanitation, bathing, and cooking. With a projected 40% population growth from 1995 to 2025, water use for food production, industry, and domestic uses will all increase and may conflict with each other.

Smakhtin et al. (2004) warn that human uses may also conflict with the water requirements of aquatic ecosystems and that water budgets needed to maintain ecological systems should be factored into water assessments and governance. Given sea-

Table 8.19. Total water withdrawal as percentage of renewable water with projections to 2025 using a business-as-usual scenario (BAU)

Region/Country	Total water withdrawal (km^3)		Total water withdrawal (km^3) as % of renewable water	
	1995	2025 BAU	1995	2025 BAU
Asia	2165	2649	17	20
China	679	846	26	33
India	674	815	30	36
SE Asia	203	287	4	5
Other S Asia	353	421	18	22
Developed	1144	1265	9	10
Developing	2762	3507	8	10
World	3906	4772	8	10

Source: Rosegrant et al. 2002.

Pressures on Rural Coasts in the Asia-Pacific Region

Table 8.20. Water consumption by sector

Region/ Country	Irrigated Land Million ha		Irrigation consumption (km^3)		Livestock consumption (km^3)		Industrial consumption (km^3)		Domestic consumption (km^3)	
	1995	2025	1995	2025	1995	2025	1995	2025	1995	2025
Asia	231.5	274.9	920.2	933.3	11.7	25.6	48.3	90.7	79.1	156.7
China	91.5	107.7	244.2	230.9	3.4	7.4	13.1	31.1	30.0	59.4
India	59.0	76.3	321.3	331.7	3.3	8.1	7.2	15.7	21.0	40.9
SE Asia	24.8	28.1	85.5	91.9	1.7	4.1	11.2	20.9	13.9	30.4
Other S Asia	29.3	34.9	163.2	169.4	1.7	3.9	1.9	4.7	7.0	16.2
Developed	82.4	86.7	271.7	276.9	15.3	18.2	94.7	113.8	58.7	68.6
Developing	293.0	354.1	1163.8	1215.5	21.8	45.2	62.2	121.4	110.6	221.0
World	375.4	440.8	1435.5	1492.3	37.0	63.4	156.9	235.2	169.2	289.6

Source: Rosegrant et al. 2002.

sonal and medium-term variability in river flow, about 20–50% of mean annual river flow is needed to maintain ecosystem functioning. Based on preliminary estimates, about 1.4 billion people live in areas where some amount of water required for ecological maintenance is appropriated for human uses, and this water withdrawal is degrading ecosystems and impairing their functions (Figure 8.9). The Yellow (Huang He) River in China typifies a stressed ecosystem because 90% of the available water is being used for human consumption. For the Asia-Pacific region, South Asia and China have the largest areas of environmentally compromised ecosystems. At basin and seasonal scales, the area of river catchments experiencing conflicts between human use and environmental water requirements will most likely increase.

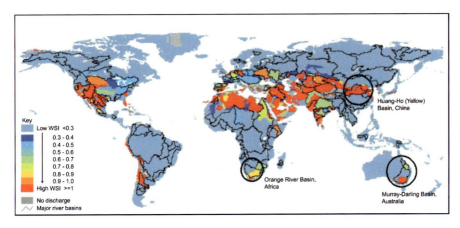

Figure 8.9. Areas in orange to red have high water stress indicators, which suggests that water needed to maintain aquatic ecosystems has been compromised (Smakhtin et al. 2004. Reproduced with permission of the comprehensive Assessment of Water Management in Agriculture)

At the level of households, domestic consumption of rural populations is determined by access to water sources. The World Development Indicators for 2005 defines access as the availability of a minimum of 20 liters per person per day or 7.3 m^3 per person per year from a water source within 1 km of the dwelling (World Bank 2005). For the 2.2 billion coastal rural inhabitants of Asia-Pacific, about 1.6 billion have water access. Of the 600 million with no access in 2003, 350 million live in East Asia and the Pacific, and the rest live in South Asia (Figure 8.10). Of those with water access, consumption levels depend whether or not households have pipe connections. In Asia, per capita domestic consumption was significantly higher (26 to 30 m^3 per person per year) for pipe-connected households than for unconnected households (17–19 m^3 per person per year) in 1995 (Table 8.21).

Falkenmark and Widstrand (1992) suggest that 100 liters per person per day (36.5 m^3 per person per year) is a minimum acceptable standard of living in developing countries, which provides for drinking, sanitation, bathing, and cooking needs. Using this recommendation as a reference, only the rural population of Southeast Asia will achieve this minimum standard by 2025. All other areas in the Asia-Pacific region will remain below the minimum standard consumption levels even by 2025. In terms of access to sanitation facilities, data from the World Bank (2005) show that rural populations in the region with access rose from 11% (225 million) in 1990 to 29% (630 million) in 2003 (Figure 8.11). Worldwide, there are 2.1 billion rural people and 633 million urban dwellers with no access to sanitation facilities, which equates to 44% of the global population in 2003.

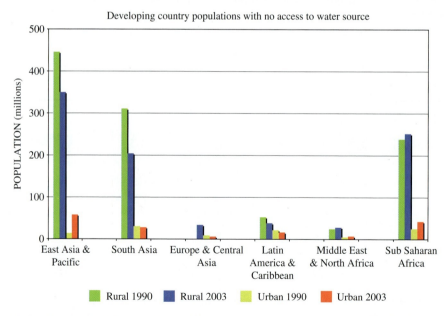

Figure 8.10. Rural and urban population with no access to a water source. Access is defined by the World Bank (2005) to be 20 liters per person per day from a source within 1 km from dwelling. This translates to 7.3 m^3 per person per year. (Data from World Bank 2005)

Table 8.21. Per capita domestic water consumption of rural populations

Region/ Country	Rural population (millions)		Per capita domestic consumption (urban + rural) (m³/person/yr)		Per capita domestic consumption of rural pipe-connected population (m³/person/yr)		Per capita domestic consumption of rural unconnected population (m³/person/yr)	
	1995	2025	1995	2025	1995	2025	1995	2025
Asia	2102	2208	24.8	36.9	27.1	29.7	17.6	18.9
China	857	778	24.5	41.2	25.7	27.6	17.1	18.4
India	679	777	22.6	30.7	26.8	27.8	17.9	18.4
SE Asia	320	324	29.9	45.6	29.6	38.2	19.2	23.6
Other S Asia	226	315	23.7	29.6	25.7	28.2	17.3	18.9
Developed	327	237	47.8	54.4	47.0	48.8	22.3	33.9
Developing	2774	3106	25.6	33.9	25.2	27.0	16.9	17.6
World	3101	3343	30.3	37.2	31.0	29.3	17.0	17.7

Source: Rosegrant et al. 2002.

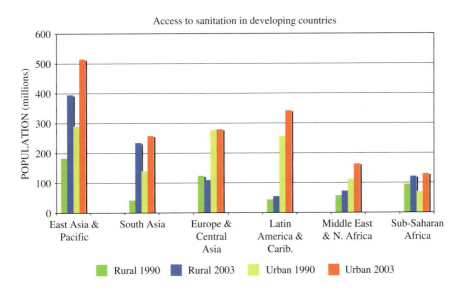

Figure 8.11. Access to sanitation facilities in urban and rural developing populations (data from World Bank 2005). Adequate facilities (private or shared, but not public) are those that effectively prevent human, animal, and insect contact with fecal material and may range from protected pit latrines to sewered flush toilets

The impacts of allocating decreasing water supply between agricultural, industrial, domestic, and environmental requirements will differentially be harsher for economically marginal populations that include the coastal inhabitants of rural Asia-Pacific. Very often, the domestic allocations are prioritised behind agricultural and industrial needs, and the environmental requirements have only recently been considered at all in water assessments. The reduction or degradation of natural resource bases on land and sea because of overfishing, eutrophication, and excessive freshwater withdrawal (among others) paints a compelling picture of economic and ecological deprivation for the rural population of the region and globally.

8.7 RURAL POVERTY: CHALLENGES AND OPPORTUNITIES

Life in rural Asia, as in most regions of the world, is inextricably linked with nature's bounty or lack thereof. As discussed in the previous sections, poverty is exacerbated by conditions where direct resource users: a) have little access to produce (e.g. high quality fish); b) receive very little for their work (agricultural GDP per capita); c) are not provided basic amenities, such as water; and d) suffer from degraded ecosystems (e.g. those with minimised function to buffer natural disasters). Many reviews on poverty agree that the rural poor is the strategic sector that must receive critical attention (but not to the exclusion of others) if millennium development goals about poverty are to be realistically achieved (IFAD 2001, 2002). Further, the analysis of poverty must include the interactions between nature's goods and services, the production of wealth from the latter, and governance that dictates use and access to these goods and services (USAID 2004, WRI 2005).

Compared to other regions, Asia has the most workers deriving livelihoods from natural resources in agriculture, forestry, and fisheries at approximately 1.8 billion, or 60% of its population (Table 8.22). About 1 billion workers, mostly farmers and fishers, live in rural coastal areas of the region, and the remaining 0.8 billion dwell inland. Data from Ahuja (1997) as cited by the International Fund for Agricultural Development (IFAD; 2002) indicate that the proportion of poor are higher in rural than in urban areas (Table 8.23). Current figures indicate that the rural poor outnumber the urban poor 3 to 1

Table 8.22. Percent of the labor force in the agriculture, forestry, and fisheries sector

Region	% of labor force
Asia & the Pacific	60
Near East & North Africa	33
Sub-Saharan Africa	62
Latin America & Caribbean	19
Developing Countries	54
Developed Countries	7
World	44

Source: WRI 2005.

Table 8.23. Share of poor in rural and urban households in Asia (%)

Country/Year	Distribution of Poor	
	Rural	Urban
Southeast Asia		
Indonesia, 1990	83	17
Malaysia, 1987	86	14
Thailand, 1992	85	15
Vietnam, 1992/93	89	11
East Asia		
China, 1995	99	1
South Asia		
Bangladesh, 1996/96	58	42
India, 1994	86	14
Pakistan, 1990/91	75	25
Pacific		
Papua New Guinea, 1996	94	6

Source: Ahuja et al. 1997 as cited by IFAD 2002.

(IFAD 2001). Applying this ratio (75%), rural coastal Asia is home to about 750 million impoverished farmers and fishers, and another 600 million who live 100 km and farther inland from the coast. In addition, there are more poor living off marginal lands than there are who live on favoured areas (those with access to irrigation and other capital inputs; Table 8.24).

This situation also roughly coincides with the 75% (1.92 billion people) of the total number of people who live on less than $2.00 a day (Table 8.25). The rural poor of coastal Asia-Pacific account for 27% of the global poor. Given that the

Table 8.24. Distribution of rural poor on productive (favoured) and marginal land

Region (no. of countries) in dataset	Total population	Total rural population	Rural population on favoured lands	Rural population on marginal lands	Rural poor on favoured lands	Rural poor on marginal lands	Average rural poverty (%)
Asia (20)	2840	2044	755	1289	219	374	29
West Asia & North Africa (40)	345	156	37	119	11	35	29
Sub-Saharan Africa (40)	630	375	101	274	65	175	64
Central & South America (26)	430	117	40	77	24	47	61
Total (126)	4145	2692	933	1759	319	631	36

Source: Population in millions; Nelson et al. 1997 as cited by Penning de Vries et al. 2003.

Table 8.25. Regional distribution of poor living in absolute poverty

Region	Millions with < $1.00 per day (% of total population)			Millions with < $2.00 per day (% of total population)		
	1981	1990	2001	1981	1990	2001
East Asia & Pacific	796 *(58)*	472 *(30)*	271 *(15)*	1170 *(85)*	1116 *(70)*	864 *(47)*
China	634 *(64)*	375 *(33)*	212 *(17)*	876 *(88)*	825 *(73)*	594 *(47)*
South Asia	475 *(52)*	462 *(41)*	431 *(31)*	821 *(89)*	958 *(86)*	1064 *(77)*
WORLD	1482 *(40)*	1218 *(28)*	1089 *(21)*	2450 *(67)*	2654 *(61)*	2735 *(53)*

Source: World Bank 2005.

most recent numbers use population figures estimated in 2003, these numbers do not include those who were above the $2.00 a day income poverty line, but who were displaced by the tsunami that struck northwest Sumatra (Indonesia), Thailand, Sri Lanka, the Maldives and southeast India on December 26, 2004. In an instant, the impoverished population of the region increased by 2 million people, and most of this population reside in coastal rural communities (ADB 2005).

In a retrospective analysis of natural resource management and rural poverty in Africa, Anderson and colleagues provide a way forward (USAID 2004). They make a compelling argument that good governance that creates accountable and transparent institutions can provide an exit out of poverty by recognising environmental limits as a guide for natural resource use, nurturing participation among resource users in decision-making processes, and creating buy-in and stewardship. The pivotal role of good institutions comes out as a significant determinant of poverty in empirical economic models that relate resource abundance, poverty, and development (Bulte et al. 2004).

Perhaps the planet is reaching a tipping point relative to its response to anthropogenic change. Perhaps also, there is a convergence of world views about how humankind can still realise sustainable scenarios. In what appears to be a shared epiphany, the World Resources Institute in cooperation with the United Nations Development Programme, the United Nations Environment Programme, and the World Bank released in 2005: "World Resources. The Wealth of the Poor. Managing Ecosystems to Fight Poverty". The book articulates pro-poor, ecosystem-based economic growth, which is more focused and directed than the vague notion of sustainability espoused by UNCED in 1992. For this strength, the 2005 book may be considered the magnum opus of the environmental movement, and, if followed through by policy at national scales, it may spell the difference on how contemporaneous societies deal with environmental change.

8.8 RESEARCH TARGETS

Rural coastal areas are a priority for further research because of the paucity of data to describe the human-environment dynamics at appropriate spatial and temporal scales. These dynamics include interactions with the adjacent upstream catchments and the marine environment offshore. Furthermore, these dynamics are embedded in patterns of urbanisation, migration, and consumption that have local, national, and global drivers.

In the rural coasts of the Asia-Pacific region, the poorest inhabitants live off the fertility of the land and the productivity of coastal waters. New knowledge should be generated to understand what scale of economies can improve the quality of human life and, at the same time, sustain the ecological functioning of ecosystems. Moreover, predicting trajectories of coastal change, as a function of globalised trade and climate change, may enable coastal societies to enhance their resilience to cope with change. Innovative governance should be targeted at enhancing pro-poor opportunities for growth and designing institutions that can nurture these vulnerable populations. LOICZ, a global change core project, provides an excellent template for pursuing integrated coastal research through its science plan and implementation strategy. Approaches like these might fruitfully be adapted at multiple scales by research and policy communities with local to international research interests (Kremer et al. 2005).

REFERENCES

Alongi, D.M. 2002. Present state and future of the world's mangrove forests. Environmental Conservation 29(3): 331–349.

ADB (Asian Development Bank). 2005. An Initial Assessment of the Impact of the Earthquake and Tsunami on South and Southeast Asia. Economics and Research Department, Asian Development Bank, 11 p.

Barbier, E.B. and Cox, M. 2003. Does economic development lead to mangrove loss? A cross-country analysis. Contemporary Economic Policy 21(4): 418–432.

Bay of Bengal News. An agenda for the BOBP-IGO. Bay of Bengal News Vol. III, No. 5–9, Jan 2002–Mar 2003, pp. 6–7.

Bay of Bengal News. Fact Sheets 1–4 (Bangladesh, Maldives, India, Sri Lanka). Bay of Bengal News Vol. III, No. 5–9, Jan 2002–Mar 2003, pp. 17–20.

Bay of Bengal Programme Inter-Governmental Organization, India. http://www.bobpigo.org/ bobp_india. htm.

Brinkhoff, T. 2005. City Population. www.citypopulation.de.

Bulte, E.H., Damania, R. and Deacon, R.T. 2004. Resource abundance, poverty and development. ESA Working Paper No. 04-03. Agriculture and Development Economics Division, FAO, 31 p.

Burke, L., Kura, Y., Kassem, K., Revenga, C., Spalding, M. and McAllister, D. 2000. Pilot Analysis of Global Ecosystems: Coastal Ecosystems. WRI, 100 p. http://marine.wri.org/pagecoastal-pub-3054.html.

Christensen, V.T., Garces, L.R., Silvestre, G.T. and Pauly, D. 2003. Fisheries impact on the South China Sea large marine ecosystem: A preliminary analysis using spatially-explicit methodology. p. 51–62. In: Silvestre, G., Garces, L., Stobutzki, I., Ahmed, M., Valmonte-Santos, R.A., Luna, C., Lachica-Aliño, L., Munro, P., Christensen, V. and Pauly, D. (eds.), Assessment, Management and Future Directions for Coastal Fisheries in Asian Countries. WorldFish Center Conference Proceedings 67,1 120 p. http://www.worldfishcenter.org/trawl/publications/assessment/pdf/Chapter-03-FA.pdf.

Delgado, C.L., Wada, N., Rosegrant, M.W., Meijer, S. and Ahmed, M. 2003. Fish to 2020: Supply and Demand in Changing Global Markets. Washington, D.C.: Penang, Malaysia: International Food Policy Research Institute; WorldFish Center, 2003. xiv, 226 pp. http://www.worldfishcenter.org/Pubs/fish-to-2020/fish-to-2020.htm.

Falkenmark, M. and Widstrand, C. 1992. Population and water resources: A delicate balance. Population Bulletin 47(3): 2–35.

FAO (Food and Agriculture Organization). 2002. The State of the World Fisheries and Aquaculture. http://www.fao.org/documents/show_cdr.asp?url_file = /docrep/005/y7300e/y7300e00.htm.

FAO (Food and Agriculture Organization). 2004a. The State of World Fisheries and Aquaculture. http://www.fao.org/documents/show_cdr.asp?url_file = /DOCREP/007/y5600e/y5600e00.htm.

FAO (Food and Agriculture Organization). 2004b. Summary of World Food and Agricultural Statistics. FAO 100 p. http://www.fao.org/es/ess/sumfas/sumfas_en_web.pdf.

Gillett, R.D. 2002. Pacific island fisheries: Regional and country information. Asia-Pacific Fishery Commission, FAO Regional Office for Asia and the Pacific, Bangkok, Thailand. RAP Publication 2002/13, 168 p.

Hampton, J., Sibert, J.R., Kleibert, P., Maunder, M.N. and Harley, S.J. 2005. Decline of Pacific tuna populations exaggerated? Nature 434: E1-E2.

Hampton, J. and Williams, P. 2003. The Western and Central Pacific tuna fishery: 2001 Overview and status of stocks. Oceanic Fisheries Programme, Tuna Fisheries Assessment Report No. 4, 54 p http://www.spc.int/oceanfish/Docs/Research/TFAR_2001_No4.pdf.

IFAD (International Fund for Agricultural Development). 2001. Rural Poverty Report 2001. The Challenge of Ending Rural Poverty. Oxford University Press.

IFAD (International Fund for Agricultural Development). 2002. Assessment of Rural Povery. Asia and the Pacific. 81 p.

Kremer, H.H., Le Tissier, M.D.A., Burbridge, P.R., Talaue-McManus, L., Rabalais, N.N., Parslow, J., Crossland C.J. and Young, B. (eds.). 2005. Land-Ocean Interactions in the Coastal Zone. Science Plan and Implementation Strategy. IGBP Report 51/IHDP Report 18, 68 p. www.loicz.org.

Langley, A., Hampton, J., Williams, P. and Lehodey, P. 2005. The western and central Pacific tuna fishery: 2003 Overview and status of stocks. Oceanic Fisheries Programme, Tuna Fisheries Assessment Report No. 6, Noumea, New Caledonia: Secretariat of the Pacific Community, 86 p. http://www.spc.int/oceanfish/Docs/Research/TFAR_2003_No6.pdf.

Lehodey, P., Bertignac, M., Hampton, J., Lewis, A. and Picaut, J. 1997. El Niño Southern Oscillation and tuna in the western Pacific. Nature 389: 716–718.

Myers, R.A. and Worm, B. 2003. Rapid worldwide depletion of predatory fish communities. Nature 423: 280–283.

Pabico, A.P. 2002. Unregulated fish pen and cage operations mess up coastal ecosystems. Philippine Center for Investigative Journalism, 17–18 June, 2002. http://www.pcij.org/stories/2002/aquaculture.html.

Pauly, D. and Chuenpagdee, R. 2003. Development of fisheries in the Gulf of Thailand large marine ecosystem: Analysis of an unplanned experiment. In: Hempel, G., Sherman, K. (eds.), Trends in Exploitation, Protection, and Research. Large Marine Ecosystems of the World 12, Elsevier Science, Amsterdam (The Netherlands), pp. 337–354.

Penning de Vries, F.W.T., Acquay, H., Molden, D., Scherr, S.J., Valentin, C. and Cofie, O. 2003. Integrated land and water management for food and environmental security. Comprehensive Assessment of Water Management in Agriculture Research Report 1. Colombo, Sri Lanka: Comprehensive Assessment Secretariat, 74 p.

Petersen, E. 2002. The catch in trading fishing access for foreign aid. Resource Management in Asia-Pacific Working Paper No. 35. Resource Management in Asia-Pacific Program, Division of Pacific and Asian History, Research School of Pacific and Asian Studies, The Autstralian National University, Canberra, 19 p. http://www.elsevier.com/wps/find/obtainpermissionform.cws_home/obtainpermissionformhttp://eprints.anu.edu.au/archive/00002284/01/rmap_wp35.pdf.

Relox, J.R. Jr. and Bajarias, F.F.A. Harmful algal blooms (HABS) in the Philippines. http://fol.fs.a.u-tokyo.ac.jp/rtw/TOP/EXabst/019JuanRReloxJr.pdf.

Rosegrant, M.W., Cai, X. and Cline, S.A. 2002. World Water and Food to 2025. International Food Policy Research Institute, 338 p.

Smakhtin, V., Revenga, C. and Doll, P. 2004. Taking into account environmental water requirement in global-scale water resources assessment. Comprehensive Assessment Research Report 2. Colombo, Sri Lanka: Comprehensive Assessment Secretariat, 32 p. http://www.iwmi.cgiar.org/assessment/-Publications/research_reports.htm.

Silvestre, G.T., Garces, L.R., Stobutzki, I., Ahmed, M., Santos, R.A.V., Luna, C.Z. and Zhou, W. 2003. South and South-East Asian coastal fisheries: Their status and directions for improved management: conference synopsis and recommendations, p. 1–40. In: Silvestre, G., Garces, L., Stobutzki, I., Ahmed, M., Valmonte-Santos, R.A., Luna, C., Lachica-Aliño, L., Munro, P., Christensen, V. and Pauly, D. (eds.), Assessment, Management and Future Directions for Coastal Fisheries in Asian Countries. WorldFish Center Conference Proceedings 67,1 120 p. http://www.worldfishcenter.org/trawl/publications/assessment/pdf/Chapter-01-FA.pdf.

Sugiyama, S., Staples, D. and Funge-Smith, S. 2004. Status and Potential of Fisheries and Aquaculture in Asia and the Pacific. RAP Publication 2004/25. Food and Agriculture Organization of the United Nations, Regional Office for Asia and the Pacific, 62 p. http://www.apfic.org/apfic_downloads/pubs_RAP/2004-25.pdf.

Talaue-McManus, L., Kremer, H.H. and Marshall Crossland, J.I. (eds.). 2001. SARCS/WOTRO/LOICZ: Biogeochemical and human dimensions of coastal functioning and change in Southeast Asia. Final report of the SARCS/WOTRO/LOICZ project 1996–1999. LOICZ Reports and Studies No. 17, ii + 277 p, LOIC, Texel, The Netherlands.

Tilman, D., Fargione, J., Wolff, B., D'Antonio, C., Dobson, A., Howarth, R., Schindler, D., Schlesinger, W.H., Simberloff, D. and Swackhamer, D. 2001. Forecasting agriculturally driven global environmental change. Science 292: 281–284.

United Nations. 2004. World Urbanization Prospects. The 2003 Revision. Data Tables and Highlights. Department of Economic and Social Affairs, Population Division, UN. www.unpopulation.org.

UNEP (United Nations Environment Programme). 2005. One Planet, Many People. Atlas of Our Changing Environment. Division of Early Warning and Assessment (DEWA). UNEP, P.O. Box 30552, Nairobi, Kenya. www.unep.org and www.earthprint.com; http://www.na.unep.net/OnePlanetManyPeople/AtlasDownload.php.

USAID (US Agency for International Development). 2004. Nature, Wealth, and Power. Emerging Best Practice for Revitalizing Rural Africa. 36 p.

Valiela, I., Bowen, J.L. and York, J.K. 2001. Mangrove forests: One of the world's threatened major tropical environments. Bioscience 51(10): 807–815.

World Bank. 2005. 05 World Development Indicators. www.worldbank.org/data/wdi2005.

WRI (World Resources Institute). 2005. World Resources 2005 – The Wealth of the Poor: Managing Ecosystems to Fight Poverty. WRI in collaboration with UNDP, UNEP, and the World Bank, 2005, 200 p. http://population.wri.org/worldresources2005-pub-4073.html.

Yadava, Y.S. 2003. Making of the Bay of Bengal Programme – InterGovernmental Organisation. Bay of Bengal News Vol. III No. 5–9, Jan 2002–Mar 2003, pp. 1–3. http://www.bobpigo.org/dnload/ BBN%20Jan02-Mar%202003.pdf.

CHAPTER 9

IMPACTS OF POLLUTANTS IN THE ASIA-PACIFIC REGION

KANAYATHU KOSHY[1], ZAFAR ADEEL[2], MURARI LAL[1], AND MELCHIOR MATAKI[1]

[1]*Pacific Centre for Environment and Sustainable Development, University of South Pacific, Fiji*
[2]*United Nations University International Network on Water, Environment and Health, Ontario, Canada*

9.1 INTRODUCTION

Over half of the world's nearly 6 billion people live within 60 kilometers of the shoreline (Morris 2004). This concentration of population is largely the result of the tremendous productivity of coastal ecosystems leading to unparalleled economic opportunities. The health of these ecosystems is vital to sustaining not only coastal communities but also human society as a whole. These natural systems – including salt marshes, mangrove forests, coastal wetlands, coral reefs, and estuaries – are under unprecedented stress from both coastal and land-based activities. In fact, municipal, industrial and agricultural waste, and run-off account for some 70 to 80% of all marine pollution (Schellnhuber 2002). Pollution of the marine environment, particularly those from land-based sources poses a major global environmental challenge for the 21st century.

Part of the problem in dealing with coastal pollution is that its sources are enormously varied, often located far away from the coastline and difficult to control. They include sewage discharges, industrial effluents, rubbish dumps, surface and ground-runoffs, and intentional dumping. Pollution from urban and agricultural runoff can travel hundreds of kilometers through river systems, ending up in the coastal environment. As the concentration of pollutants increases in our oceans and seas, the effects are becoming increasingly serious both for human beings and for the marine environment. For example, we now have higher levels of carcinogens and toxins (ciguatera) in the fish we eat, and there are an increasing number of beaches designated off limits because of water-borne contaminants and associated poor water quality (Livingston 2002).

Coastal and marine pollution is a global problem, though the severity and type of pollution vary from country to country and region to region. In the developing world, the construction of basic sewage treatment facilities and the enforcement of rules on industrial and commercial effluent often do not keep pace with the tremendous economic and population growth being experienced in many coastal cities.

Environmental pollution can have direct and indirect linkages to poverty. Poor people are the most vulnerable to losses of basic food supplies for subsistence or income earning due to contamination of land or marine resources. They are also vulnerable to harmful pollutants and the associated health problems. Pollution can reduce land value and productivity due to contamination, and may have direct effects on fish and other marine resources. Export markets are also potentially at risk as many developed countries continue to tighten their monitoring and control of contaminated foods.

There are many social aspects to marine and coastal environmental pollution. At a very basic level, poorer neighborhoods are often the "preferred" locations for polluting industries and for rubbish dumps. The residents of these neighborhoods are most at risk from the pollutants, due to their generally poorer health status compared to other groups. Women and children are more at risk than men from many pollutants. For example, most organic chemicals accumulate in body fats and the accumulation rates tend to be higher for women than for men because of their generally higher fat levels. Those same pollutants can be readily transferred from mothers to the unborn children or via breast milk to babies. Infants and young children are especially vulnerable to the effects of pollutants such as mercury and lead, which can affect both mental and physical development.

Effective pollution prevention brings many benefits. In the manufacturing and energy sectors this can include reductions in raw material costs (including fuel) and increased processing and operating efficiencies. Reducing packaging will reduce the cost of packaged goods. The marketing advantages of "pollution-free" products are now being widely recognized, especially in the tourism sector and for organic agricultural produce. Waste recovery and recycling operations represent potential business opportunities for both the public and private sectors.

9.1.1 Coastal Impacts in the Pacific Islands

Coastal zones in the Pacific are among the most pristine in the world and with the exception of Papua New Guinea, a significant proportion of the Pacific islanders live by the coast. Coastal zones are dynamic since they are at the interface of land and sea and thus exposed to continuous perturbations. Prior to the introduction of western forms of agricultural practices and the growth of industries, the people within the islands experienced clean and unpolluted coastal areas. The state of the coastal zones in the Pacific has changed significantly in the last few decades with the radical changes emanating from newer agricultural practices, education systems, consumerism, and industrial development, all of which have introduced numerous types of chemicals in varying quantities into the Pacific Island environment. With increasing stresses, usually driven by increases in population, urbanization, and economic activities, it is plausible that severe perturbations may cause irreversible changes to the coastal zones of the Pacific Islands in the future.

The populations of Small Island States (SIDS) are acutely vulnerable to environmental degradation, climate change, overexploitation of fisheries resources, land-based pollution, and natural disasters. Moreover, they share a number of disadvantages, including a limited population, a narrow range of available resources, excessive dependence on international trade, and vulnerability to global developments. The main types of pollution within the region are shipping-related, sewage, hazardous chemicals, hazardous wastes, and solid waste. The region's coastal and marine resources are also threatened by the introduction of exotic invasive marine species, shipwrecks, marine accidents and spills, ships' waste including ballast waters, and antifouling paints on vessels. Increasing quantities of solid waste, poor control of chemicals imported into the region and the lack of capacity to manage potential chemical pollutants and sewage are primary problems in the Pacific region.

The limited land area of many of the islands combined with a lack of appropriate technology for waste recycling and safe disposal has resulted in a proliferation of plastics, paper, glass, metal, and also drums of hazardous chemicals. Much of this rubbish breaks down slowly and leaches into the soil and into drinking water. The foul-smelling organic wastes usually attract vectors of diseases. Piles of household rubbish are accumulating on beaches and in mangrove swamps (Figure 9.1 below). The issue of plastics is thought to be a priority pollution threat in this region today; the occurrence of plastic bags in the ocean is increasing and it is known that the ingestion of only a few plastic bags can kill juvenile cetaceans and turtles.

The Pacific Ocean is home to a wide range of large marine animals including mammals such as whales, dolphins, porpoise, dugongs, and marine turtles. Maintaining a healthy population of these animals is essential to sustaining oceanic productivity. The diversity of these marine creatures is recognized as a fundamental element of Pacific Islands' culture and heritage. Over half the world's known species of whales are found in this region. There are six marine turtle species that feed and migrate through the Pacific waters. The Pacific Island region is a globally significant area for marine turtle breeding and migration. The coastal populations of the Pacific islands have exploited marine turtles for their meat, eggs, shell, and oil for centuries. The latter half of the 20th century has been marked by catastrophic declines of sea turtle populations throughout the Pacific region (UNEP 1999). The types of pollution that may impact whales and dolphins in the region include chemicals (heavy metals and excess nutrients); plastics (ingestion), and sewage. Sewage discharge may cause nutrient enrichment and possible habitat destruction; it may also introduce disease vectors, heavy metals, and pesticides. A few cases of impact have been recorded from this region, including disease issues in dolphins adjacent to sewage discharge. At this stage it is not considered a significant issue for the region but ongoing monitoring is required.

Figure 9.1a and b. Solid waste in form of plastics covering water body and roadside in Tuvalu (a: Nick Harvey, b: Koshy Kanayathu)

The major industries in the small island states are agriculture, tourism, forestry, mining, and fisheries related. Each of these industries generate wastes – some are by-products of the activity, whereas some are a necessary part of the product stream. By-product wastes are generally the result of poorly managed operations and include siltation (from mining and land clearing of marginal forests for agricultural activities), oil pollution (used oil from machinery and from accidental spills), poisons (from pest control), and miscellaneous plastic trash (old fishing gear, plastic sheeting, drums, and bags). Production wastes include organic wastes from food processing, and chemical wastes (from oil palm refineries, mining processes and timber treatment).

Hazardous chemicals and nutrient pollution find their way into the inshore estuarine and marine environment via effluents, dumps, storm runoff, sewage, and wind-blown dust. This is especially damaging to coastal marine nursery areas like sea grass beds, coral reefs, and mangrove forests. While many of these effluents cause local environmental stress, siltation (Figure 9.2), oil pollution, poisons, and plastic trash contribute to extensive damage to inshore marine environments.

Chemical pollution has not yet been considered a significant issue in this region. Nonetheless, persistent organic pollutants, which include several pesticides, polychlorinated biphenyls and dioxins, mainly produced in the Northern Hemisphere, are potentially a significant threat to cetaceans and other marine organisms. These compounds affect the hormonal system and can cause low fertility and birth defects. They are transmitted through atmospheric deposition into the ocean and by run-off from land, particularly agricultural chemicals (e.g. pesticides). Pollution due to toxic waste is a dire threat – toxic substances flowing into coastal waters may be the biggest threat to marine life today. Excessive amounts of pollutants, such as nitrogen from agriculture, sewage and lawn chemicals have, in fact, the potential to transform the natural systems into to a new state, totally changing marine ecosystems.

Discharge of pollutants into the terrestrial and marine environments usually represents wasted resources. For example, smoky vehicle exhausts mean that fuel is being wasted through inefficient combustion. Some of the materials thrown away as rubbish

Figure 9.2. Land-ocean interaction in Sigatoka (Fiji) with siltation and agrochemical run off into reef areas (Photo: Koshy Kanayathu)

represent lost resources when they could be used in other ways, for example using green waste as compost. The recovery of some of these materials can have direct economic benefits, like collecting aluminum cans to resell and using waste oil as a fuel substitute. Tourism, a key money-earner for some Pacific island countries, is also starting to be affected by the spread of litter. The coastal regions of the Pacific Island have a competitive advantage over other tourist destinations because of the reputed beauty of its lagoons and beaches. But that advantage is shrinking under the weight of solid waste piling up around shorelines and waterways.

Pacific Island countries now list the prevention of pollution as their major environmental concern as it remains one of the major threats to sustainable development in the Pacific islands region. The increase in the sources and extent of pollution are threatening the Pacific islands' efforts to maintain healthy societies, to stimulate development and new investment, and a sustainable future for its people.

Pollution prevention needs to be addressed predominantly at national and local levels, although there are also some regional and global aspects. At a national level, there are currently only a very few Pacific countries and territories with specific environment acts and associated regulations. There are even fewer with laws and regulations that deal with specific aspects of pollution, such as waste management. Most authority in this and other related areas is derived from rather outdated and fairly generic legislation, such as the Public Health Act. Progress has been made in recent years in developing national policies and strategies to address issues such as waste management in some Island countries. However, much work remains to be done to transform these into active programs that can achieve real on-the-ground solutions. A related issue is often the lack of any clear and unified approach within the government systems, with responsibilities being spread across a number of agencies.

The critical issues regarding the inter-linkages between environmental pollution and the coastal zones of the Pacific Islands are:
- Loss of biological diversity;
- Threats to freshwater resources;
- Degradation of coastal environments;
- Climate change and sea-level rise; and
- Land and sea based pollution.

From the list of recognized environmental problems (UNEP 1999), the coastal zone is directly or indirectly referred to in all of them, signifying the importance it has to Pacific Islanders. The emerging pressures originating mainly from the shift to semi-industrialization necessitates serious measures that should be taken to ameliorate or mitigate the negative impacts of chemical pollutants in the coastal zones of the Pacific Islands Countries.

9.1.2 Chemical Pollution in the Pacific

In dealing with the chemical pollution of coastal zones in the Pacific, it is important to acknowledge that all chemicals found in the Pacific are imported and none is manufactured internally apart from the reformulation of some chemicals from imported raw materials. Chemical reformulation is usually restricted to some agricultural products (fertilizers, pesticides), petrochemical fuels, and sanitary reagents. The scale of chemical pollution in the Pacific may therefore be considered as insignificant in the global context; however, localized pollution from industries (e.g. mining, fuel spillage, and agricultural plantations) and the vulnerability of the environment to medium chemical perturbations requires appropriate attention and action. Given that coastal zones are at the interface of the land and sea, much of the chemicals including solid and liquid wastes eventually find their way to the coastal zones through surface runoff, storm water outlets, industrial waste outlets, streams, and rivers. Furthermore, given the lack of appropriate chemical management practices and disposal facilities, some of the chemicals are even intentionally disposed of in coastal zones.

Assessing chemical pollution in the coastal zone of the Pacific Islands often remains a challenge because of the processes of dilution and dispersion, which are predominant in the vast coastal zone where there is constant interchange between the land and the sea on the one hand and the lagoon and the ocean on the other. Furthermore, the land-use processes responsible for non-point pollution are very difficult to identify and assess, given their diffuse nature and the enormity and variance in terms of location and quantity of pollutants. Moreover, normal pollution regulatory measures (usually derived from point source pollution) are seldom relevant for non-point sources.

9.1.3 Coastal Impacts in the East Asian Region

Damage to coastal ecosystems poses a direct threat to human survival in many parts of the East Asian region. This is primarily because coastal resources are an important source of both food and income in this region, particularly for low-income groups. It is important to note that fish and seafood are primary sources of animal protein in this region. By the mid-1990's, the level of consumption of fish and other seafood in this region exceeded the global averaged per capita seafood consumption. This is because fish prices are relatively cheap in Asia, compared to other sources of animal protein, such as beef and pork (Vincent et al. 1997). Any impacts on fisheries put poorer coastal communities at higher risk, because it is more difficult for them to afford alternative sources of protein and maintain steady income.

The East Asian fisheries are one of the most important economic activities in the region. This sector – mostly marine fisheries – provides a high level of employment opportunities for the population. The number of people working in fisheries varies among countries, ranging from 1,600 people in Brunei Darussalam (Silvestre and Pauly 1997) to over four million in Indonesia (ADB 1997). The bulk of employment is in the various fishing activities that take place in shallow waters and around coastal shelves. Furthermore, aquaculture in particular can play a significant role in development for poor communities (Edwards 2000).

Pollution in the coastal zones can potentially affect the health of both the coastal populations and the ecosystems. The awareness of these issues among the coastal communities and policy makers is gradually improving in the region. For example,

Table 9.1. Major pollutants in East Asian countries

Type of pollutants	Indonesia	Malaysia	Singapore	Philippines	Thailand
Oil	x	x	x	x	x
DDT, pesticides, organochlorines	x	x		x	
Heavy metals	x	x		x	
Organic and biological pollutants, fertilizers	x	x	x	x	x

Source: Hidayati 2000.

a regional report on the health impacts of land-based coastal pollution describes a growing awareness of the risks, and human and economic costs of infectious diseases related to bathing in contaminated water and the consumption of contaminated seafood (GESAMP 2001). Recent research on the status of mangroves, coral reefs, seagrass, and mussel beds in Indonesia, the Philippines, and Thailand shows that harm to one coastal ecosystem may also have direct or indirect impacts on the other.

9.1.4 Chemical Pollution in East Asia

Domestic and industrial wastes are the most serious land-based pollutants in this region (Table 9.1). These can be broadly classified as organic and biological pollutants, agricultural wastes (such as rubber, palm oil, tapioca processing wastes, animal excrement, and chemical fertilizers), and industrial wastes (primarily from textile and paper industries). Sewage and industrial outfalls, rivers, land runoff, and transport of chemicals through the atmosphere are believed to be responsible for most of the marine pollution in this region.

Takada (2001) presented the benchmark monitoring of the major chemicals present in coastal pollution around urban areas. These chemicals were attributed to sewage, detergents, and petrol products. For East Asia, the identified priority is attention to pollution from sewage (GESAMP 2001).

9.2 SYNOPSIS OF COASTAL PROJECTS IN THE ASIA-PACIFIC REGION

An intergovernmental conference held in Washington D.C., USA from 23 October to 3 November 1995 adopted a Global Programme of Action (GPA) for the Protection of the Marine Environment from Land-Based Activities with a view to prevent the degradation of the marine environment from land-based activities, by facilitating the realization by States of their duty to preserve and protect the marine environment. United Nations Environment Programme (UNEP) was designated as the Secretariat of the said GPA and was mandated to undertake the following (UNEP/GPA Coordination Office,

2000), as coordinator and catalyst of environment activities within the United Nations system and beyond:
- Promote and facilitate the implementation of the Programme of Action at the national level;
- Promote and facilitate the implementation at the regional, including sub regional level through, in particular, a revitalization of the UNEP Regional Seas Programme; and
- Play a catalytic role in the implementation at the international level with other organizations and institutions.

The South Pacific Region launched the region's participation in the GPA with projects and activities (SPREP, 2000) which include:
- Preparation of A Strategic Action Programme for the International Waters of the Pacific Region in 1998;
- Pacific Pollution Prevention Programme (PACPOL);
- UNITAR/IOMC National Profiles to Assess the National Infrastructure for the Management of Chemicals Project;
- Management of Persistent Organic Pollutants in the Pacific;
- Development of the Hazardous Waste Management Strategies in Pacific Island Countries Project; and
- Pacific Regional Waste Awareness and Education Programme.

Subsequently, a workshop was organized in Apia, Samoa in October 1999 by the Secretariat of the Pacific Regional Environment Programme with assistance provided by the UNEP/GPA Coordination Office. The key objectives of this workshop were:
- To review the objectives of the GPA and its implications for the region;
- To identify possible elements of regional framework strategies with special reference to recommended approaches by sources category;
- To consider the development and implementation of national programmes, including the assistance required and available for this purpose through the organization supporting the GPA; and
- To review and amend a draft regional program of action to addresses land-based activities.

A report on land-based sources and activities affecting the marine, coastal and other associated water resources was prepared as the main background document for the workshop. The report provided useful information to assist the governments of the region, both individually and collectively, in their efforts to protect the marine environment and achieve sustainable development of their coastal areas. The report identified and assessed the problems related to each country and the region and outlined basic steps for remedial action as well as effective environmental management to prevent future degradation from the identified land-based sources.

Again, a regional review was undertaken of the production, use, environmental impacts, and environmental transport of the group of chemicals known as persistent toxic substances (PTS) in the Pacific Islands region. The review was undertaken to provide a measure of the nature and comparative severity of damage and threats posed at national and regional levels by PTS and to identify immediate international action needed to protect human health and the environment through measures which will reduce and/or eliminate the emissions and discharges of twelve persistent organic pollutants (POPs) listed in the 2001 Stockholm Convention and other pollutants with similar characteristics. This work, financed by the Global Environment Facility (GEF) through a global project with co-financing from the Governments of Australia, France, Sweden, Switzerland, and the United States of America, led to the production of a comprehensive report on the status of PTS in the Pacific Region (including 22 countries and territories) in 2002 produced within the framework of the Inter-Organization Programme for the Sound Management of Chemicals (IOMC).

The United Nations University (UNU) in partnership with APN organized an international workshop focused on Persistent Organic Pollutants (or POPs) in September 2003. The purpose of the workshop was to address key issues in the interpretation of the threats posed by environmental levels of POPs through the examination of existing standards, guidelines and toxicity assessments for POPs in the environment. The workshop also aimed to build the capacity in the East Asian region to undertake monitoring of POPs through a monitoring network that had been established by UNU. This network has been collecting data in the region for a number of years.

The UNU network in East Asia has been conducting environmental monitoring activities, including POPs monitoring, for more than eight years. The regional sampling programme is the only one of its type existing for the analysis of water and sediments across the region. This workshop was a valuable opportunity to continue discussions amongst the scientists in the region, giving due consideration to the potential for harmonization of monitoring activities and threat assessment in the region.

The workshop was a step in the direction of being able to fully understand the extent and severity of POPs pollution in the Asia-Pacific region. At the same time, it initiated a policy dialogue on ways to ameliorate the situation through the use of data-based early-warning mechanisms for identifying and reacting to severe POPs threats in the region.

9.3 ORGANIC COMPOUNDS

Organic pollutants have been operationally defined to include agrochemicals, ship-based discharges, oil, dissolved organic carbon (DOC), dynamite used for fishing and mining, industrial discharges, and persistent organic pollutants (POPs).

9.3.1 Agrochemicals

The agricultural economy plays a significant role in almost all countries and territories of the region, with the possible exception of Nauru. Much of the agricultural activity in the Pacific islands is associated with subsistence farming and the economy, whereas major commercial agricultural operations exist in the East Asian region.

In the East Asian region, agricultural activities play an important role in increasing pollution around urban areas where farming is intensified, involving the use of large quantities of fertilizers. This, together with other organic pollutants from agricultural activities causes eutrophication and Harmful Algal Blooms (HAB's) in the coastal zones. For example, China is the world's largest consumer of synthetic nitrogen fertilizers (Kraemer et al. 2001). Yeung describes how this has led to the occurrence of 'red tides' – a type of HAB – along the Chinese coast (Yeung 2001). During 2000 there were a total of 28 red tide events on the Chinese coasts (Hu 2001). Similar effects have also been recorded in Indonesia, Thailand, Brunei Darussalem, Singapore, Vietnam (Azanza 1998), and the Philippines. An analysis of agrochemical usage in East Asia is shown in Table 9.2.

A most striking example of pesticide usage in the East Asian region is that of DDT (dichlorodiphenyltrichloroethane). Although its use has been banned in almost all the countries, DDT has been widely monitored and the most clearly observed in the region to date. It has been used in the region both as an agrochemical, and also for vector control. Recent studies undertaken for GEF/UNEP to assess the status of POPs (Choi et al. 2002, Ibrahim et al. 2002) show that DDT should be considered a priority POPs substance for assessment in Asia with regard to sources, concentrations, ecotoxicological and human effects. It has been observed at significant levels in regional coastal waters (Figure 9.3).

While much agricultural activity in the Pacific region is subsistence farming, there are some commercial agricultural activities. The sugar industry in Fiji, the coffee industry in Papua New Guinea and the squash farms in Tonga are notable examples. The pollutants arising from agricultural areas are primarily nutrients, pesticides, and sediments and are derived from the application of agricultural chemicals, erosion of exposed soils with naturally occurring nutrients, and runoff from piggeries and other areas with concentrated animal wastes. Runoff and wastes from piggeries have also been identified as a major pollutant source by the countries of the region.

Table 9.2. Analysis of agrochemical usage in East Asia

Country	Rice fields (1,000 ha)	Aquaculture areas (10^3 ha)	Fertilizer use (ton/year)	Pesticide use (ton/year)
Cambodia	1,835	No data	>40,000	No data
China	3,425	2,476	3,636,685	>89,423
Indonesia	4,966	243	5,670,117	28,706
Malaysia	No data	7	No data	No data
Philippines	1,236	20	181,084	No data
Thailand	8,613	No data	No data	No data
Vietnam	1,500	No data	110,250	No data
Total			>9,638,136	118,129

Source: Talue-McManus 2000.

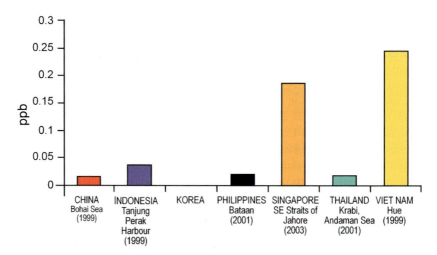

Figure 9.3. Maximum observed levels of DDT seawater in the East Asian coastal areas (King 2003)

The qualitative understanding and anecdotal evidences provide the significance of the effects of agricultural pollution on the marine environment. However, no quantitative estimates are available on the nutrients and agricultural chemicals loading in runoff from agricultural areas and the resultant concentrations in marine, or even fresh water. The available statistics in the use of agricultural chemicals suggests that the use of these chemicals is small compared to other regions; however, the inconsistencies in the quality of current management, use, and disposal practices have led to environmental concerns in localized areas (SPREP, 2000). Anecdotal evidence of eutrophication and turbidity in several areas of the region are cited as evidence of the significance of this source of pollution.

Table 9.3 (adapted from SPREP, 2000) presents the available data for fertilizer and pesticide use in the South Pacific Region. Data from several countries and territories was not available. In some cases, the data is out-dated. Some of the data were obtained from Stone (1990) who calculated quantities from monetary totals. The original sources for the Stone data are not known. The use of agricultural chemicals in the region generally does not appear to be changing dramatically; thus the data serve as reasonable estimates of pesticide use.

Agrochemicals (fertilizers and pesticides) have been used extensively and in some cases intensively in agricultural activities geared towards cash economies; for example, in the coconut and sugar industries and in commercial root crop farming. There is very little data on the extent of coastal pollution by agrochemicals.

PTS pesticides have been used in the past throughout the region, although the level of usage has been generally low by world standards. The primary uses were in crop production, termite control, general household and public health applications, and for vector control. Malaria is a significant problem in the Solomon Islands,

Table 9.3. Argicultural Chemical Use in the Region (tonnes)

COUNTRY	FERTILIZERS					PESTICIDES								
	Nitrogen Based	Phosphates	Potassic	Other	TOTAL	Insecti-cides	Fungi-cides	Herbi-cides	Rodenti-cides	Mineral oils	Fumi-gants	Mollusci-cides	Other	TOTAL
American Samoa														
Cook Islands[2]					0.00	3.65	2.44	2.44						8.53
FSM														
Fiji[2]	12.00	23.00	18.00		53.00	283.00	24.00	198.00	198.00					703.00
Kiribati					2.8									9.6
Marshall Islands														
Nauru														
Niue[5]							1.20	1.00			0.10			2.30
Palau														
PNG					20.55	143.00	2.10	467.00					6.00	618.10
Samoa[10]	150.00	150.00	150.00		14018.00	0.63	0.087[10,11]							
Solomon Islands[7]		2238.00	160.00		2398.00	283.00	5.00	205.00					4.00	497.00
Tonga[8]	430.04	0.24			430.28	0.98	10.91		0.08			0.85		12.82
Tuvalu		2.00	5.00	17.00	24.00	1.00	0.30		1.00				2.00	4.00
Vanuatu[9]						1.00		10.00					30.00	41.30
TOTAL	592.04	2413.24	333.00	17.00	16943.89	738.06	46.44[10]	885.54[11]	199.08	0.00	0.10	0.85	42.00	2164.25[13]

Source: UNEP/GPA Coordination Office & SPREP 2000.

Notes: Data from (1) Mowbray (1988); (2) United Nations (1989) for fertilisers; Mowbray (1988) for pesticides; (3) Mowbray (1988); (4) Tiaeke, N. Kiribati Environment Unit, MESD National Profile of Chemicals Management Infrastructure (Undated, 1999?); (5) United Nations (1990); (6) Pitcairn High Commissioner (1992); (7) Pesticide data from Mowbray (1988); fertiliser data from Stone (1990); (8) Foliaki (1991); (9) Albert William, Vanuatu representative to RPA Meeting (1999); Mowbray (1988) and Taylor (1991); (10) Convard (1993); (11) *ibid*.
Additional small quantities: (12) + 0.585 kilolitres; (13) + 43.56; + 44.585 kilolitres.

Table 9.4. PTS pesticides listed in the Pacific Regional Agro-Pesticide Index (ARSAP/CIRAD/SPC 1994), as being available for sale and/or approved for use within the region

Pesticide	Area available for sale and/or approved for use
Aldrin	Solomon Islands
Atrazine	Fiji, French Polynesia, New Caledonia and Tonga
Chlordane	French Polynesia, Samoa
DDT	Solomon Islands
Dieldrin	Solomon Islands
Endosulphan	Cook Islands, Federated States of Micronesia, French Polynesia, New Caledonia and Vanuatu
Heptachlor	French Polynesia
Hexachlorocyclohexane (as Lindane)	French Polynesia, Kiribati, New Caledonia and Solomon Islands
Organomercury compounds	Solomon Islands (as phenylmercury acetate)
Organotin compounds	French Polynesia, Guam, New Caledonia, Vanuatu and Wallis & Futuna (as azocyclotin, cyhexatin and/or fenbutatin oxide)
Pentachlorophenol	Fiji

The following PTS pesticides are currently used within the countries indicated:
Atrazine: Fiji (10 t/a), Tonga (10 kg/a), Vanuatu (25–30 kg/a).
Endosulphan: Fiji (2 t/a), Vanuatu (15–20 kg/a).
DDT is known to be used in the Solomon Islands for malaria control. Annual usage is believed to be of the order of 5 to 10 tonnes per year, although this has not yet been confirmed.

Vanuatu PNG, and New Caledonia. However, spraying for mosquitoes is also practiced in most other countries for the control of dengue fever.

In 1994, a number of PTS pesticides were listed in the Pacific Regional Agro-Pesticide Index (ARSAP/CIRAD/SPC 1994), as being available for sale and/or approved for use within the region (Table 9.4).

Agricultural activities primarily represent non-point sources of pollution, increasing the difficulty in the identification and quantitative assessment of the problem. Changing agriculture practices is increasing the potential for runoff, which brings with it increased nutrient loading of the surface water. In some areas of greater agricultural chemical use the runoff may also carry persistent organic pollutants to the marine area. The effects of these contaminants in the marine environment are not yet well understood. Pollutant loading to the marine environment from agricultural chemicals only has been estimated using the empirical assumption that about 5% of all applied chemicals will reach the marine environment. This pollutant loading is summarized in Table 9.5.

9.3.2 Ship Based Discharges

The ship based discharges can, in broad terms, have one of three sources of origin, namely, (i) the waste associated with maintenance and operation of the vessel (e.g. lubricating oil, fuel sludges, paint chips, and used engine components), (ii) domestic wastes generated by passengers and crew (e.g. food waste and associated

Table 9.5. Argicultural Chemical Pollutant Loading to the Marine Environment (tonnes)

COUNTRY	FERTILIZERS					PESTICIDES								
	Nitrogen Based	Phosphates	Potassic	Other	TOTAL	Insecti-cides	Fungi-cides	Herbi-cides	Rodenti-cides	Mineral oils	Fumi-gants	Mollusci-cides	Other	TOTAL
American Samoa														
Cook Islands						0.18	0.12	0.12						0.43
FSM														
Fiji	0.60	1.15	0.90		2.65	14.15	1.20	9.90	9.90					35.15
Kiribati					0.0031									
Marshall Islands														
Nauru														
Niue							0.06	0.05			0.01			0.12
Palau														
PNG					1.03	7.15	0.11	23.35[3]					0.30	30.91
Samoa	7.50	7.50	7.50		22.50	0.03[1]	0.04[2]		0.04	12.50	0.05			12.66[4]
Solomon Islands		111.90	8.00		119.90	14.15	0.25	10.25					0.20	24.85
Tonga	21.50	0.01			21.51	0.05	0.55		0.004			0.04		0.64
Tuvalu		0.10	0.25	0.85	1.20	0.05			0.05				0.10	0.20
Vanuatu					6.00	0.05	0.02	0.50					1.50	2.07
TOTAL	29.60	120.66	16.65	0.85	174.79	36.91[1]	2.34[2]	44.28[3]	10.00	12.50	0.06	0.04	2.10	108.25[4]

Source: Convard 1993; adopted from SPREP, 2000.
Notes: All quantities are formulation quantities.
1) + 0.02 kilolitres   3) + 2.18 kilolitres
2) + 0.029 kilolitres   4) + 2.23 kilolitres

packaging, sewage, stationery, and printed material), and (iii) cargo-associated wastes (e.g. hold sweepings, packing materials, pallets, drums, containers, and oil tank residues).

These wastes can also be categorized according to type for the purposes of regulation and management, with the principal classifications being garbage, oil, sewage, and hazardous materials. Most oily wastes are in the liquid phase, except for oil filters and oily rags, which are solid wastes. The oily wastes can be further divided into two broad categories, namely, concentrated oil wastes (e.g. used lubricants and hydraulic oil, contaminated fuel oil, and oil sludge), and oily mixtures largely in a water medium (e.g. oily bilge water, tank washings, oil-contaminated ballast water). Oily wastes may also contain a range of impurities (e.g. detergents, degreasers, engine additives, and greases).

Much of the garbage generated in ships and boats is analogous to that generated in a residential setting, namely domestic wastes such as food and associated packaging, paper, cardboard, disposable products, and other consumer items. This garbage type is generated in both commercial and recreational vessels and the amount produced is generally a function of the tasks undertaken by the vessel, the number of people onboard and the duration of the voyage. Food waste and associated packaging form a significant proportion of the garbage generated in ships. In practice, food waste can be difficult to manage by virtue of the quantities involved and its putrescible nature. Food wastes of overseas origin and associated packaging also present a quarantine risk. Within the Pacific islands region, garbage sourced from overseas vessels which contain foodstuffs is generally treated as quarantine waste.

Some elements of the garbage waste stream are materials which can be recycled, such as aluminum cans, paper, cardboard and certain plastics. The partially damaged shipping containers, and wooden pallets, may be suitable for reuse. Some garbage generated in vessels is of a hazardous and/or toxic nature, including dry and wet cell batteries, pressure pack containers, and receptacles containing residues of noxious substances such as greases, oils, solvents, paints, adhesives, and engine additives. Medical wastes, with associated biological and sharps hazards, are also produced in ships, particularly those carrying many people onboard, such as cruise liners and large warships. A significant amount of cargo-associated garbage, particularly packaging wastes (e.g. damaged containers, broken pallets, empty drums, dunnage, steel strapping, used lining, and packing materials) and cargo residues such as deck and hold sweepings, spillages are produced by merchant vessels. The garbage from fishing vessels include used nets and trawl gear, marker buoys, bait boxes, and the plastic packing straps used to bind these boxes. Litter and debris from commercial fishing activities has been recognized as a persistent and widespread problem in oceanic and coastal zones of the Pacific island countries.

Provided the discharge is beyond the minimum mandated distance from nearest land, MARPOL 73/78 does permit most types of garbage to be disposed to sea. However, plastics or toxic or hazardous materials can not be discharged to sea. Given the geography of many Pacific island states, with many small islands and islets, and rocks and reefs exposed at low-water, the mandated distances for disposal of garbage may be much further out to sea than might otherwise be expected. Under the circumstances, coastal shipping may rarely be in a position where even food waste, with the least stringent disposal criteria, can lawfully be discharged to sea.

A range of noxious and hazardous wastes such as toxic, flammable, explosive, corrosive, poisonous, radioactive, or infectious in nature, or materials such as Persistent Organic Pollutants which otherwise require special treatment are also likely to be generated by the shipping vessels. These are most commonly in solid or liquid form, but could also be gaseous. Special, hazardous and noxious wastes are frequently mixed with general garbage and often appear to be innocuous. The vessel-generated wastes typically also include batteries, pressure pack containers, greases, oils (packaged), solvents, acids, paint, paint chips, adhesives, and engine additives. Medical wastes, with associated biological and sharps hazards, need special handling and disposal as these present an additional quarantine risk.

Special wastes may also include animal-related wastes such as urine and faeces, and carcasses in the case of livestock carriers. Special wastes in the form of putrescible catch residues, bycatch, processing wastes, or bulky fishing gear are most common types found in fishing vessels and require special handling. Certain cargoes, such as bulk or packaged liquid chemicals (e.g. empty CCA [wood preservative] containers), and fertilizers produce cargo and packaging residues requiring special handling and disposal.

Sewage is typically considered as human excreta directed into urinals and toilets, but is defined by the IMO to also include drainage from onboard medical premises and spaces containing living animals. In some boats this may include greywater and dishwater that is drained into common holding tanks. Some sewage is discharged from vessels in raw, untreated form, while other vessels are fitted with treatment systems that variously macerate and/or dose the effluent with chemicals or otherwise sanitize it, often with chlorine. More elaborate ship waste management includes biological treatment systems.

Shipping activities bring foreign chemicals and species into East Asian coastal ecosystems. Some other forms of pollution that may be carried on ships are required to be cleaned off by some ports such as Singapore, before entry. This often means that they are often cleaned away in coastal waters elsewhere, causing pollution in those locations. The effects of international shipping on water quality include high oil and grease contents, and phosphates from the handling of phosphate rocks, degradation of sediment, contamination of soil and air, and the generation of vibrations, noise and other types of wastes.

Table 9.6 lists the potential demand for waste reception arising from ships normally using Pacific Island Ports. Port reception facilities for ship-borne wastes are of fundamental importance for Pacific Island countries and there is a need for improvement of these facilities (Figure 9.4). Such facilities should also include efficient waste disposal systems; and they should be isolated from the general public because of possible diseases or pests associated with the wastes. Although the focus of the reception and subsequent management of ship-generated waste

Table 9.6. Estimated Rates of Potential Demand for Waste Reception Arising from Ships Normally Using Pacific Island Ports

Vessel Type	Indicative No. Persons Onboard	Indicative Displacement (t)	Sludge and Waste Oil[1] m³/day (at sea before arrival)	Oily Bilge Water[2,3] amount per ship visit (m³)	Garbage[1] kg/day (at sea before arrival)	Sewage[4] m³/day (in port)
Merchantmen[3]	18	3,000–20,000	0.18	n/a	27	1.3
Tankers[3]	15	2,000–20,000	0.18	n/a	22	1.0
Cruise Liners[3]	600–1,500	10,000–20,000	0.27	n/a	1,800–4,500	42–105
Inter-island Traders	15–20	100–250	0.05	5	22–30	0.4–0.6
I/island Ferries (large)	600	1,500	0.05	10	900	n/a
Inter-island Ferries	100	250	0.05	2	150	n/a
Tourist Charter Boats	10–20	n/a	0.01	n/a	5–10	n/a
Warships (very large)[3]	1,000–6,000	20,000–100,000	0.18	n/a	1,700–10,200	50–300
Warships (large)[3]	200	2500	0.18	n/a	340	10
Warships (small)	20	100–250	0.01	5	26	1.0
Fishing (oceanic)	18	250–1,000	0.02	10	32	0.7
Fishing (mothership)	18	2,000–4,000	0.05	10	50	0.7
Fishing (local)	2–5	n/a	0.005	n/a	2–4	n/a
Local workboats	2–5	n/a	0.01	0.05	2–4	n/a
Yachts (itinerant)	3	n/a	n/a	n/a	1.5	0.06
Local craft (day trips)	2	n/a	n/a	n/a	1	n/a

[1] Estimates are indicative only and assume all waste is retained onboard for disposal ashore (including food waste) without any treatment (e.g. incineration, compaction, shredding).
[2] Does not include tank washings or non-segregated ballast water.
[3] Older ships not fitted with IMO approved pollution control equipment may need to discharge to shore that oily bilge which is produced while alongside/at anchor.

upon the ship – port interface, effective management of this waste stream is a continuum of measures, of which the ship – port interface is but one component. Current arrangements for the management of ship-generated waste in the Pacific islands region are piecemeal and of varying quality, ranging from effective and comprehensive in some ports to virtually non-existent in others. Given the intent of MARPOL 73/78 and the technical and economic factors applying in the region, a cooperative regional approach has been identified as the most effective manner in which to minimize the pollution of the marine environment by ships. In order to improve ship waste management, it is essential that the Pacific island Countries accede to and properly implement MARPOL 73/78. Accession to MARPOL

Impacts of Pollutants in the Asia-Pacific Region 249

Figure 9.4. Suva Harbour with ship-based oil pollution and small scale industrial discharge present (Photo: Koshy Kanayathu)

73/78 carries obligations and responsibilities for signatories, the most important of which is arguably the requirement to provide adequate port waste reception arrangements.

Waste management is a major issue for Pacific Island States. For many reasons, including lack of technical expertise, land availability, and cultural factors, waste is often not managed in an environmentally acceptable manner. Advances in the management of ship-generated waste can only be accomplished in concert with improvements in the overall management of wastes in the region.

9.3.3 Oil Pollution

The effects of releases of oils (hydrocarbons) into the marine and coastal environment has been an issue of environmental concern for many decades. From originally being related mainly to pollution from sea-based sources such as shipping and offshore (accidental as well as operational discharges), the problem of oil pollution of the marine environment of Pacific Island countries has gradually grown to encompass a very wide range of both land-based (coastal and inland) and sea-based sources within most sectors of society.

Accidental oil spillages are frequent in the Strait of Malacca (Ismail 2001) and between 1987 and 1997, 89 oil spills were documented off the coast of Vietnam (Nguyen 1999). While use of organotin compounds in antifouling paints is banned in locations like Japan and Hong Kong, they are still used elsewhere and can be found in the waters of Tokyo Bay and the Straits of Malacca as a result of their transport through foreign vessels (Hashimoto et al. 1998).

The International Convention on Oil Pollution Preparedness, Response and Cooperation, 1990 (OPRC 90) is focused upon providing cooperative, regional responses to combat oil pollution in the event of a spill. A fundamental tenet of OPRC 90 is the reduction of risk of oil spill incidents, with a mechanism of obliging Parties to OPRC 90 to also implement the requirements of MARPOL 73/78.

Regional and National programmes could include components such as targets, timetables, and sector-specific measures according to the precautionary principle and the principles of best available techniques (BAT), best environmental practice (BEP), and integrated pollution prevention and control (IPPC). Programmes could also include fiscal and economic incentives and measures, including voluntary agreements, to encourage reductions in emissions and discharges of oils, the recycling of used lubricating oils, and fuel-use efficiencies. Programmes could also comprise the provision of reception and recycling facilities for oily wastes, the development of plans and measures to prevent accidental releases of oils, particularly from refineries (e.g. in PNG), storage and waste reception facilities, and the capacity to respond to such accidents, the establishment of cleaner production programmes in cooperation with industry, and means to ensure the effective implementation of the programme of action.

9.3.4 Dissolved Organic Carbon (DOC)

DOC is naturally produced in coastal areas by phytoplanktons via exudation and cell lysis (Lancelot 1984, Lee and Wakeham 1988, Libes 1992, Lee and Henrichs 1993). However, output from rivers, streams, creeks, surface and ground water run-off, storm water, and direct discharge of sewage can also contribute to the DOC levels in coastal areas (Robards et al. 1994). DOC plays an important role in the biogeochemistry of any aquatic system, as it is the component of organic carbon that is cycled through organisms (Thurman 1985, Stumm and Morgan 1981). DOC provides energy and carbon for metabolism in heterotrophic bacteria and for some species of phytoplankton which can subsist heterotrophically on dissolved organic substrates (Riley and Chester 1971, Volk et al. 1997). On the other hand, DOC in an aquatic system is normally oxidized to carbon dioxide by bacteria, fungi, protozoan and animals present in water (Libes 1992), consequently, excess DOC may impose considerable demand on the oxygen budgets of water bodies (Jackson 1988, Robards et al. 1994). Furthermore, dissolved organic forms of micronutrients like phosphates and nitrates, may also be valuable sources for autotrophic growth (Kennish 1997). Consequently, excess DOC may enhance the likelihood of eutrophication in a water body.

Impacts of Pollutants in the Asia-Pacific Region 251

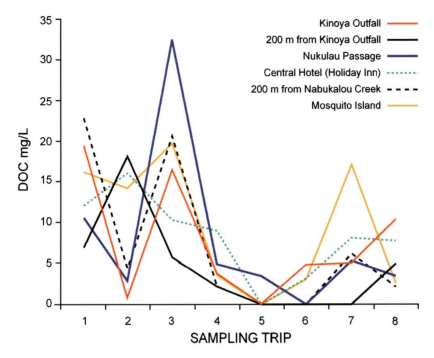

Figure 9.5. Baseline DOC study (Mataki 1999)

A baseline DOC study by Mataki (1999) of six sites on the shoreline of the Suva lagoon showed that land-based discharges significantly influenced the concentration of DOC in the coastal areas of Suva lagoon (Figure 9.5). The average concentration of DOC measured per site ranged from 0 to 33 mg/L. The study also showed that surface and groundwater run-off, discharges from creeks, storm water outlets and small rivers are also significant conduits of terrestrial DOC input to the coast. Furthermore, rainy weather could also enhance DOC export to the coast. Apart from the study of Mataki (1999), very little data on DOC is available through out the Pacific region.

9.3.5 Use of Dynamite for Fishing

Reefs were once used to test atomic bombs. Extensive areas were flattened and contaminated by radioactive substances. For example, the Bikini-Atoll (1946) was the location of 23 atmospheric atomic bomb tests from 1946 through 1958. Today, coral reefs are often dynamited to harvest small fish. Although it is illegal, it is practiced in many countries worldwide and is a major threat to coral reefs.

Destructive fishing practices, including the use of dynamite and poison are still practiced in East Asia, particularly in Indonesia, the Philippines, and Thailand. Mining coral as a source of building materials and lime is an important problem in Indonesia and the Philippines. As a result of over fishing, virtually all of the Philippines reefs, and 83% of Indonesia's reefs, are at risk. The area with the highest coral damage in Thailand is around Phuket Island, mainly due to increased tourism activities.

The Pacific Islands region is no exception. These destructive fishing method uses bottles filled with explosives made from potassium nitrate (a common fertilizer). The explosives create an underwater shock wave that ruptures fish swim bladders so they float to the surface where fisherman can easily collect them. Not only fish are killed but also other inedible reef animals. Often a reef area is dynamited twice, first to kill the smaller fish, then again to kill the larger predators attracted by the carnage. But the most damage is caused by the destruction of the corals by the blast, reducing the surrounding area to lifeless rubble. These areas don't recover very fast, because larvae have difficulty in finding places to take hold in the rubble. Fish will reproduce to fill the environment – coral is the environment.

9.3.6 Industrial Discharges

The growing industrial activity in the region provides the potential for economic expansion and increased environmental impacts. Industrial pollution is a major concern throughout the Asia-Pacific region. Particular examples, such as that involving mercury caused the well-known disaster in Minamata, Japan, during the 1950's and 1960's. Since then, the development of industry has been recognised as a major source of coastal pollution in East Asia. According to Clarke (1998), wastewaters from heavy industries and mining may cause damaging pollution from the discharge of toxic substances including heavy metals (lead, mercury, cadmium, etc), radioactive elements, acids, Polyaromatic hydrocarbons (PAH's), and innumerable other toxic industrial chemicals such as Polychlorinated biphenyls (PCB's).

There are many specific examples throughout the East Asian region that can be cited. For example, increases in heavy metal content in Jakarta Bay have been observed (1997) as a result of industrial waste (Mahbub and Kuslan 1997). Pollution concentration in marine animals are useful in determining accumulation of these pollutants in the food chain. Examples of these studies include work on Hong Kong's porpoises by Parsons (1998), in which trace metals were found, and the study on several porpoise and whale species in which high levels of organochlorines were detected (Minh 2001). The latter work also provided valuable information about the transfer of these pollutants from mothers to newborns through placenta and milk.

In the global context, Pacific Island Countries have practically very small industrial base except mining operations. Some countries in the region have limited medium size industrial activity which has resulted in environmental degradation in recent years. Cook Islands, FSM, Kiribati, Niue, Tuvalu, RMI, and Palau have small industrial bases related to tourism only such as processing of coconuts, fish, and fruit. The Cook Islands industrial base includes a small brewery. CNMI has a fairly significant garment factory base. Yap, in the FSM, also has a small garment factory employing about 350 persons, though many of these are foreigners. There are also printing shops, laundries, boat building, and other small commercial activities that utilize solvents and other hazardous chemicals which are usually disposed of in domestic waste-water drains or pits in the ground. These chemicals degrade very slowly and there is a potential for accumulation, even if the annual generation is small.

American Samoa, Solomon Islands, Tonga, Vanuatu, and Samoa have small industrial bases with larger manufacturing facilities of a more diverse nature and/or larger facilities for other industries such as breweries and fish canning plants. Several countries also have small to medium sized abattoirs and meat processing facilities. Slaughtering and meat processing activities range from household activities to small all-manual slaughterhouses to the larger more modern facilities, such as those found in Vanuatu.

Fiji and Papua New Guinea have medium-sized industrial bases too. Papua New Guinea has a comparatively larger industrial base associated with mining activity. The major industrial complex in Papua New Guinea is primarily copper and gold mining, but also includes smaller scale industries in Port Moresby and other urban centres. The industrial complex in Fiji has a variety of manufacturing and industrial activities ranging in size from small family-owned operations to large exporting facilities. Nauru may be considered to have a medium sized industrial base limited only to phosphate mining.

Except for the solid waste and small quantities of waste water, the marine environmental impact from the limited industrial activities in the region is not well quantified. Nonetheless higher wastewater flows can be expected where bottling and cleaning with water occurs. The pollutants that reach the marine environment, either through direct discharge or leachate from landfills, include BOD, nitrogen, phosphorus, and solids. Oil and grease pollution could also result from the storage of fuels and lubricating oils associated with these processes. The total loading and concentrations of BOD, suspended solids, nutrients, solvents, oil and grease, and cleaning agents discharged from breweries, sugar refineries, breweries, edible oil production and fish canning plants are reported to be relatively higher.

POPS and other toxic chemicals result from power and energy production, and other smaller scale industrial operations including printing, photo developing, metal plating, auto repair, paint factories, metal fabricating, manufacture of cleaning products. The industries located in the region's urban areas contribute substantially to marine pollution problems, particularly the potential cumulative effect of heavy metals, solvents, and other hazardous materials from these operations. For example, several studies have documented the introduction of organic chemicals, heavy metals, and toxics into Suva Harbour in Fiji which are largely of industrial origin (Convard 1993, Morrison 1991, Morrison 1992, Naidu et al. 1991). Similar pollution has been observed in marine waters near other regional urban centres (Morrison 1990, Gangaiya and Green 1991).

The waste streams from most industrial facilities have little or no treatment. The most common type of treatment provided for almost all facility types is simple sedimentation. The increase in the presence of several contaminants in the marine environment has been observed in recent years which have the potential to disturb the coral reef environment including reef fisheries and the pelagic fisheries. Many countries with canneries or plans for opening new ones also wish to promote tourism. Ill-managed or heavily polluting cannery operations are not compatible with the development of a successful tourism industry.

Table 9.7 summarizes the available quantitative information on industrial pollutant loadings reaching marine waters, excluding mining wastes. The information is structured so that the process from which they originated and the country that they are located in identifies the pollutant loadings. Many of the more hazardous and toxic wastes such as solvents and heavy metals associated with the medium scale industrial bases of Fiji and Papua New Guinea, are, however, not well represented in here. For example, the summary does not include most of industrial pollutant

Table 9.7. Industrial Pollutant Loadings by Country and Industry (for Industries where information was available to use rapid assessment method)

COUNTRY	Process	B.O.D.	S.S.	Oil	N	P	Other
		t/yr	t/yr	t/yr	t/yr	t/yr	t/yr
American Samoa	Fish Canning	4,53	179,18	64,71	255	167,3	
Cook Islands	Beer Production	No data					
Fiji	Beer Production	12,9	18,7				
	Fish Canning	8,18	6,35	4,52	25,63		
	Sugar milling	263,81	124,58				
	Food Production		0,333	0,061		0,91	
	Brewery	129,7	187,1				
	Edible Oils	95,9	94,8	107,9			
	Soap Manufacturing	0,14	0,04	0,03	0,001		
	Paint Manufacture			0,04			Pb 0.04
	Battery Manufacture	0,001	0,006	0,03			
	Bulk Fuel Storage	0,001	0,007	0,032			
	TOTAL	510,632	431,916	112,613	25,631	0,91	0
Kiribati	None						
Marshall Islands	No data						
Nauru	No data						
Niue	Slaughter House	0	0	0	0	0	
Palau	No data						
Papua New Guinea	Edible oil	246,6	974,6	765,3			
	Brewery	48,9	8				
	Sugar Milling	213,44	100,8				
	TOTAL	508,94	1083,4	765,3	0	0	
Samoa	Soft Drinks						
	Slaughterhouse						
	Beer Production	63,7	10,42				
	TOTAL	63,7	10,42	0	0	0	0
Solomon Islands	Slaughter House	9	1,12	1,04	1,4	0,1	
	Fish Processing	14,1	9,09	6,185	17,3		
	Edible Oil	490,5	484,6	553,6			
	Food Manufacture						
	Soft Drinks						
	TOTAL	513,6	494,81	560,825	18,7	0,1	0

Tonga	Fish Canning	0	0	0	0	0	0
Tuvalu	none						
Vanuatu	Beer Production	211,7	34,63				
	Soft Drinks Prod.	126	88,2				
	Slaughterhouse	152,69	101,99	98,03	117,21	42,72	
	Milk Production	57,7	16,6				
	Fish Processing	0	0	0	0	0	
	TOTAL	548,09	241,42	98,03	117,21	42,72	0
TOTAL		**2 149,5**	**2 441,1**	**1 601,5**	**416,5**	**211,0**	**0,0**

Source: Convard 1993.

Note: PNG also has other industries, including oil palm processing, abattoirs, coffee pulping, food processing, and meat canning.

sources identified by Cripps (1992) for Suva Harbour due to the lack of production data and/or loading factors as per WHO Methodology.

In Fiji, sugar milling is a critical part of the industrial activity. The suspended solids and BOD loading to the marine environment are quite high; visual observations in Lautoka confirm this. Breweries are also increasing in prominence in the industrial sector of the region. Sediments, BOD, and heated water are the primary constituents of concern for this industry. This industry can be relatively innocuous and simple treatment technologies provide adequate treatment, if adequately sized. Shifting effluent discharges in non-sensitive areas with good circulation, and preferably below thermocline levels, can minimize marine pollution.

South Pacific Tourism Organisation (SPTO) data clearly demonstrate that tourism is an important industry for the region, and since the 1970s, visitor arrivals have grown here steadily. Not accounting for the decline in visitor arrivals during the coups of 1987 and 2000, the average rate of growth is close to 6% or even higher. The total number of visitor arrivals (and hence tourism earnings) reached a record high of around 431,000 in 2003. In Fiji, tourism is now the highest foreign exchange income earner ahead of sugar. The economic benefits of tourism are clear, however, the threats of the tourism industry to the environment, particularly coral reefs and water quality are numerous and significant. Sound management of tourism solid and liquid waste generation and its associated physical alterations to the coastal environment is vital in order to allow sustainable development of this important economic sector. The potential to destroy the same high quality resources that gave rise, and presently maintain this industry is great.

9.3.7 Persistent Organic Pollutants (POPs) and Persistent Toxic Substances (PTS)

The introduction of the group of chemicals known as persistent organic pollutants (POPs) and persistent toxic substances (PTS) into the environment and resulting effects is a major issue that gives rise to concerns at local, national, regional, and global scales. The nature and scale of the threats to the environment and its resources posed by past and present uses of POPs and PTS both within the Pacific Islands region and beyond are considered serious. Table 9.8 provides a list of existing waste quantities of POPs and the number of contaminated sites in the Pacific Region.

The 12 chemicals covered under the Stockholm Convention on Persistent Organic Pollutants (POPs) are aldrin, chlordane, DDT, dieldrin, dioxins and furans, endrin, hexachlorobenzene, heptachlor, mirex, polychlorinated biphenyls (PCBs) and toxaphene. Information on endosulphan, hexachlorocyclohexanes (HCH), phthalate esters, polyaromatic hydrocarbons (PAHs), pentachlorophenol, and organolead, organotin, and organomercury compounds are also relevant. Other chemicals that need to be considered here are, atrazine, chlordecone, hexabromobiphenyl, polybrominated biphenyl ethers, chlorinated paraffins, octylphenols, and nonylphenols. However, no data were available for any of these chemicals, although this does not preclude their existence within the region.

Polychlorinated Biphenyls (PCBs) were almost universally used as transformer oils prior to the 1980s, and it can be assumed that this was so throughout the Pacific region. Conversely, transformers and transformer oils imported in more recent times are expected to be free of PCBs, although the only countries with specific import controls are the US and French territories. A survey by SPREP (Burns et al. 2000) identified significant stockpiles of old transformers and transformer oils throughout the region, and some of these were shown to contain PCBs. Other possible sources of PCBs are old capacitors, lighting ballasts, hydraulic fluids, and various building products.

Table 9.8. Existing Waste Quantities of POPs in the Pacific Region (SPREP 1999)

Waste Type	Quantity
Waste Oil	180 tonnes
Potentially PCB Contaminated Oil	135 tonnes
Waste Bitumen	330 tonnes
Waste Timber Treatment Chemicals	160 tonnes
Waste Fertilisers	87 tonnes
Waste DDT	10 tonnes
Waste Pesticides (not including DDT)	47 tonnes
Buried Waste Pesticides	11 tonnes
Waste Medical Drugs	21 tonnes
Miscellaneous Special Wastes	38 tonnes
Oil Contaminated Sites	26 sites
Bitumen Contaminated Sites	9 sites
Hydrocarbon Contaminated Groundwater Lens	11 sites
Pesticide Contaminated Sites	21 sites
Buried Waste Pesticides Sites	7 sites
Timber Treatment Chemical Sites	7 sites
Miscellaneous Contaminated Sites	7 sites

Dioxins, furans and other complex organics are formed as unintentional byproducts from most combustion processes, along with trace amounts of PCBs, PAHs and hexachlorobenzenes. The possible sources include motor vehicles, oil-fired power stations, industrial fuel use, rubbish burning and the combustion of trees, wood, and other biomass. No data is available on the emissions of these chemicals within the region. However, some simple estimates of the order of the emissions can be calculated using published emission factors. The results are generally quoted in dioxin toxic equivalents (TEQs), which provide a simple method of comparing the relative potential impacts of different emission sources (UNEP Chemicals 2001). A preliminary assessment of dioxin emissions for Samoa was reported by Cable (2001). These estimates were made using the emissions factors recommended by UNEP Chemicals, combined with local estimates for factors such as petroleum fuel consumption, and biomass and rubbish burning. The work identified motor vehicle emissions, rubbish burning, and forest or scrub fires as the three most significant sources.

There is no information available on the use within the region of chemicals such as the chlorinated paraffins, polybrominated biphenyls and diphenyl ethers, or phthalate esters. Nor are there any established systems in the region for obtaining this information. These chemicals are known to have been used in a wide variety of industrial, commercial, and consumer products but, as such, their presence in the products is usually not clearly identified.

There is no manufacture in the region of any of the PTS chemicals covered in this report, although many of them are known to have been used in the region. The current usage of PTS pesticides in the region is low, and should be eliminated over the next 10 years or so. Existing stockpiles of PCBs and PTS pesticides should also be eliminated over the next few years. There is no evidence of PCBs being actively used in the region although small quantities are believed to still exist in a few in-use transformers. No information could be obtained on the use or otherwise of organolead and organomercury compounds.

Numerous hot spots have been identified, consisting mainly of stockpiles of hazardous wastes and obsolete chemicals, pesticides, and transformer oils. Over 100 contaminated sites were identified, of which 54 were assessed as needing major remediation work. These sites include PCBs, buried pesticides, pesticides storage, timber treatment, and rubbish dumps. Significant efforts will be required for remediation of these sites. Generally, there is a lack of data on the emissions of dioxins, furans, and other complex organics from combustion processes and other sources in the region. Some estimates of dioxin emission have been made for some of the countries, on the basis of existing fuel use data.

The amount of data available on environmental levels of PTS in the region is extremely limited. Of those samples analyzed, the majority have been of environmental media (air, water, sediment, and marine organisms used as pollution indicators). Hardly any data exists for levels in humans (plasma, milk, fat). Data on drinking water and food are also very limited. Of the data available, a large number of samples had detectable levels of PTS, owing both to local usage and global transport, especially by wind currents. PTS were recorded in some samples for which there is no record that that particular chemical was ever imported into that country. This could indicate either illegal entry or environmental transport. In general, concentrations are relatively low for most samples. There are a few samples, however, especially of sediments from urban areas that would lead to a classification as contaminated sites in developed countries and warrant remediation. There are also contaminated areas in Micronesia due to past military activities that have impacted marine food samples.

Overall the highest concentrations of PTS tend to have been found for DDT and its derivatives, especially in Papua New Guinea and Solomon Islands where DDT is used to control malarial mosquitoes, and PCBs, which have been used as electrical oil insulating material and often disposed of in a haphazard manner. Organolead and organotin levels are also high in some areas, probably due to their use in gasoline and marine paints respectively. Table 9.9 lists the chlorinated compounds and PCBs found in marine food samples examined in some of the Pacific Islands.

Toxicology of PTS in the region is in its infancy. Very few toxicological or exposure investigations have been conducted in the Pacific Islands and even fewer relate to PTS. The only body burden studies found in the region were conducted in Samoa in 1979 and the Northern Mariana Islands in 2000. Information from the neighboring country of Papua New Guinea identified DDT in mothers' milk with one sample above 3 mg/kg (ppm).

In the absence of comprehensive human health studies and body burden data, chemical contamination in food sources has been assessed, although this data is also very limited. Examination of some of the more widely utilized PTS around the region, gives an indication of the impact that use has had on food sources. The highest levels of DDT in marine foods were found in Fiji shellfish, but once again, the sampling was extremely limited. In Papua New Guinea, DDT levels in oysters were considerable. Consistent with use, PCBs have been detected at notable levels in imported foods and in marine foods, especially in former United States trust territories contaminated by electrical oil residues. In 1979, Samoan fish were found to have the highest levels of HCHs and heptachlor in the region. Unfortunately, it was a small sampling and there is no recent data. Most lead and mercury levels were in the range found in developed countries (mean values around 200 µg/kg wet weight).

While toxicological studies in the Pacific Islands have been limited and human health risks have not yet been effectively assessed, some data has been gathered, which reveals no serious problems. More work is needed to effectively assess the impact of PTS on human health in the Pacific Islands. In small island countries, the rule of thumb is that chemical pollutants in the land means pollutants in the sea and thus the biota.

Impacts from PTS to ecosystems of the Pacific could have a major effect on the economic base of the Pacific Islands. The fisheries, agriculture, timber, and tourism industries in the region could be heavily impacted by contamination of PTS.

Table 9.9. Chlorinated Pesticides and PCBs in Marine Food (ng/g)

Location Food Year (samples)	HCB	HCHs	Heptachlor	Aldrin	Chlordanes	DDTs	Mirex	Dieldrin	Endrin	PCBs
Fiji shellfish Morrison et al. 1994	<0.1–0.1	<0.3–0.2	<0.4–0.3	<0.13	<0.5	5.0–52	<0.19	0.4–0.7	<0.5	<5.6
Tonga shellfish Harrison et al. 1994	<0.4	<0.8	1.0–2.3	<0.7	<0.7	2.1–2.2	<0.10	<0.1–0.6	<0.5	<1.1
Solomon Is. fish Kannan et al. 1992	<0.01–0.05	0.2–1.6	<0.01	<0.1–0.5	<0.01–2.1	0.07–0.83	-	0.1–2.5	-	1.2–11
Solomon Is. crab/oyster Kannan et al. 1994	<0.01–0.05	0.3–0.9	0.04–0.15	<0.1–2.1	<0.01–0.8	0.3–1.4	-	0.1–0.7	-	5.9–16
Guam mollusc Denton et al. 1997	-	-	-	-	-	-	-	-	-	1.2–47
Guam fish Denton et al. 1997	-	-	-	-	-	-	-	-	-	0.1–85
Solomon Is. fish Kannan et al. 1995	0.01–0.06	0.23–1.9	-	0.1–1.0[a]	0.11–1.6	0.91–24	-	0.1–1.0[b]	-	0.66–15
Samoa shellfish Govt of Samoa 1993		<0.1–0.55			<0.01–0.58	5.6–54				0.6–7.0
Guam fish USEPA 2001	<0.10–3.3	<0.01–8.1	<0.5–2.0	<0.9–5.3	<20–22600	11–17900	<1.3–140	<1.7–4600	<0.6–48	26–80160
Guam fish USEPA 2001	<0.18–0.3	<0.01–4.2	<0.12–0.5	<0.1–3.8	<0.01–33.1	<0.01–17.3	<0.01–0.7	<0.11–11.0	<0.13–1.5	0.29–8348
"Acceptable" levels[b]	1.8	2.3	0.32	-	8.4	8.6	-	0.18	88	1.5

[a] Agency for Toxic Substances and Disease Registry (ATSDR) unrestricted fish consumption limits.
[b] Sum of aldrin and dieldrin.

While no published ecotoxicological data are found in the Pacific, some of the work done on non-migratory sea-otters on the Pacific coast of the United States (Jarman et al. 1996) indicated that PCBs were found in higher concentrations in the Aleutian Islands. While local military sources may have contributed to these levels, there are also indications that transboundary movement of PCBs has also contributed to the levels of PCBs found. While there have been no specific transboundary studies in the Pacific Islands, there are a number of theories that have been developed regarding transboundary movement of PTS in and out of the Pacific region. Some of these theories are supported by studies conducted in the surrounding continents and work carried out relating to climate change issues.

Transport mechanisms in the Pacific include some typical means, as well as regionally specific features including the freshwater lens under many islands, highly porous substrata, the possibility of significant contributions from imported foodstuffs, and large fish movements contributing to PTS transport. Information on contaminant concentration and pathways of transport in the Pacific Islands is rare. DDT's continued illegal use in some Pacific Island Countries may be contributing to recent loading of PTS into the environment and transport out of the region. Extrapolations from other regions indicate that there may be transboundary movement of contaminants into the Pacific region and that there are a few special situations of major significance in this region. More work on integrating environmental chemistry with other components of the regional contamination assessment is required.

9.4 INORGANIC COMPOUNDS

9.4.1 Effects of Mining Operations and Mine Tailings Discharges

Mining is usually a high impact activity with detrimental environmental impacts.

PNG, New Caledonia, Nauru, Solomon Islands, and Fiji are the major mining centres in the sub-region. Four materials are mined in the Pacific region: phosphate, nickel, copper, and gold. Phosphate mining primarily occurs in Nauru. Both copper and gold are mined in Papua New Guinea and gold is also mined on a small to medium scale in Fiji, Solomon Islands, and Vanuatu. PNG also produces petroleum and natural gas, mostly from off-shore wells. Mining in all these countries results in unavoidable localized environmental damage. Regulations attempt, with varying degrees of success, to mitigate damage from mine tailings, processing fumes, and siltation of streams and rivers. In New Caledonia and PNG, for example, minerals are taken by strip mining in mountainous areas. The more rugged the terrain, the more practical difficulties in preventing massive siltation of waterways. Prior to the 1980s there were few, if any, environmental precautions taken with mining activities. Siltation of waterways and coastal areas was common. Even after regulations were enacted the practicalities of mine operation in rugged terrain often precluded effective environmental protection. For example, siltation settlement ponds at the Ok Tedi gold mine in PNG were destroyed by an earthquake but the mine was still allowed to operate. As a consequence, sediments caused serious pollution of the Fly River damaging coastal gardens and fisheries.

The environmental impacts of mining in the region, particularly those in Papua New Guinea have been reviewed and are reported in literature (Carpenter and Maragos 1989, Brodie et al. 1990, Hughes 1989). While most of the mining activities have been without negative environmental impact, it has been reported that the phosphate mining operation discharges little waste to the marine environment (Morrison 1992). Misima mine produces a daily discharge of approximately 20,000 tonnes of soft waste rock and 15,000 tonnes of tailings per day. The tailings are washed to recover 75% of the process chemicals and then mixed with seawater and discharged at a depth of 75 to 100 meters on the outer edge of the coral reef. Operators of the Ok Tedi Mine have admitted that environmental degradation has been even greater than anticipated. Discharges from PNG's Ok Tedi mine discharges to the Fly River and are carried to the marine environment. Sediment loading to the Ok Tedi and Fly Rivers has exceeded the rivers' capacity to transport sediment. This has resulted in flooding and ecosystem changes (tree species changes) along riverbanks.

Mining operations have contributed to substantial socioeconomic and financial benefits from royalties, infrastructure development, employment, training, medical, and other services over the years. None-the-less, these benefits must be weighed against the inevitable environmental degradation. Prevention of environmental degradation through planning, design, and implementation of strategic environmental control measures must accompany economically necessary mining activities.

The excavation of coral, sand, and aggregate from the region's coastal areas is also an integral part of the mining activities in the Pacific as this is the only real source of construction materials for many of the smaller islands. A review of non-living resources in the Pacific in preparation for the SAP consultative meetings was conducted (Howarth 1997) that also included a review of aggregate mining sources in the Pacific.

The three principal locations of aggregate and sand mining in the Pacific include quarries, riverbeds, and the coastal zone. At locations such as a beach, reef flat or lagoon, mining by dredging, front end loader, or manual operation is likely in the coastal zone. Table 9.10 suggests that the coastal mining occurs in most countries of the region and is the only source of aggregates (other than importation) for atoll islands in general and particularly in the atoll countries of Kiribati, Marshall Islands, and Tuvalu.

In addition to aggregate mining in the coastal zone, since 1962 coral sand has been dredged from the reef in Laucala Bay, Suva, Fiji for use as the major ingredient in the manufacturing of Portland Cement by Fiji Industries Limited. This operation

Table 9.10. Source of Aggregate in Pacific Island Countries

Country	Quarry Source	River Source	Coastal Source
American Samoa[3]	X		X
Cook Islands	X		X[1]
FSM	X		X
Fiji	X	X	X[2]
Kiribati			X
Marshall Islands			X
Nauru	X		X
Niue			
PNG	X	X	
Samoa	X		X
Solomon Islands	X	X	X
Tonga	X		X
Tuvalu			X
Vanuatu	X		X

Source: Howarth 1997.

[1] Beach mining on Rarotonga is now banned.
[2] Includes dredging for carbonate reef sand for Portland Cement manufacture near Suva.
[3] American Samoa data added to Howorth data.

supplies cement to Fiji as well as several others countries in the region. Quantities mined over the past 5 years have averaged 70,000 cubic metres per year, at a value of almost FJD$1 million per year (Ohman and Cesar 2000). Where quarry or river materials are available as an alternative to a coastal source the former is usually of better quality and hence more expensive.

The problems associated with coastal aggregates mining include two major issues, (i) the proximity to shore often enhances the possibility of erosion, and (ii) the extraction rates are far exceeding the natural replenishment rates, which makes it unsustainable. The threats associated with reef flat dredging also include the following:
- Heavy turbidity of the water in the working area increases the risk of degrading large portions of the reef and seagrass beds;
- At seabed level, the dredge workings produce craters, which complicates restoration of the area at a later date;
- Modification of the hydrologic regime, including poor flushing of semi-enclosed bodies of water;
- Potential risk of introducing ciguatera poisoning by promoting the spread of microscopic algae on the suspended fines; and
- Aesthetically displeasing remains of old workings, and associated rusting equipment.

The problems associated with mining in the coastal zone could perhaps be avoided through alternative source areas further offshore such as in lagoon areas where the risk of inducing erosion is negligible, though other environmental problems are not eliminated altogether. SOPAC has carried out surveys during the past few years in Tonga (Tongatapu and Vavau'u), Tuvalu (Funafuti), Marshall Islands (Majuro), and Pohnpei (Federated States of Micronesia) to identify potential alternative source areas.

The commercial oil and gas production also leads to environmental pollution and degradation. In 1996, the value of oil exports from Papua New Guinea just exceeded 1 billion Kina (approximately 0.4 billion US$). Information on the environmental impacts of the PNG petroleum industry is, however, not available. Outside of PNG, Fiji and Tonga have the best potential, mostly offshore, for hydrocarbons, though largely unexplored as yet.

9.4.2 Effect of Sewage Discharges

Pollution created by large volumes of sewage is also a threat to coastal ecosystems. Currently, about 50% of the world's urban sewage is directly discharged into the sea or nearby water bodies (Clarke 1998). Yeung (2001) examined the pollution of South Asia's coastline, caused by lack of sewage treatment in particular. Nearly all near-shore waters along the urbanized coastline in South Asia are polluted with high bacteria counts, making seafood caught unfit for human consumption (Hinrichsen 1990). Hidayati (2001) attributed this to the high population densities in these areas, and found that the highest proportion of sewage pollution was reported in the Strait of Malacca, near Bangkok in the upper Gulf of Thailand, and in the metropolitan areas of Manila and Jakarta. Table 9.11 indicates population and estimated BOD generation and removal in selected South China Sea countries.

From the perspectives of eutrophication, sewage is the most significant source of marine pollution throughout the Pacific region. Sewage discharges give rise to a host

Table 9.11. Population and estimated BOD generation and removal in selected South China Sea countries

Country	Population in South China Sea Subdivisions	Proportion of population in cities (%)	Population growth rate (%)	BOD generated (10^3 t/yr)	BOD removed by sewage treatment (10^3 t/yr)
Cambodia	1,985	89	2.7	36.2	No treatment
China	59,694	35	1.6	1,089.4	<109
Indonesia	105,217	48	2.9	1,920.2	364
Malaysia	10,336	15	3.3	188.6	53
Philippines	23,633	27	2.1	431.3	149
Thailand	37,142	0	1.4	677.8	89
Vietnam	75,124	3	1.6	1,371.0	No treatment
TOTAL	313,131	>27	1.4	5,714.5	655

Source: Talue-McManus 2000.
Note: Population recalculated to 1996 using growth rate shown, obtained using a weighted mean method. Estimated BOD produced using 0.05 kg/person/day.

of problems such as oxygen demand and the protection of human health in the vicinity of untreated or incompletely treated discharges. The sheer rate and ubiquity of discharge has, in recent years, overwhelmed the capacity of many coastal areas in the Pacific. About half of all sewage is dumped into coastal waters of the Pacific Islands in its original form. No efforts are made to disinfect the sewage (for reasons of microbial contamination) or to remove especially harmful pollutants. Nearly every Pacific Island nation has identified critical environmental and public health problems resulting from the disposal of human excrement. These have included algal blooms and eutrophication in lagoons, dying reefs, contaminated drinking water wells and outbreaks of gastro-intestinal disease, Leptospirosis and cholera. The causes of this pollution include overflowing latrines and privies, water-sealed toilet, septic systems of inadequate capacity, piggeries and sewage treatment plans as well as the complete lack of sanitation facilities in most of the coastal villages in rural settings.

In the Marshall Islands and Fiji, signs of eutrophication resulting from sewage disposal are evident adjacent to settlements, particularly urban centers. A number of studies have found sewage pollution to be adversely affecting coral reefs in the FSM, especially when discharged into lagoons with low circulation. Untreated sewage contains bacteria, parasites, and viruses. The greatest impact is on the fringing reef, which is used for catching small fish and shellfish. Shellfish concentrate bacteria and viruses from sewage. Consuming raw or partially raw shellfish can lead to transmission of viral diseases. Also since shellfish are at the bottom of the food chain, other fish species are affected. Untreated sewage also contains high levels of nutrients, which stimulate massive algal growth leading to decreased oxygen levels. These anoxic conditions can result in death of fish and reef, which in turn decreases fish stock. Only ecologically engineered solutions can provide sound sewage disposal and sanitation, without wasting valuable freshwater resources in the region.

9.4.3 Effect of Wood Treatment: CCA Leaching

In the South Pacific, increasing quantities of hazardous waste due to rapid economic growth and urbanization, the lack of controls on imported chemicals, and the lack of capacity to manage a range of pollutants have been issues of serious concern. Per capita consumption of imported and processed goods on some islands in the region may soon rival that of developed countries and about 80% of hazardous wastes result from imports. The approximate existing waste quantities of persistent organic pollutants in the region include almost 160 tons of waste timber treatment chemicals among others. In Samoa and the Solomon Islands and perhaps other countries timber treatment operations have now been abandoned. Site contamination with copper – chromium – arsenic is the inevitable result.

CCA, Chromated Copper Arsenate, a form of arsenic, was first used in the early 1970s. It is a wood preservative that is estimated to extend the life of wood structures such as outdoor decking, fences, and boat docks almost fivefold. Although CCA is infused deep into wood fibres under very high pressure, the arsenic seems to be leaching out.

Arsenic is a known carcinogen and is highly toxic. The potential health risks from any carcinogen depend largely on the length of exposure and dose of the chemical received. Human exposure and toxicity limits during contact with CCA-treated wood or surrounding soil could be harmful if leaching of arsenic occurs (depends on the pressure at which wood has been treated). Poor plant CCA waste retention and abandonment of old timber treatment facilities have contributed to the seepage of CCA lazed run-off to the coast. An abandoned timber treatment plant near Port Vila (Vanuatu) located less than 50 meters from the coast had about 70,000L of CCA stored in tanks with obvious signs of disrepair. There was evidence of CCA within the perimeter possibly linked with intentional and careless handling.

9.4.4 Heavy Metals & TBT

Heavy-metal contamination is becoming a severe problem because of industrial effluents and land-based waste being dumped into the coastal waters of the Pacific Ocean. Excessive levels of metals in the marine environment can affect marine biota and pose risks to consumers of seafood. Heavy metal compounds are found in many of the materials and processes of regional industrial activities and to a lesser extent agricultural activities. Lead, cadmium, and chromium, which were used for anti-algae and fungi properties, are found in older paints and in anti-fouling paints for marine craft and structures. More toxic, and of documented environmental degradation in the Pacific, is the use of organotins, such as TBT in anti-fouling paints. Metals and antifouling compounds kill the larvae and young stages of all marine life. This even happens at extremely low concentrations. Sandblasting of ships in preparation for painting has also resulted in the release of heavy metals to the marine environment. In the Marshall Islands, heavy metals have been detected in marine sediments and shellfish (CIA 2003).

Organotin compounds have been widely used as anti-fouling coatings on the hulls of ships and boats. There is only limited information available on the use of these chemicals within the region, but this seems highly likely. The most significant uses would have been at the numerous small boat yards and boat repair facilities around the region. This use of organotin compounds has now been prohibited in some parts of the world. However, the international moves for a global ban have not yet been successful, and continued use should be expected throughout most of the region. It should also be noted that once organotin compounds have been applied to the hull of a vessel there is a continuous slow release of the chemicals into the surrounding water. As a result, any vessels operating within the region can act as a source of organotin compounds, if these have been used for treatment of the hull. This can be a significant source of contamination in ports and harbours.

Contamination of mussel beds has been observed in the East Asian region as part of the Asia Pacific Mussel Watch (APMW) programme. This programme was implemented by Ehime with support from the Government of Japan and cooperation with UNESCO-IOC and UNU. The programme found increasing levels of contamination, and suggested that further monitoring of other species is needed. During the monitoring of organotin and organochlorine compounds as part of the mussel watch programme, sediments and biological samples revealed widespread contamination along the coastal waters of Asian developing countries (Tanabe 2000).

The distribution of organotins detected in these studies were found to correlate with the level of industrialization of the surrounding areas. In Korea, pollution from Polycyclic Aromatic Hydrocarbons (PAH's) was investigated in harbour areas, and 'hotspots' located close to industrial areas (Yim et al. 2000). Also, relatively high concentrations of organotin compounds were found at locations with intensive boating activities and coastal aquaculture facilities, implying the usage of tributyltin (TBT) as a biocide in antifouling paints used on boat hulls and in marine aquaculture.

A study carried out by Maata (1997) in several harbors in Fiji where shipping ports and slipways are located, revealed a serious problem of Tri-n-butyltin (TBT) contamination (Table 9.12). It can be inferred from this study that other shipping ports in the Pacific region may also suffer from such contamination and that appropriate measures need to be taken to counter such high levels of contamination.

Mining activities are major contributors to the region's environmental load of heavy metals. Other industrial operations such as foundries, metal plating and other processes involving combustion and use of petroleum products are also contributors. Batteries may also form a significant source of lead and mercury if their disposal is not properly managed.

9.4.5 Fuel Products

Leaded petrol is the only known major source of organolead compounds in the region. Petrol sold in the Pacific islands is sourced entirely from refineries outside the region and as a result, changes in the use of leaded petrol tend to mirror developments in the supplier countries. Thus leaded petrol was phased out of the US territories in the early 1980s and from most of the independent states during the 1990s.

Table 9.12. TBT level in Fiji Harbors

Sites	TBT $ngg^{-1} \times 10^3$
Walu Bay (Suva)	72–360
Lautoka	15–53
Savusavu	16–40
Labasa	35–89
Laucala Bay (Suva)	17–55

Source: Maata 1997.

9.4.6 Fertilizers

Apart from the smaller atoll countries in the Pacific like Kiribati, Marshall Islands and Tuvalu most of the other Pacific Island countries have agro-based economies where fertilizer is used in commercial and small scale agricultural activities (sugarcane, coffee, coconut, Oil Palm, vegetables, and rootcrops). However, inorganic fertilizer use in the Pacific is dominated by Papua New Guinea and Fiji especially in their coffee, oil palm and sugar industries. Fiji imports on average about 50,000 tons of fertilizers annually (Graham 2004). In other Pacific countries, inorganic fertilizer use is relatively low and limited only to plantation crops, such as squash and watermelon in Tonga; banana, citrus and pineapple in Cook Islands; banana in Samoa and cocoa and coconut in the Solomon Islands. There is a lack of information on the adverse effects of fertilizers on the environment except for a few localized cases such as the suspicion in Samoa that fertilizer applied in plantations near water catchment areas may have contributed to recent occurrences of Red Tide. Estimates by UNEP/GPA Coordination Office & SPREP (2000) suggested that in the Solomon Islands fertilizer input to the marine environment is 120 tonnes annually. However, these estimates were obtained from rapid assessment methods which do have attendant difficulties. They do, however, indicate the scale of the nutrient input to the coastal zone as a result of agricultural activities. The Pacific's tropical climate, where rainfall typically averages around 2 meters annually, renders coastal zones susceptible to receiving high levels of nutrients as a consequence of surface run off from plantations and farms.

9.5 NUCLEAR WASTE POLLUTION

The main source of radioactivity in the Pacific ocean can be attributed to worldwide fallout from atmospheric nuclear weapons testing. Measurements of selected artificial radionuclides in the Pacific Ocean were first conducted in the 1960's where it was observed that fallout radioactivity had penetrated the deep ocean. Problems from radionuclides in the region are also associated with past military activity. There is no known dumping of nuclear wastes in the region at this time. However, historically many of the islands were used for nuclear testing, including Bikini and Eniwetok in the Marshall Islands, Christmas Island in Kiribati, Mururoa in French Polynesia, and Johnston Atoll (United States). Biki and Eniwetok atolls were used for US atom-bomb tests for nearly two decades (1950s and 1960s) with radioactivity lasting for 100 years. On September 5, 1995 a French test nuclear bomb was detonated in the Moruroa atoll lagoon about 750 miles southeast of Tahiti in the South Pacific (177 nuclear tests have been exploded at Moruroa and Fangataufa atolls since 1966, the two most recent being on September 5, and October 1, 1995). In general, levels of fallout radionuclides in the surface waters of the Pacific Ocean seem to have decreased over the past 4 decades and are now more homogeneously distributed. Resuspension and the subsequent deposition of fallout radionuclides from previously deposited debris on land has become an important source for the surface ocean waters. Radionuclides may also be transported from land to sea in river runoff – these transport mechanisms are more important in the Pacific Ocean where large quantities of river water and suspended sands/fluvial sediments reach the coastal zone. Another unique source of artificial radionuclides in the Pacific Ocean is derived from the slow resolubilization and transport of radionuclides deposited in contaminated lagoon and slope sediments near U.S. and French test sites. Low level radionuclides sources such as those from medical waste (x-rays) have not been quantified. However, this is not believed to be a significant source.

Subsequent to nuclear testing in the Pacific by France at Moruroa Atoll in 1996, the shipment of plutonium and high-level nuclear wastes and proposals to dump nuclear waste on isolated atolls continue to threaten the region. Ships transporting radioactive nuclear material (MOX – plutonium and uranium mixed oxide fuel) to Japan continue to sail through the Pacific Ocean. There are plans for up to 30 shipments over the next 15 years. In the case of a severe accident leading to the breach of nuclear casks being carried as cargo on these ships, the impact on the health, environment and economies of Pacific Island communities would be devastating. Growing environmental and social threats posed by the transportation and importation of hazardous and radioactive wastes in the Pacific region re-iterates the urgency for Pacific Governments to ratify the Waigani Convention (Convention to Ban the Importation into Forum Island Countries of Hazardous and Radioactive Wastes within the South Pacific Region) and ensure its enforcement as soon as possible.

9.6   EFFECTS OF INDUSTRIAL EFFLUENTS (FISH PROCESSING & CANNING, AND OTHER SMALL SCALE OPERATORS)

Fish canneries exist in several countries and several new ones are also planned for the region. Effluents from canning factories consist mainly of organic-rich waste from the processing of fish. In 1993 these canning factories were estimated to have contributed to the marine environment approximately: BOD (26.77 tonnes), nitrogen (297.93 tonnes), phosphorus (167.3 tonnes), and suspended solids (194.75 tonnes). Canneries may be an appropriate industry for the region given its pelagic resources. Care must be taken, however, to plan for and implement adequate waste treatment. Concerns over localized pollution from the tuna factories in the Solomon Islands and Fiji have been raised by residents and Government officials. However, there is a lack of concrete and readily available data on the composition and quantities of effluents discharged by these factories that should be addressed. Furthermore, regular effluent monitoring should be enforced although this is often complicated by the lack of manpower and weak industrial, environmental and public health legislation. In addition to the tuna canning factories, there are also a variety of small to medium sized food processing industries which discharge effluents to the coastal zone. For a few countries like Fiji where liquid waste including sewage is treated in a centralized sewage treatment plant, the industries usually apply preliminary treatment to their effluents before flushing into the sewage treatment lines. Although, direct discharge into creeks and the coastal zone via streams and rivers is also practiced.

9.7 CAPACITY BUILDING

The Pacific Region currently has very limited capacity to manage PTS. There is a need for increased monitoring capacity, improved regulations, management structures and enforcement systems, and perhaps most importantly more people in the region with the skills, knowledge and experience to implement and utilize the above. There are also significant needs in the area of technology transfer, especially in relation to alternatives to the use of PTS and other possible reduction measures.

DDT, PCBs, dioxins, furans, PAHs and the organometallic PTS are considered to be the highest priority PTS for the region. The region has priority needs in education, training and community awareness and participation, and for chemical management systems, technology information and research.

9.8 LEGISLATIVE FRAMEWORK

There are a number of multilateral environmental agreements relating to the coastal chemical Pollution in the South Pacific Region. The relevant ones are listed below:
- Convention on the Prevention of Marine Pollution by Dumping of Wastes and Other Matter, 1972 – London Convention;
- International Convention for the Prevention of Pollution from Ships, 1973 as amended by the Protocol of 1978 relating thereto – MARPOL Convention;
- Convention for the Protection of the Natural Resources and Environment of the South Pacific Region and Related Protocols, 1986 – SPREP Convention;
- Basel Convention on the Control of Transboundary Movements of Hazardous Wastes and their Disposal, 1989 – Basel Convention;
- Convention to Ban the Importation into Forum Island Countries of Hazardous and Radioactive Waste and to Control the Transboundary Movement and Management of Hazardous Waste Within the South Pacific Region, 1995 – Waigani Convention;
- The Rotterdam Convention on the Prior Informed Consent Procedure for Certain Hazardous Chemicals and Pesticides in International Trade, 1998 – Rotterdam Convention; and
- Stockholm Convention on Persistent Organic Pollutants, 2001 – Stockholm Convention.

All member countries of the South Pacific Regional Environment Programme (SPREP) in the Pacific Island region also have a number of legislations relating to the prevention of chemical pollution (varies from country).
- Public Health Act;
- Poisons and Drugs Act;
- Pesticide Act; and
- Environment Act etc.

9.9 OBSERVATION/MONITORING CAPACITY

Land based activities such as agriculture, industry, urbanization, and mining disturb essential natural processes in the coastal waters and could be potentially detrimental to environmental quality and the social and economic well being of the countries in the Pacific Islands region. Sewage, sedimentation, and agricultural activities are primary threats to critical species and habitats as well as non-living resources. The decision-makers need an inventory of all relevant data on sources and strength of the key pollutants in developing pathways to address the issue and to respond to imminent threats. The ultimate root cause of the imminent threats is deficiencies in management, which need to be grouped in terms of governance and technical understanding. It is particularly important that all strategic information are compiled in an appropriate form and made available to decision-makers, resource managers, and communities to evaluate costs and benefits of proposed management alternatives.

9.10 EXCLUSIVE ECONOMIC ZONE AND POLLUTION

The Pacific Ocean occupies 180 million square kilometers, half of the earth's sea surface and more than a third of the Earth's surface. Scattered in the western half of this immense area are 200 high islands and 2,500 low islands or atolls, which make up the 22 countries and territories of the Pacific Islands.

The region's unique geographical characteristics have helped shape the cultural traits of its people. Arriving first to the region, Melanesian ancestors settled in the high islands of the Western Pacific. Faced with abundant resources and a complex topography, Melanesian communities developed largely isolated from one another, leading to a multiplicity of languages and cultural traits. In contrast, the resource-poor islands of Polynesia and Micronesia provided the impetus for sea travels and expansion into the outer edges of the Pacific Ocean.

In this sea of islands where the ocean exceeds landmasses by an average factor of 300 to 1 (Table 9.13), the people of the Pacific have developed a unique relationship with the ocean that has shaped their sense of place, their economies, and their culture. For them, the ocean is both a shared resource and a source of isolation. It helps define the ways communities communicate and are governed, and it continues to be a source of cultural significance and inspiration.

The relation that Pacific Island people have with the ocean is dualistic. The vast offshore areas, the deep ocean represents the frontier, a region of underexploited resources of high economic and strategic value. Yet for most Pacific Islanders, it is the coastal areas surrounding their islands that provide the food, income, culture, and recreation that are so important to the Pacific way of life.

Coastal areas in the Pacific are increasingly threatened. Overfishing, pollution, mining, and poor coastal planning are leading to the depletion of fisheries and to coastal degradation, undermining the livelihood of coastal communities. The

Table 9.13. Pacific Islands-Land and Ocean Areas

Country or Territory[a]	Land area (square kilometers)	EEZ[b] (square kilometers)	Ratio of Ocean to Land Area
Samoa	2,934	120,000	41
Solomon Islands	29,785	1,340,000	45
Vanuatu	12,189	680,000	56
Fiji	18,376	1,290,000	70
New Caledonia	19,103	1,740,000	91
Guam	549	218,000	397
Tonga	696	700,000	1,006
Palau	500	629,000	1,258
French Polynesia	3,521	5,030,000	1,429
Niue	258	390,000	1,512
American Samoa	197	390,000	1,980
Wallis and Futuna	124	300,000	2,419
Marshall Islands	720	2,131,000	2,960
Northern Marianas	475	1,823,000	3,838
Fed. States of Micronesia	702	2,978,000	4,242
Kiribati	726	3,550,000	4,890
Cook Islands	180	1,830,000	10,167
Nauru	21	320,000	15,238
Tokelau	12	290,000	24,167
Tuvalu	26	900,000	34,615
Pitcairn	5	800,000	160,000
Total	**91,099**	**27,449,000**	**301**

[a] Papua New Guinea is not shown as its large land mass is typical of other Pacific Islands.
[b] The Exclusive Economic Zone (EEZ) is the 200-mile limit sea area surrounding coastal states. Within this area, the Pacific Islands have exclusive rights to exploit their natural resources. Where states have not declared EEZs, or where the main fisheries area did not correspond exactly to the EEZ, data were modified or estimated as appropriate.

decline of mangroves and coral reefs is increasing the islands' exposure to cyclones and storm surges. Pacific Island governments can no longer afford a policy of inaction. The degradation of coastal areas is imposing significant economic and social costs, leaving coastal communities in need of urgent assistance. Managing the use of coastal areas is a major challenge for Pacific Island countries in current times.

Neither governments nor communities can manage coastal areas on their own. The distances involved and the existence of customary marine tenure in many islands make it virtually impossible for government-only efforts to succeed. At the same time, communities need help in accessing the technical advice they may require to manage their coastal areas, and in addressing problems such as pollution and dredging that cannot be handled at the local level. The challenge will be to use well the comparative strengths of communities, governments, and other stakeholders (such as NGOs), and to develop a common goal for the management of coastal areas that uses each partner to its best advantage.

The deep ocean presents challenges and opportunities of a different kind. Chief among them is the management of tuna fisheries in the Central and Western Pacific, the most important tuna fishing ground in the world. Because tuna are highly migratory, their management requires close regional collaboration. Pacific Island countries and distant water fishing nations have just concluded negotiations on a new regional convention to manage the tuna resources of the Western and Central Pacific. In contrast with past arrangements, distant water fishing nations would be a full member of the commission. This is likely to influence the outcome of critical decisions such as the allocation of total allowable catch which could affect the benefits that Pacific Island derive from tuna fisheries for years to come. The need for the coastal states to carefully review the available options and strengthen their collaboration cannot be over-emphasized.

Another emerging challenge in the offshore areas of the Pacific is seabed mining. After a long period of hiatus, there has been a recent resurgence in investors' interest in seabed minerals. Several applications for exploratory licenses have been made and are presently being considered. Given the potential scale of these operations, it is urgent that Pacific Island countries adopt appropriate offshore mineral policies.

Under the Law of the Sea Convention, some Pacific Island countries had only until 2004 to extend maritime claims beyond the 200 mile EEZ, by delineating their continental margin. The three key challenges mentioned above – management of coastal areas, regional collaboration on tuna management, and regulation of seabed mining – are the most urgent issues currently faced by Pacific Island countries in ocean management. Many other challenges and opportunities could emerge in the future. The Pacific Ocean has long been an area of strategic importance for national, regional, and external interests, and they are expected to continue to be a major shaping force in the future.

9.11 CONCLUSIONS AND RECOMMENDATIONS

The Asia-Pacific region encompasses an extremely wide range of physical, climatic, and ecological diversity yet all of the islands rely on the marine environment for their existence. The coastal and marine environment houses its population and the most important economic activities occur in the coastal areas. Land based activities such as agricultural and industry, urbanization, and mining disturb essential natural processes. A number of activities and specific contaminant sources have been identified as potentially detrimental to environmental quality and the social and economic well being of the countries in the region.

The decision-makers need to identify and develop ways to address root causes and to respond to imminent threats of coastal pollution. The data gaps represent a lack of strategic information in an appropriate form available to decision-makers, resource managers, and communities to evaluate costs and benefits of proposed management alternatives. The data gaps, however, do not prevent action from being taken, as there are substantial anecdotal evidence to confirm the threats and are an important reminder to take a precautionary approach. The prevention of environmental problems costs less than their remediation.

The following overarching threats have been identified from land-based sources of pollution to the marine environment:
- Pollution from land-based activities; and
- Physical, ecological, and hydrological modification of critical habitats.

Pollution from land-based activities threatens water quality, critical habitats and sustainable use of resources. These linkages require comprehensive measures to address the concerns effectively. The recognition of the need for integrated management is essential both on technical and on practical resource considerations. Integrated management is necessitated by the linkages and the limited resources of the individual countries and the region as a whole do no allow disconnected management programmes. Sewage, sedimentation, and agricultural activities are identified as primary threats to critical species and habitats as well as non-living resources. The root causes of these threats include, but are not limited to:
- Issues of policy, regulation and enforcement;
- Technical capacity;
- Data gaps;
- Inadequate infrastructure;
- Economic valuation of the resources;
- Land tenure;
- Resource pricing;
- Loss of traditional management systems;
- Development pressures;
- Lack of community awareness and education;
- Lack of community involvement;
- Lack of planning; and
- Other management issues.

Importantly, the ultimate root cause of the imminent threats is deficiencies in management, which are grouped in terms of governance and understanding.

9.12 REFERENCES

ADB (Asian Development Bank) (1997) Report and Recommendation of the President to the board of directors on Prepared Loans to the R.I. for the Coastal Community Development and Fisheries Resource Management Project, ADB, the Philippines

Azanza RV (1998) Harmful Algal Bloom Management Programs in Southeast Asia. Paper presented to IOC/WESTPAC International Scientific Symposium–Role of Ocean Sciences for Sustainable Development, Okinawa, Japan, 2–7 February, 1998, Intergovernmental Oceanic Commission Workshop Report, No. 148, UNESCO

Brodie JE, Arnould C, Eldredge L, Hammond L, Holthus P, Mowbray D, Tortell, P (1990) State of the Marine Environment in the South Pacific Region. UNEP Regional Seas Reports and Studies No. 127 and SPREP Topic Review No. 40, South Pacific Regional Environmental Programme, Apia, Western Samoa

Burns T, Graham B, Munro A, Wallis W (2000) Management of Persistent Organic Pollutants in Pacific Island Countries. South Pacific Regional Environment Programme (SPREP), Apia, Samoa

Cable W (2001) Preliminary Dioxin Inventory of Samoa. In UNEP Chemicals Proceedings: Sub-Regional Awareness Raising Workshop on the Prior Informed Consent Procedure, Persistent Organic Pollutants and the Basel and Waigani conventions, 2–6 April, 2001, Cairns, Australia

Carpenter RA, Maragos JE (1989) How to Assess Environmental Impacts on Tropical Islands and Coastal Areas. Environment and Policy Institute, East-West Centre, Hawaii

Central Intelligence Agency (2003) The World Fact Book. New york: Bartleby.com

Choi K, Grosheva E, Sakai S, Shibat, Y, Suzuki N, Wang J, Zhou H, Leung A, Wong MH (2002) Regionally Based Assessment of Persistent Toxic Substances – Central and Northeast Asia (Region VII). Proceedings of the Workshop on Environmental Monitoring of Persistent Organic Pollutants (POPs) in East Asian Countries, 2–4 December 2002, Tokyo, Tsukuba, Japan, Ministry of the Environment, Japan, pp. 132–139

Clarke JR (1998) Coastal Seas: The Conservation Challenge. Blackwell Science, Oxford, United Kingdom

Convard N (1993) Land-Based Pollutants Inventory for the South Pacific Region. South Pacific Regional Environmental Programme (SPREP) Reports and Studies Series No. 68, SPREP, Apia, Western Samoa

Cripps K (1992) Survey of the Point Source of Industrial Pollution Entering the Port Waters of Suva. For Ports Authority of Fiji, January 1992, 74pp + appendices

Edwards P (2000) Aquaculture, poverty impacts and livelihoods. ODI Natural Resource Perspectives 56: 4 pp

Gangaiya P, Green DR (1991) Water Quality in the Monasavu Reservoir and Wailoa Dam - 1991. INR Environmental Studies Report No. 58, University of the South Pacific, 13pp

GESAMP 71 (2001) Protecting the Oceans from Land-based Activities. UNEP, Nairobi, Kenya

Government of Western Samoa Public Works Department (1993) Apia Sewerage Project: Water Quality and Biological Studies, unpublished report

Graham B (2004) Fiji National Profile of Chemicals Management Infrastructure. Department of Environment, unpublished report

Hashimoto S, Watanabe M, Noda Y, Hayashi T, Kurita Y, Takasu Y, Otsuki, A (1998) Concentration and Distribution of Butyltin Compounds in a Heavy Tanker Route in the Strait of Malacca and in Tokyo Bay. Marine Environmental Research 45(2): 169–177

Hidayati D (2000) Coastal Management in ASEAN Countries: The Struggle to Achieve Sustainable Coastal Development. UNU, Tokyo, Japan

Hinrichsen D (1990) Our Common Seas: Coasts in Crisis. Earthscan Publications, London, United Kingdom

Howorth R (1982) Erosion in the SOPAC Region. Proceedings of the Eleventh Session of CCOP/SOPAC, pp 122–125

Howarth R (1997) Review of Non-living Resources and Threats in the Pacific Region. Prepared for International Water SAP consideration

Hughes PJ (1989). The Effects of Mining on Island Environment. UNDP Regional Workshop on Environmental Management and Sustainable Development in the South Pacific

Hu Tao (2001) China's Coastal Environmental Management. Proceedings of the 10th NorthEast Asian Conference on Environmental Cooperation, 16–19 October, Korea

Ibrahim Md. S, Jacinto G, Connell D, Leong LK (2002) Regionally Based Assessment of Persistent Toxic Substances Region VII – Southeast Asia and South Pacific. Proceedings of the Workshop on Environmental Monitoring of Persistent Organic Pollutants (POPs) in East Asian Countries, 2–4 December 2002, Tokyo, Tsukuba, Japan, Ministry of the Environment, Japan, pp 140–148

Ismail A (2001) Studies on Hazardous Chemicals along the Straits of Malacca and their Ecological Effect. Paper presented at the 3rd UNU-ORI Joint International Workshop for Marine Environment, Coastal Ecology, Nutrient Cycles and Pollution, 21–26 October 2001, Otsuchi, Japan

Jackson AG (1988) Implications of High Dissolved Organic Matter Concentrations for Oceanic Properties and Processes. Oceanography Magazine: 28–33

Jarman WM, Bacon CE, Estes JA, et al. (1996) Organochlorine Contaminants in Sea Otters: The Sea Otter as a Bio-Indicator. Endangered Species Updated 1996, 13pp

Kannan K, Tanabe S, Williams RJ, Tatsukawa R (1994) Persistent Organochlorine Residues in Foodstuffs from Australia, Papua New Guinea and the Solomon Islands: Contamination Level and Human Dietary Exposure. The Science of the Total Environment 153: 29–49

Kannan K, Tanabe S, Tatsukawa R (1995) Geographical Distribution and Accumulation Features of Organochlorine Residues in Tropical Asia and Oceania. Environmental Science and Technology 29: 2673–2683

Kannan K, Tanabe S, Tatsukawa R, Williams RJ (1995) Butyltin Residues in Fish from Australia, Papua New Guinea and Solomon Islands. International Journal of Environmental Analytical Chemistry 61: 263–273

Kennish MJ (1997) A Practical Handbook of Estuarine and Marine Pollution. CRC Press, New York

King C (2003) Capacity Development for Monitoring Major Persistent Organic Pollutants (POPs) in East Asian Waters: Examples of UNU Monitoring Activities in East Asia. Proceedings of the UNU International Workshop Monitoring of POPs in the East Asian Hydrosphere, 1–2 September, 2003, United Nations University, Tokyo, Japan

Kraemer A, Choudhury K, Kampa E (2001) Protecting Water Resources: Pollution Prevention, Thematic Background Paper. International Conference on Freshwater, Bonn, Germany

Lancelot C (1984) Extracellular Release of Small and Large Molecules by Plankton in the Southern Bight of the Northsea. Estuarine Coastal and Shelf Science 18: 65–77

Lee C, Henrichs SM (1993) How the Nature of Dissolved Organic Matter Might Affect the Analysis of DOC. Marine Chemistry 41: 105–120

Lee C, Wakeham SG (1988) Organic Matter in Seawater: Future Research Challenges. Marine Chemistry 39: 95–118

Libes SM (1992) An Introduction to Marine Biogeochemistry. John Wiley & Sons Ltd, New York

Livingston RJ (2002) Trophic Organization in Coastal Systems. CRC Press, 408 pp

Maata M (1997) TBTs in Fiji Marine Sediments, Ph.D. thesis, University of the South Pacific, Suva, Fiji

Mahbub and Kuslan (1997) Pemda DKI Jakarta, Studi Potensi Kawasan Perairan Teluk Jakarta. Paper presented on "Pemantauan Kualitas Lingkungan Jakarta tahun 1996" to Workshop on Jakarta environmental quality monitoring, Jakarta, 12 March 1997

Mataki M (1999) Online Photochemical Oxidation and Flow Injection Conductivity Determination of Dissolved Organic Carbon in Estuarine and Coastal Waters, M.Sc. Thesis, USP Chemistry Department

Minh T Binh (2001) Contamination by Persistent Chlorinated Endocrine Disrupters in Cetaceans from the North Pacific and Asian Coastal Waters. Presented at The 3rd UNU-ORI Joint International Workshop for Marine Environment Coastal Ecology, Nutrient Cycles and Pollution, 21–26 October, 2001, Otsuchi, Japan

Morris N (2004) Living at the Coast. The Creative Company, ISBN: 1583404864

Morrison RJ (1990) Assessment and Control of Pollution in Coastal and Open Ocean Areas of the SPREP Convention Region. Unpublished paper presented at the Third SPREP Intergovernmental Meeting, Noumea, New Caledonia, 24–28 September, 1990

Morrison RJ (1991) Pacific Atoll Soils: Chemistry Mineralogy and Classification. Atoll Research Bulletin 339: 1–25

Morrison RJ (1992) Soil and Physical and Chemical Limitations to Increased Agricultural Production on Atoll Soils. In: Chase RG (ed.) A Review of Agricultural Development in the Atolls. University of the South Pacific, Apia, Western Samoa

Naidu SD, Aalbersberg WGL, Brodie JE, Fuavao V, Maata M, Naqasima MR, Whippy P, Morrison RJ (1991) Water Quality Studies on some Selected Pacific Lagoons. UNEP Regional Seas Reports and Studies No. 136, UNEP, Nairobi, 99 pp

Nguyen Chu Hoi (1999) Issues and Activities Related to the Seas and Marine Environment in Vietnam. Integrated Coastal Zone Management and Non-living Marine Resources Development in Asia and the Pacific. UNESCAP, Vol. 4, UN, New York, USA

Ohman MS, Cesar HSJ (2000) Costs and Benefits of Coral Mining. In: H.S.J. Cesar (ed.) Collected Essays on the Economics of Coral Reefs. CORDIO: Kalmar, Sweden

Parsons ECM (1998) Trace Metal Pollution in Hong Kong: Implications for the Health of Hong Kong's Indo-Pacific Hump-Backed Dolphins (Sousa chinensis). The Science of the Total Environment 214(1–3): 175–184

Riley JP, Chester R (1971) Introduction to Marine Chemistry. Academic Press, New York

Robards K, McKelvie, DI, Benson LR, Worsfold JP, Blundell JN, Casey H (1994) Determination of Carbon Phosphorous, Nitrogen and Silicon Species in Waters. Analytica Chimica Acta 287: 147–190

Silvestre G, Pauly D (eds) (1997) Status and Management of Tropical Coastal Fisheries in Asia. ADB and ICLARM. ICLARM Conf. Proc. 53, 208 pp

SPREP (1999) Management of Persistent Organic Chemicals in Pacific Island Countries - Draft. SPREP. Apia, Samoa

Schellnhuber Hans-Joachim (2002) World in Transition. James & James/Earthscan, ISBN: 1853838527

Stumm W, Morgan J (1981) Aquatic Chemistry: An Introduction Emphasizing Chemical Equilibiria in Natural Waters. John Wiley & Son Ltd, Brisbane

Takada H (2001) Environmental Monitoring in South East Asia Using Molecular Markers. Paper presented to 3rd UNU-ORI Joint International Workshop for the Marine Environment, Coastal Ecology, Nutrient Cycles and Pollution, 21–26 October, 2001, Otsuchi, Japan

Talue-McManus L (2000) Transboundary Diagnostic Analysis for the South China Sea. EAS/RCU Technical Report Series No. 14., UNEP, Bangkok, Thailand

Tamata B, Thaman B (2000) Water Quality in the Ports of Fiji. IASA Environ. Studies Report C103, IAS, USP, 71 pp

Tanabe S (2000) Asia-Pacific Mussel Watch Progress Report. Marine Pollution Bulletin 40(8): 651

Thurman EM (1985) Organic Geochemistry of Natural Waters. Martinus Nijhoff/DR W. Junk Publishers, Boston, USA

Vincent JR, Ali RM and Associates (1997) Environment and Development in a Resource-Rich Economy: Malaysia Under the New Economic Policy. Harvard Institute for International Development, Harvard University Press

Volk JC, Volk BC, Kaplan AL (1997) Chemical Composition of Biodegradable Dissolved Organic Matter in Streamwater. Limnology and Oceanography 42: 39–44

UNEP (1999) Regional Environmental Outlook. Compiled by G. Miles. UNEP, Nairobi, Kenya

UNEP Chemicals (2001) Persistent Organic Pollutants (POPs) in Pacific Island Countries (PICs). Proceedings of the Sub-Regional Awareness Raising Workshop on the Prior Informed Consent Procedure, Persistent Organic Pollutants, and the Basel and Waigani Conventions, 2–6 April 2001, Cairns, Australia

UNEP/GPA Coordination Office & SPREP (2000) Overview on Land-based Pollutant Sources and Activities Affecting the Marine, Coastal, and Freshwater Environment in the Pacific Islands Region. Regional Seas Reports and Studies No. 174

WHO (1993) Guidelines for Drinking Water Quality. Volume 1 Recommendations. WHO, Geneva

Yeung Yue-man (2001) Coastal Megacities in Asia: transformation, Sustainability and Management. Ocean & Coastal Management 44: 319–333

Yim Un-Hyuk, Jae-Ryong Oh, Sang-Hee Hong (2000) Status of PAH Contamination in Coastal Korea: Results of 1999 and 2000 Korean Mussel Watch Program. Paper presented to 2nd UNU-ORI Joint International Workshop for the Marine Environment, Coastal Ecology, Nutrient Cycles and Pollution, 3–8 December 2000, Otsuchi, Japan

CHAPTER 10

LANDSCAPE VARIABILITY AND THE RESPONSE OF ASIAN MEGADELTAS TO ENVIRONMENTAL CHANGE

COLIN D. WOODROFFE[1], ROBERT J. NICHOLLS[2], YOSHIKI SAITO[3], ZHONGYUAN CHEN[4] AND STEVEN L. GOODBRED[5]

[1]*School of Earth and Environmental Sciences, University of Wollongong, NSW 2522, Australia*
[2]*School of Civil Engineering and the Environment and the Tyndall Centre for Climate Change Research, University of Southampton, Southampton SO17 1BJ, UK*
[3]*Geological Survey of Japan, AIST, Central 7, Higashi 1-1-1, Tsukuba, 305-8567, Japan*
[4]*Department of Geography, East China Normal University, Shanghai 200062, China*
[5]*Department of Earth and Environmental Sciences, Vanderbilt University, Nashville, TN 37235, USA*

10.1 INTRODUCTION

Deltas, occurring at the mouths of river systems that deposit sediments as they enter the sea, are some of the most dynamic sedimentary environments. They contain a long, and often economically significant, sedimentary record of their response to past episodes of climate and sea-level change. Geological investigation of these deposits, and the processes controlling sedimentation, provide insights into the response of deltas to environmental change, which in turn may offer rational and cost-effective strategies for the sustainable management of natural resources and land use in these dynamic systems in the face of future environmental change.

This chapter examines the megadeltas associated with the mouths of the nine largest rivers in south, southeast, and east Asia, the Indus, Ganges-Brahmaputra-Meghna (GBM), Irrawaddy (Ayeyarwady), Chao Phraya, Mekong, Red (Song Hong), Pearl (Zhujiang), Changjiang (Yangtze), and Huanghe (Yellow) rivers. Most of these are influenced by the seasonal Asian monsoon, with headwaters in the Himalayan-Tibetan massif (see Chapter 4.1), and carry large, highly-seasonal sediment loads to the sea (Table 10.1). There are extensive low-lying sedimentary plains associated with the mouths of each of these systems, termed 'megadeltas', which appear particularly vulnerable to impacts as a result of any change in sea level (McLean and Tysban 2001) and other global change

Table 10.1. Principal characteristics of the large rivers of Asia including estimated pre-disturbance and present suspended sediment load

Megadelta	Catchment area[1] km² × 10³	Mean annual discharge[1] m³ s⁻¹	Mean annual discharge[2] m³ s⁻¹	Mean annual sediment[3] load t × 10⁶	Pre-human sediment discharge[2] kg s⁻¹	Post-human sediment discharge[2] kg s⁻¹	Max tidal range m	Mean wave height m	Trapping efficiency[2]	Regulation by dams[1] %
Indus	1082	6564	3333	385	9593	3686	~3	~2	0.54	13
GBM	1667	22102	40025	1402	40534	46287	>4	~1.2	0.02	8
Irrawaddy	414	11953	20501	260	16331	8239	~3	~1	0.00	1
Chao Phraya	179	961	987	11	452	256	~2	~0.2	0.26	76
Mekong	806	15900	15029	150	2551	2531	~4	~0.9	0.17	3
Red	171	3900	2487	130	1039	4119	~4	~0.7	0.00	3
Pearl	409	10700	8732	69	1547	1427	~3	~0.7	0.35	31
Changjiang	1722	29460	29583	480	109444	49504	>4	~1	0.67	12
Huanghe	945	1990	1438	1080	3237	931	1.2	~1	0.97	51

[1] Data from Nilsson et al. (2005).
[2] Data from Syvitski et al. (2005).
[3] Data from Métivier and Gaudemer (1999).

Asian Megadeltas and Response to Environmental Change 279

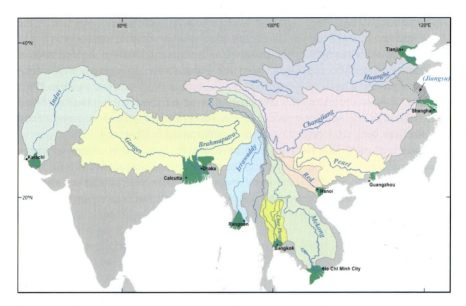

Figure 10.1. The large rivers of Asia and megadeltas that have developed at their mouths (landward margins demarcated by the Holocene maximum transgression shoreline), and associated cities/megacities

(Kremer et al. 2005). Although smaller Asian deltas also share many characteristics with these deltas, the megadeltas are of special concern because they support large rural populations, and are all associated with at least one large and growing city (Figure 10.1).

The Asian megadeltas have responded to a variety of stimuli as they have evolved through the Holocene. They have become increasingly shaped by human influence, both through land-use change within the catchments and on the deltas themselves. The productive low-lying deltaic plains have been largely cleared of their natural vegetation, and exploited for agriculture, silviculture, aquaculture, or settlement (Stanley and Chen 1996). The economic future of these deltas in the face of global climate change, particularly sea-level rise, is inextricably linked with their environmental well-being (Milliman et al. 1989). It is the aim of this chapter to reassess and re-emphasise the interrelationships between geological evolution of deltas and sustainable economic development of the plains that characterise them. There is particular emphasis on the nine Asian megadeltas, but the principles of this analysis will be of general utility to the more numerous smaller deltas in the Asia-Pacific region as well as other megadeltas elsewhere in the world.

10.2 RECENT GEOLOGICAL EVOLUTION OF MEGADELTAS

The primary control on the development of the megadeltas of Asia is the large, highly-seasonal sediment load the rivers carry as a result of relatively rapid weathering and transport associated with active tectonism in the world's highest mountain range

(Milliman and Meade 1983). The region plays a disproportionately large role in the global land-sea flux of sediment as a result of substantial summer monsoonal rainfall and spring snowmelt caused by the seasonal northwards migration of the intertropical convergence zone. The big rivers, from the Indus to the Huanghe, carry annual suspended sediment loads of 100 to 1000 million tonnes (Milliman and Syvitski 1992; see Table 10.1). It is primarily this catchment-derived sediment that has built the broad plains, although tide and wave processes can also effectively rework sediments from the delta front back onto the delta shoreline, as discussed in Chapter 4.

Evolution of each delta has been controlled directly by sea-level change as shown schematically in Figure 10.2. Megadeltas have developed at the mouths of these rivers during the Holocene for several reasons. Previous sea-level cycles, comprising alternate transgression (landwards movement of the shoreline) and regression (seawards

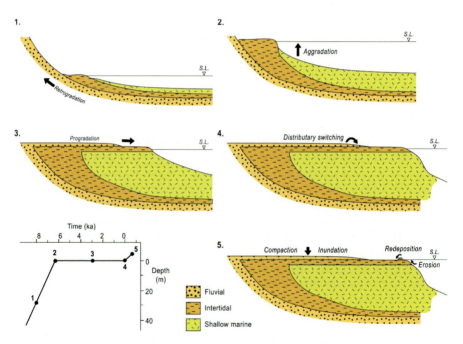

Figure 10.2. A schematic reconstruction of the response of Asian megadeltas to sea-level change during the Holocene. The inset shows a highly generalised pattern of sea-level change for the past 10,000 years based on sea-level reconstructions in the Asia-Pacific region, and the cross-sections show the evolving stratigraphy of intertidal and shallow subtidal sediments associated with delta evolution. The sea-level diagram has been generalised to comprise 1) a rapidly rising postglacial sea level at which stage deltas aggraded; 2) stabilisation of sea level around present level, termed maximum transgression, which occurred around 6000 years ago; 3) subsequent progradation of the shoreline – the deltas became increasingly complex during this stage; 4) present situation with extensive prograded delta plains; and 5) hypothetical situation if sea level rises in the future. Note that future sea-level rise does not result in the same response as during postglacial sea-level rise, because there are now extensive near-horizontal delta plains (in some cases compacted to below modern sea level) that the sea will inundate and rework

movement of the shoreline), have formed low gradient plains that have provided suitable space, termed 'accommodation' space in the geological literature, for the accumulation of an extensive depositional wedge of sediment. Sand and mud have partially filled this accommodation space, forming near-horizontal plains, as a result of deceleration and stabilisation of sea level. The surface morphology of individual Asian megadeltas has formed during the past 6000 years while sea level has been relatively stable around its present highstand. There are structural controls that constrain the shape of some of the deltas; for example, the Irrawaddy fills a trough between the Araken mountain range and the Shan plateau, and similarly the Chao Phraya occupies a structural depression in the central plain of Thailand. Inheritance from previous Quaternary landscapes also constrains the shape of modern megadeltas, with outcrops of Pleistocene terraces formed at previous sea-level highstands conspicuous as higher surfaces within, or flanking, the delta plains. Similar Pleistocene deposits underlie the modern deltas and have determined the accommodation space that has been available for deposition during the Holocene. Transgressive sediments underlie each of the deltas as a result of a fairly consistent pattern of postglacial sea-level rise across the region, but the various Asian megadeltas appear to have followed increasingly individual pathways over the past 6000 years. These two phases, the transgression associated with sea-level rise and progradation of delta landforms during the mid and late Holocene, are described below.

10.2.1 Transgression

At the peak of the last glaciation the sea was at least 120 m below present, with some evidence in the Indo-Pacific region for it having reached as much as 140 m below present (Emmel and Curray 1982, Yokoyama et al. 2001). Postglacial sea-level rise resulted in inundation of the underlying Pleistocene surface. The position of the former shoreline is indicated by mangrove deposits, which record rapid flooding across the Sunda shelf (Hanebuth et al. 2000). In common with deltas worldwide, Holocene sedimentation began approximately 8,000 years ago as sea level inundated this underlying surface, typically encountered in drillholes at around 20–30 m below sea level (Stanley and Warne 1994). The shoreline is recorded particularly by deposition of organic muds associated with mangrove forests in intertidal environments; for example, a pattern of landward migration of this narrow mangrove-fringed shoreline is implied by radiocarbon dating of the base of cores throughout the central plain of Thailand (Somboon 1988). This transgressive basal sedimentary unit can also be detected on other deltas in the region (Hori et al. 2001, 2002, 2004, Tanabe et al. 2003a, b, c). The exception appears to be the GBM system, which is discussed below, where inundation of the prior surface occurred 9000–10000 years ago at depths of up to 70 m below present sea level (Umitsu 1993, Goodbred 2003).

Around 7000–6000 years ago, the sea level reached a level close to present, with several lines of evidence suggesting that it was slightly above present (1–2 m) throughout the region (Geyh et al. 1979, Woodroffe and Horton 2005). Deceleration, and subsequent stabilisation of sea level marked a change from aggradational to progradational sedimentation (Chen and Stanley 1998). At this time the delta shoreline extended to its most landward position, marking the inner extent of the delta; for example, in the central plain of Thailand the shoreline was north of the location now occupied by the city of Ayutthaya (Tanabe et al. 2003a), and in the case of the Mekong the shoreline was near the location now occupied by the city of Phnom Penh in Cambodia (Tanabe et al. 2003b). At the peak of the transgression rivers entered into the bayhead of estuarine embayments. Since that time there has been a regressive sedimentary pattern with the continued build-out of sediment seawards (Figure 10.2).

10.2.2 Progradation of Delta Plains

The Asia-Pacific region is distant from present polar ice or formerly extensive ice sheets which accumulated during the last glaciation. As a consequence the relative sea-level curve for the past 6000 years for most of the region is similar to the eustatic curve driven by ice-melt. Most melt of glacial ice terminated around 6000 years ago and for that reason sea level has shown little variation over recent millennia. In fact, in southeast Asia, radiocarbon dates indicate that the sea was relatively higher than present around 6000 years ago as a result of subtle hydro-isostatic adjustments to redistribution of mass on the earth's surface. A series of oscillations have been inferred since then with periods during which sea level appeared higher than present around 5000, 4000 and 1500 years BP, interspersed with lower phases (Tjia 1996, Woodroffe and Horton 2005).

Deltas have prograded over this period. The resulting stratigraphy is shown schematically in Figure 10.2. During the period of sea-level rise up until 6000 years ago the mangrove fringe and associated intertidal environments migrated landwards. Since sea-level stabilisation, the shoreline has prograded with intertidal sediments such as mangrove muds and peat being deposited over nearshore shallow marine sediments. During this mid and late Holocene shoreline regression and progradation of the plains, distributary channels have extended seawards such that their hydraulic efficiency has decreased and in many cases they have switched to a shorter route to the sea and a new locus of deposition (see Chapter 4.3.4 and 4.4.2).

10.3 DELTA MORPHOLOGY

The Holocene delta plain of each megadelta is defined as those areas which formed as a result of progradation. The landward margin of each can be defined as the limit of the mid and late Holocene delta plains. This maximum transgression demarcates the approximate position of the shoreline around 6000 years ago when sea level appears to have reached its present highstand in the region. The landward margin of several deltas has been mapped during recent geological studies on a number of river systems (for example, the Huanghe, Saito et al. 2001, the Changjiang, Chen and Stanley 1995, Hori et al. 2002, the Pearl, Li et al. 1991, the Red River, Tanabe et al. 2003c, the Mekong, Ta et al. 2002a, b, and the Chao Phraya, Tanabe et al. 2003a). We are unaware of any mapping of this maximum transgression for either the Indus or the Irrawaddy, and have consequently adopted an extent as mapped by Wright et al. (1974). The extent that these authors show appears verified using digital terrain models (DTM) based on Shuttle Radar Topography Mission (SRTM, described below). SRTM analysis of those deltas for which the maximum transgression has been mapped confirms that this margin coincides with a marked change in gradient. In the case of the Ganges and Brahmaputra Rivers a large part of Bangladesh, as well as parts of West Bengal, north of the Indian city of Calcutta, comprises the GBM delta plain (Morgan and McIntire 1959). Examining mass balances for the GBM has shown that the volume of sediment deposited within the delta in the period 10,000-8,000 years BP, termed the hypsithermal, equates to more than twice the volume presently brought down the rivers. The delta was aggradational from 11,000-7,000 years BP, implying that the rivers supplied as much as 2.5 times their present load (Goodbred and Kuehl 2000a, b, Goodbred 2003). The maximum transgression occurred around 23 °N in West Bengal (Bannerjee and Sen 1988), and coincides with an elevation of around 3 m (Allison 1998a). The extent of Holocene deposition for this system is based on a generalised margin inferred from the studies described by Allison et al. (2003), together with SRTM DTM data.

10.3.1 Plains Topography

Shuttle Radar Topography Mission (SRTM) elevations, obtained by Synthetic Aperture Radar (C/X band, single pass radar interferometry), are available from the National Aeronautics and Space Administration (NASA) in a geographical projection at several resolutions (http://edcsns17.cr.usgs.gov/srtm/index.html). SRTM-1 is at 1 arc second, but the research-grade SRTM-3 was used in this study, averaged at 3 arc second, corresponding to a cell size of around 90 m on the ground. The SRTM data contain elevations to the nearest metre, with the sea surface set to zero. Data were masked using a shapefile of each delta extent defined as described above. Delta margins are imprecise because water levels rise and fall, land is eroded and deposited and in places is highly irregular. It is important to emphasise that many elements of delta surface topography encompass much variability within this 90 m cell size, for example, the ridge-basin relief on meander scroll plains (Brammer 1996). Table 10.2 summarises the proportion of the elevations for each of the delta plains in a series of elevation slices, enabling a broad overview of variation in elevation of the plains, and providing insight into the complex mosaic of delta topography. The nature of the shuttle-derived topography is such that it cannot be regarded as precise; the returns for the Huanghe (especially for Jiangsu) and the Mekong both contain noise, backscatter from the sea surface, and data voids. For this compilation, in order to give a broad overview of the topography of the plains, voids or outlier returns were discarded. The results for those finite values ranging from 1–20 m above sea level are summarised in Figure 10.3 and Table 10.2. Some general trends are described below for the individual deltas.

The summary of the proportions of different deltas in each of a series of broad elevation classes indicates considerable variability between deltas. The SRTM DTM

Table 10.2. Relative elevation of megadelta plains as derived from Shuttle Radar Topography Mission (SRTM) 3-arc DTM. Percentages are for cell distribution within 0–20 m range

Megadelta	<2 m	2–4 m	4–6 m	6–8 m	8–10 m	>10 m
Indus	21.0	23.2	13.7	11.0	8.9	22.2
GBM (N of 23 °N)	0.3	2.6	8.7	15.1	16.4	56.9
GBM (S of 23 °N)	9.2	25.2	27.2	17.0	10.8	10.6
Irrawaddy	4.0	28.1	29.1	13.3	8.2	17.3
Chao Phraya	12.4	35.8	33.8	13.1	3.5	1.4
Mekong	45.3	35.7	13.6	4.0	1.1	0.3
Song Hong (Red)	24.8	37.8	20.7	9.0	4.3	3.4
Pearl	37.5	21.9	13.1	7.9	5.2	14.4
Changjiang	13.7	42.8	33.4	7.81	1.5	0.8
Huanghe	24.1	31.3	24.2	13.7	4.9	1.8
[Jiangsu]	53.8	29.3	10.1	3.8	1.8	1.2

for the Indus indicates that there is a general increase in elevation with distance from the coast. The Holocene history of this delta appears to have consisted of a series of delta lobes to the east of the present channel with progressive abandonment as the river has avulsed westwards (Holmes 1968, Kazmi 1984). The Indus delta does not reflect this delta lobe chronology in its relatively gradual shore-normal topographic gradient, in contrast to other deltas, for which DTMs show considerable variability in surface topography. For example, in the case of the GBM there are many natural levées flanking existing and former channels (Umitsu 1985). These play an important part in channelling flow, are significant for human transport and livelihood, and influence the pathway of flood waters and the persistence of flooding. They can be clearly seen in the SRTM DTM, and coalesce to form a fluvial landscape over much of the portion of the delta north of 23°N, meaning the northern section is considerably higher in elevation (Table 10.2). Furthermore, through the complex interaction of a series of factors, including subsidence, compaction and the reduced sedimentation in backwaters cut off from direct inundation, there can be extensive areas of plains that lie below flood levels (Stanley and Hait 2000). For example, the low-lying, actively-subsiding Sylhet basin, in places only 2 m above sea level, is clear in the DTM of the GBM system, and can be flooded by water that is up to 6 m deep in the wet season. The elevations cannot be used directly to infer the extent of flooding because of the variability of tidal range, which in places exceeds 5 m and the complex flood-water surfaces that vary from year to year. Nevertheless, the SRTM DTM indicates the low-lying areas flanking the former distributaries of the Ganges.

A particularly large proportion of the Mekong appears especially low-lying (note, many SRTM data voids were excluded from this delta). This is also the case with the Pearl, and the Huanghe. The abandoned former delta of the Huanghe river, along the coast of Jiangsu, contains some spurious STRM noise, but also appears especially low-lying (Table 10.2). The Irrawaddy by contrast comprises low-lying plains that extend well inland towards the apex of the delta (Figure 10.3). The significance of

Asian Megadeltas and Response to Environmental Change

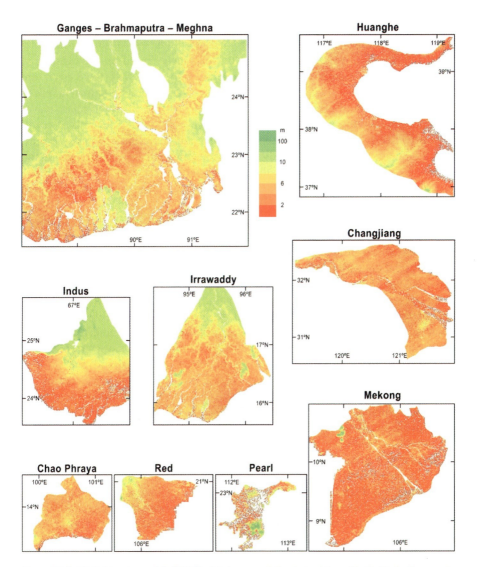

Figure 10.3. Digital terrain models (DTM) of Asian megadeltas derived from Shuttle Radar Topography Mission (SRTM) data

these elevations needs to be considered in relation to the elevation of flood surfaces, whether these are high tide levels, storm surge levels, or river flood levels in the wet season. However, none of these surfaces will be horizontal, so actual flooding remains a highly localised phenomenon.

Depositional coasts can be viewed in terms of the dominant processes that influence them, particularly in relation to river, wave, and tide energy. Deltas have traditionally been classified into river-, wave- and tide-dominated deltas (Wright and Coleman 1973, Galloway 1975, Wright 1985, Suter 1994), summarised in a ternary diagram in which individual deltas are placed based on the balance of processes at their margins. Many of the deltas associated with large rivers are 'river-dominated', but the degree to which any of these factors dominate needs to be viewed relative to other processes, and it is possible to differentiate wave-dominated deltas from tide-dominated deltas. Wave-dominated sections of coastline are typically swash-aligned with a sequence of shore-parallel sandy ridges; tide-dominated sections are muddy and characteristically contain tapering and meandering tidal creeks fringed with mangrove forests. River-dominated sections of a delta are indicated by former distributary channels (paleochannels) and meander scroll bars.

The processes operating within megadeltas have become more clearly discriminated as each has evolved during the Holocene, either as a result of spatial separation of the processes that shape them, or with discrimination in time. The subaerial delta plain can be divided into an upper and lower deltaic plain. The upper deltaic plain is dominated mainly by river processes. River domination is characterised by fluvial sediment supply and meandering and avulsing channels flanked by levées that are elevated above the general surface of the plains. The lower deltaic plain is usually within the zone of tidal influence, indicated by tapering and meandering tidal channels, or it may be dominated by waves in which case it is usually characterised by shore-parallel ridges of sand and shell. Asian megadeltas may contain different sectors dominated by different processes, as can be seen in the Red River delta (Figure 10.4). The western sector is wave dominated with shore-parallel beach ridges, and the eastern sector is tide-dominated with mangrove-fringed tidal creeks. The apex of this delta is river-dominated (Figure 10.4a), with the levées flanking active and former distributaries clearly distinguishable from SRTM DTMs (Figure 10.4b). These topographic trends can still be seen, although the delta plain is heavily settled with intensive land use, as apparent on the JERS image (Figure 10.4c).

The Indus is situated on a coast that experiences high wave energy (Wells and Coleman 1984), and the Huanghe, emptying into the Bohai Sea with minimal tides, experiences wave reworking of the shore (Saito et al. 2000). In the case of most of the other Asian megadeltas, wave energy is relatively low, and the abandoned delta is tide-dominated. The delta can be divided into active sectors and abandoned sectors, as shown by Wright et al. (1974), and tide-domination becomes particularly apparent along the margin of the abandoned delta. The GBM shows the characteristics of an active delta plain, the Meghna Delta plain to the east is river-dominated, and an abandoned delta plain (or 'moribund' delta), the Gangetic Tidal plain to the west is tide-dominated. The distributaries draining into the Sundarbans, which is a mangrove reserve covering most of the Gangetic Tidal plain, carry less than 4% of the flow, and the plain is dominated by tides (Figure 10.5). There is an intricate network of tidal creeks. These are prominently tapered, and diverge into sinuous channels that meander though the mangrove

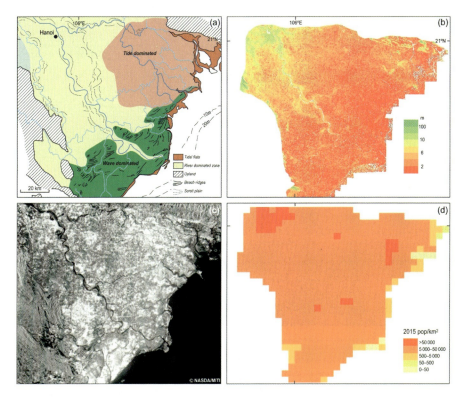

Figure 10.4. The example of the Red River Delta; (a) an interpretation of subaerial delta geomorphology which shows distinct sectors of the delta dominated by different processes (based on Mathers and Zalasiewicz 1999); (b) DTM derived from SRTM; (c) JERS radar imagery of the delta showing the complexity of land-use, and (d) gridded population density for 2015 based on estimates of population increase – note the relatively coarse scale of grid cell

forests (it is likely that the dense canopy of these mangrove forests has been detected by SRTM radar along the central portion of the delta front, west of the Haringhata River and south of Khulna, to record the uncharacteristically high returns in contrast to the surrounding low-lying tidal plain seen in Figure 10.3). This abandoned part of the plain, no longer receiving direct river-borne sediment, but continuing to subside in response to the loading of the shelf by sediment, is actively receding. For example, 100 km^2 has been lost from the Sundarbans in the past 30 years, with erosion concentrated on the southern shores of islands (Allison and Kepple 2001), in contrast to the Meghna Delta plain which has prograded (Allison 1998b).

The stratigraphy and Holocene progradation of the Mekong Delta has become clearer as a result of a series of recent studies. Sandy ridges have been formed episodically in the eastern side of the delta over the past 3000 years (Nguyen et al. 2000). Stratigraphy and chronology suggests that the delta has changed from a more

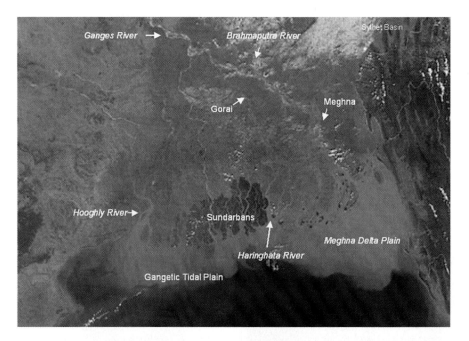

Figure 10.5. The Ganges-Brahmaputra-Meghna (GBM) Delta. This MODIS image shows the river-dominated Meghna Delta Plain and the abandoned, tide-dominated Gangetic Tidal Plain. Turbidity is high at the mouth of the Meghna as a result of river discharge and sediment supply, and also high in the tide-dominated Sundarbans and Hooghly River as a result of tidal resuspension of sediment

estuarine and tide-dominated system, prograding seaward at around 30–35 m a^{-1}, to a more wave-dominated delta, as the offshore gradient became steeper, prograding at 11 m a^{-1} (Tanabe et al. 2003b). This represents an intrinsic threshold within the development of the delta. After the delta has infilled the initial extensive accommodation space, it has built into deeper water consequently developing a steeper shoreface (Ta et al. 2002a). Other intrinsic thresholds in delta development include the transition from mangrove forests to freshwater vegetation as the plains accrete vertically beyond the level at which they are influenced by saline or brackish water. Pollen diagrams from several of the deltas in the region record this transition (Woodroffe 1993).

10.3.2 Distributary Switching

Distributary switching occurs when, after some time of building seaward, a delta distributary has become over-extended and may adopt a shorter, alternative course to the sea. It is generally an intrinsic change triggered by the stage of delta development, although floods or earthquakes can also induce distributary avulsions. The Huanghe shows the most extreme of these avulsions (see Figure 10.1); it adopted a course discharging into the Yellow Sea in North Jiangsu in 1128 AD, and then readjusted to again flow into the Bohai Sea in 1855 AD (Saito et al. 2000). The muddy coast of the abandoned Hunaghe delta in Jiangsu has been rapidly eroding at a decelerating rate, from 147 m a^{-1} in 1855 to rates of around 27 m a^{-1} in the 1980s, and 20 km and 1400 km^2 of delta plain has been lost since 1855 (Li et al. 2004).

Distributary switching appears to have occurred repeatedly in the GBM system with active distributaries migrating eastwards with the successive abandonment of distributaries such as the Hooghly and Gorai (Figure 10.5). Levées persist as topographic irregularities in the system, and SRTM DTM shows that levées remain as high ground adjacent to abandoned paleochannels (Umitsu 1985). SRTM DTMs also show prominent levées, giving considerable local variability to the topography for other deltas, for example the Red River, whose paleochannels can be seen in Figure 10.4c.

10.3.3 Delta Morphodynamics at Different Time Scales

It is clear that there are a range of processes in operation and that different processes operate over different time scales. This is summarised in Figure 10.6; insight into past geological processes can be most effectively reconstructed on millennial time scales, due to the control on rates of change provided by stratigraphy and radiometric dating. Thus the pattern of sea-level rise and the aggradation of deltas up to around 6000 years ago, which occurred synchronously throughout the region, can be studied by coring through the delta and deciphering the sedimentary sequence within cores. By contrast, the details of development of the plains over the past 5000 years are less clearly understood and require more detailed coring at a variety of sites, using a range of techniques. Dating of beach ridge sequences is beginning to constrain the pattern of coastal build-out (Ta et al. 2002a). Similarly the fluid dynamics and processes of sedimentation on short time scales (<years) are reasonably understood; application of dating techniques such as ^{210}Pb and ^{137}Cs has clarified recent change in the GBM and Changjiang (cf. Kuehl et al. 1989, Chen et al. 2004), but there remains a gap between these dating techniques and radiocarbon timescales, resulting in uncertainty about patterns of change at decadal to century time scales. This constrains the effectiveness of forecasting change, with the way in which a delta will evolve becoming increasingly uncertain into the future.

Associated with the substantial sediment input has been subsidence of parts of the delta. Localised subsidence and associated consolidation, dewatering, and compaction of sediments were identified as major issues for the GBM by Milliman et al. (1989) and for the Changjiang estuarine depositional sink (Chen and Stanley 1993, 1995, Chen et al. 2000). Under natural circumstances the delta may have some propensity to offset this local subsidence by deposition of fluvial sediment, but this is considerably reduced when flows are constricted within embankments and levées. As illustrated below (see Figure 10.8) the pattern of subsidence is likely to vary spatially, as is the degree to which it is offset by the availability and accretion of sediment.

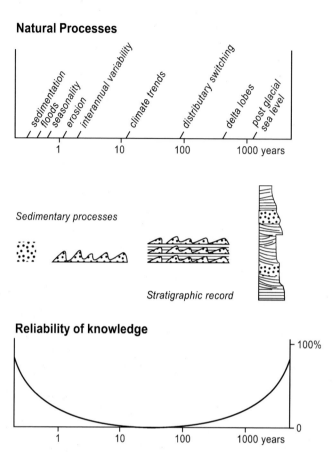

Figure 10.6. Temporal variability in pattern, process, and prediction; at short time scales sediment processes are described by the physics of fluid dynamics, at long time scales (of millennia) stratigraphic analysis of drillholes and radiocarbon dating provides an overview of changes, but the decadal to century record of sedimentary processes is generally not completely recorded at any site and reliability of knowledge, and hence ability to predict, at these time scales remains inadequate (modified from de Groot 1999)

10.4   THE HUMAN LANDSCAPE

The expansion of low-lying Holocene deltas appears to have coincided with, and is regarded as a trigger for, the appearance of urban centres of population in semiarid areas, because these societies could be supported by agriculture on the productive, seasonally flooded plains (Stanley and Warne 1993). By contrast, it has been suggested that the tropical Asian deltas were less favoured for human settlement as a result of their inaccessibility because of seasonal flooding and malaria (Büdel 1966). However, the extent of early human use of the Asian megadeltas may have been underestimated. There are known to have been civilisations associated with the

Indus Valley (Harappan period, around 6000 BP) and the Changjiang (Majiabang culture, around 7000 BP). In these cases, as with the deltas of the Nile and Mesopotamia, occupation began around the time that the deltas started prograding. In view of the prolific sedimentation and rapid vegetation growth, evidence of human use of the Asian deltas has not been well-preserved; archaeological remains are less obvious, less well-preserved, and less studied in these monsoonal, well-vegetated, and highly-active deltas. An exception is the Changjiang delta, where the earliest human occupation and agricultural cultivation was dated to around 7000 years BP on the southern delta plain (Stanley and Chen 1996, Chen and Stanley 1998). Numerous well-preserved Neolithic sites coincide with delta initiation (Chen et al. 2005). Preliminary prehistoric evidence has also been reported from the Mekong Delta extending back at least two millennia, prior to the rise of the Angkor civilisation (Higham and Lu 1998), and further evidence may remain to be discovered from the other wetter Asian deltas.

Associated with occupation of the river catchments, and the evolving deltas, human impacts have been both direct and indirect. Direct impacts result from exploitation of resources within each delta, including widespread land-use change to agriculture, silviculture, and aquaculture in the rural areas, and more recent urbanisation. Indirect impacts have occurred through land-use and water-management changes within the catchment (Nilsson et al. 2005). The 20th century saw an escalation of human impacts across Asian megadeltas, and a further intensification seems inevitable as populations continue to increase and the Asian economies expand through the 21st century. On the delta plains, land use varies from subsistence agriculture particularly for rice, but also for other crops such as jute, to more intense aquaculture, such as shrimp farms, to the growth of major cities. Along the delta front of most megadeltas, extensive areas of the coastal fringe, supporting mangrove forests or saltmarshes have been destroyed to make way for short-lived shrimp farms, many of which do not appear sustainable. The deltas are important for local fish production, and major offshore fisheries are sustained by nutrient output from the distributaries.

Indirect human impacts have led to substantial changes in sediment load on many of these systems, simultaneously increasing sediment supply by enhanced erosion on some rivers and reducing it through retention in dams on others (Syvitski et al. 2005). The most extreme example is the Huanghe River, called the Yellow River because of its enormous sediment load that results from human clearance of vegetation in the loess plateau and consequent soil loss, increasing suspended load 2–10 times over the past 2000 years (Ren and Shi 1986, Milliman et al. 1987, Jiongxin 2003). Similar land-use changes have increased sediment loads at least 2 times on the Changjiang and Red River (Saito et al. 2001), and have led to enhanced loads on other rivers (see Chapter 4).

The spread of wet-rice cultivation transformed these deltas into the rice bowls of Asia, once mastery of the floods was established. Flooding has been controlled to some extent, through engineered levées and embankments and with the spread of irrigation. Exclusion of natural flood waters from many areas of the delta plains has had some less desirable consequences, such as compaction and acid sulphate soil development, which are examined below. The redirection of nutrient-rich waters and their sediments have also had major implications for the productivity of deltas (see Chapter 4).

Patterns of flooding are not the same from year to year, but vary widely with climatic factors, and may show gradual trends as a result of human influence. For example, increased flooding has been observed in Vietnam, as a result of a combination of upstream deforestation, heavy rains, sea-level rise, and blocking of lagoonal inlets and river mouths (Thanh et al. 2004). Flooding can be severe when river floods coincide with spring tide or surges. The August-October floods on the Mekong were especially extensive in 1991, 1994, 1996, and 2000, with severe consequences for rice production (Wassman et al. 2004) and waves broke dykes on the Red River delta in 1955 and 1996. Increased saline intrusion has been noted in tide-dominated distributaries of many of the deltas (Aung 1993, Douglas 2005).

Delta fronts along many of these megadeltas are undergoing a lowering of the surface and a landward retreat of the margin. Whereas in many instances this may be the result of natural processes in the abandoned delta plain, it has been exacerbated by human action. A further problem can arise where groundwater is extracted as is occurring for many of the big cities. Rapid subsidence associated with groundwater extraction was resulting in subsidence of parts of Shanghai at rates of up to 28 mm a^{-1}, but this has been reduced to 3–4 mm a^{-1} by regulating withdrawal (Han et al. 1995). Nevertheless, central Shanghai now lies 2–3 m below storm surge levels, and is dependent on the embankment built around the city to an elevation believed to correspond to the 1 in 1000 year flood level (6.9 m above the Wusun Datum, Chen and Wang 1999). This contrasts with dykes on the Pearl River which are built for the 1 in 20 year flood level, and which do not have freeboard for sea-level rise. Similar restraints on groundwater extraction may slow subsidence rates within other cities on Holocene deltaic sediments such as Bangkok and Hanoi. Much of the erosion along the Gulf of Thailand may result from the relative sea-level rise due to compaction following ground water pumping (Phienwej and Nutalaya 2005). As the morphodynamic behaviour of a delta is responsive to human intervention, an understanding of the distribution of population and their activities within each delta is necessary. The population characteristics of the megadeltas are described below; regional trends in population growth and urbanisation, and globalisation, are described in more detail in Chapter 7.

10.4.1 Delta Populations

The Asian megadeltas now support enormous populations. In part this reflects the burgeoning populations of the two most populous nations on earth, China and India (see Chapter 7.2), but it is clear from gridded maps of population density that people within those countries are disproportionately concentrated within the river valleys and on the deltas of the large rivers (Small and Nicholls 2003). The extensive subaerial delta plains of the Ganges and Brahmaputra rivers have a surface area in excess of 110,000 km², and almost the entire population of Bangladesh, comprising over 140 million people, as well as many millions in West Bengal in India, live on the delta plains built by these rivers. Excluding the uplifted fault blocks comprising the Barind Tract and the Madhupur Terrace on which much of Dhaka lies, the population of the GBM delta exceeds 129 million.

Table 10.3 lists the populations of the other deltas calculated using the delta margins defined by the maximum Holocene transgression, and using the gridded population density estimates. The gridded population datasets are based on census returns distributed within administrative units, using a mass-conserving algorithm (Tobler et al. 1997). The updated and expanded GPW-3 (beta version), produced by the Center for International Earth Science Information Network (CIESIN), is adjusted to UN estimates of country total populations. These data include population estimates for 1990, 1995 and 2000, and an estimated population for 2015 using country-specific growth rates. Population densities have been interpolated assuming population to be uniformly distributed across each administrative unit. Data are gridded using 2.5 arc minute cells (4.6 km at equator). The geographic projection of these data, using latitude and longitude, means that the size of the cell on the ground varies with latitude, and population density figures need to be weighted for area using a further grid from CIESIN (http://sedac.ciesin.columbia.edu/gpw). Coastal grid cells have been given an area based on the proportion of that cell that is land. Spatial uncertainties in population distributions are generally larger than boundary uncertainties, and an example of the vagueness of the landward boundary is illustrated by the details of the Red River outlined by Tanabe et al. (2003c, see their Figs. 2 and 3) who portray slightly different boundaries in consecutive figures of the delta. Delta populations for 2000 and 2015 shown in Table 10.2 were derived using ArcGIS raster calculator.

Table 10.3. Estimates of population within the Holocene deltaic plains of megadeltas, based upon GPW-3, 2.5 arc minute gridded population of the world (CIESIN) for 2000 and 2015

Megadelta	Area km²	Population 2000	Population 2015	Increase (%)
Indus	19800	3058500	4425100	+45
GBM	115600	129931100	166217000	+28
Irrawaddy	31500	10591700	12163600	+15
Chao Phraya	11600	11485600	16487900	+44
Mekong	37900	15754200	19039800	+21
Song Hong (Red)	9900	13293900	16063400	+21
Pearl	5900	9846400	27166900	+176
Changjiang	15600	25945700	33147500	+28
Huanghe	25100	14060400	16614100	+18
[Jiangsu]	30300	19930700	14978400	−25

Note: Area of delta plain has been determined from gridded population cell count, and is only approximate as it does not take into account areas that are covered by water, including major distributaries.

After the GBM system which stands out by its size and population, the next largest delta population consists of nearly 26 million and is associated with the Changjiang. The population on the Holocene plains deposited at the present mouth of the Huanghe consists of around 14 million, but the coast of Jiangsu on the abandoned Huanghe, adjacent to the Changjiang, has nearly 20 million. The Indus has the smallest population with around 3 million, and the other deltas each have 10–15 million. It is noteworthy that several of these estimates appear lower than some previous figures, for example, those reported for Vietnamese deltas by Thanh et al. (2004), which presumably results from our stricter definition of the delta as defined by the Holocene maximum transgression.

The gridded population data are only an approximation of the spatial distribution of people within the deltas. Proportional allocation population estimates have generally been shown to be more reliable in denser population areas, such as the deltas, because there are usually more collection districts and the topographic expression of delta margins also often coincides with administrative boundaries. However, the spatial resolution of census data does pose a fundamental constraint on the conclusions that can be drawn and is a limitation on calculating population in the near-coastal zone (see Figure 10.4d). Extrapolations of populations based on GPW2015 show a scenario of population at a sub-national resolution for 2015 (Table 10.3). These should not be considered projections because they are determined assuming continuation of recent geometric demographic patterns of growth between the 1990 and 2000 censuses, adjusted at national level to UN 2015 population projections. They are presented only to emphasise the often dramatic patterns of continued growth. Urbanisation is the most significant contributor to growth (as described in Chapter 7.3); rural populations are expected to remain constant as illustrated by Jiangsu, where there is no major city, and where population is projected to decrease.

Although the gridded population data enable updated estimates of populations for each delta based on administrative units, it is important to recognise that people within the deltas are generally in highly nucleated settlements, and the spatial resolution of the gridded cells does not capture the concentration of settlements along higher ground, such as levées and beach ridges as can be seen in the case of the Red River delta in Figure 10.4. The scale of administrative units on which data are based is a major constraint, which although not known for the Red River delta, averages 25 km for Vietnam as a whole. Nevertheless this approach was preferred over alternatives, such as the modelled Landscan (Oak Ridge) or the global city lights dataset, each of which has their own constraints. As discussed below (see Figure 10.8), natural processes operate at different spatial scales, as do socio-economic processes; for example, land tenure is often very fragmented, and individual holdings may involve properties of 1–2 ha spread across small plots in several dispersed places. It is critical to emphasise that, in common with the topographic summary above, the population data are not at sufficient resolution for detailed hazard analysis, or local vulnerability assessment. The rural-urban distribution of population is a major issue in these deltas; it is the subject of further

Table 10.4. Actual and emerging megacities associated with Asian megadeltas; their growing populations (1975, 2000 and 2015 projected: after UN/DESA, 2002)

Large city	Megadelta	Relative location	Population (millions)		
			1975	2000	2015
Karachi	Indus	Delta margin	4.0	10.0	16.2
Calcutta	GBM	Delta margin	7.9	13.1	16.7
Dhaka	GBM	Mid-delta terrace	2.2	12.5	22.8
Rangoon	Irrawaddy	Delta margin	1.8	4.4	6.3
Bangkok	Chao Phraya	Mid delta	3.8	7.4	9.8
Ho Chi Minh City	Mekong	Delta margin	2.8	4.6	6.3
Hanoi	Red	Delta apex	1.9	3.8	5.2
Guanzhou	Pearl	Mid delta	3.1	3.9	4.2
Shanghai	Changjiang	Mid delta	11.4	12.9	13.6
Tianjin	Huanghe	Delta margin	6.2	9.2	10.3

development of the global gridded population data set using a GIS layer of urban extents, as part of a CIESIN project called GRUMP, but present data do not allow resolution of this issue, which is discussed below, and in Chapter 7, based on more traditional assessments of city population data.

10.4.2   The Emergence of Megacities

Table 10.4 shows the rapid growth of several actual and emerging megacities in the region (cities of more than 8 million people, Nicholls 1995), and their projected growth to 2015. Almost all the large urban agglomerations in the Asia-Pacific region are on the coast (see Figure 7.1 in Chapter 7), and many of the cities occur in one of several settings within, or adjacent to, the megadeltas. Shanghai, Guangzhou, and Bangkok are each located on the Holocene plains in mid delta. Dhaka lies in mid delta, but sits largely on the Madhupur terrace, an uplifted section of mid-Holocene delta deposits (Brammer 1996), meaning it is a little more elevated than the adjacent settlements. Also on the margin of this same delta is Calcutta, on the Hooghly River, a former distributary of the Ganges. Karachi, Ho Chi Minh City, and Rangoon are similarly on the coastal margin of their respective deltas, whereas Tianjin is on the inland margin of the Huanghe. It is often the case that significant cities develop at the apex of the delta. Such is the case for Hanoi. Phnom Penh occurs near the apex of the Mekong Delta, whose margin has been considered to coincide with the border between Vietnam and Cambodia, mainly because Quaternary mapping of plains in southern Cambodia has not been undertaken; and other smaller cities also occupy such a position, such as Zhengjiang on the Changjiang. Populations of cities have been obtained from UN/DESA 2002. Collectively, the projected population of these cities for 2015 exceeds 125 million. The city of Dhaka is the most rapidly expanding in the world and is projected to expand by a further 71% by 2015, to exceed 22 million people. Megacities are considered in greater detail in Chapter 7.

10.5 RESPONSE TO CLIMATE CHANGE

Anticipated impacts associated with climate change and concomitant sea-level rise involve coastal erosion, inundation of low-lying areas, saltwater intrusion, habitat loss, and flooding by river and storm surges. On Asian coasts these have implications for human populations, including displacing people, requiring land-use changes, and complicating water management, navigation, and waste management (Paw and Thia-Eng 1991). Regional forecasts of climate change, based on intercomparison of model-based projections using the latest IPCC emission scenarios provide only broad indications of what might occur. Scattergrams of the range of model predictions for the regions South Asia, Southeast Asia, and East Asia, plotting temperature change against precipitation change, have been prepared by Ruosteenoja et al. (2003). These show that for Southeast and East Asia all models predict that temperatures will be warmer and that it will be wetter during each of the seasons through the 21st century. For southern Asia, models predict that it will be warmer, but they show a greater spread of predictions about rainfall, including some model predictions that show it decreasing.

Present-day patterns of sedimentation are dynamic and changeable, showing substantial variations spatially and temporally; and are already impacted by human adjustments. Future changes should be anticipated, although the consequences may not become apparent until some unanticipated outcome alerts coastal planners to a problem. Even without sea-level or rainfall changes, the deltas can be expected to continue to evolve and to show geographical and year-to-year variability. The challenge is to disentangle human impacts from natural adjustments, particularly intrinsic changes within the system itself. This becomes harder still where some of the physical constraints on the delta may be indirectly influenced by human factors, obscuring the magnitude of natural processes. Whereas erosion occurs because waves reach the shore having been generated by offshore wind, this is the proximal cause, but the underlying issues may be human activities in terms of changes to sediment budgets, subsidence, or other causes. In this section, four issues are considered; the impact of sea-level rise, vulnerability to flooding, the effect of dam construction on sediment supply to megadeltas, and the role of extreme events.

10.5.1 Sea-level Rise and Inundation

Average global sea-level rise over the second half of the 20th century has been 1.8 ± 0.3 mm a^{-1} (White et al. 2005), and sea-level rise of the order of 2–3 mm a^{-1} is considered likely during the early 21st century as a consequence of the greenhouse effect. However, the regional expression of this is geographically variable, and as previously stressed must be considered in conjunction with highly variable rates of subsidence within each megadelta. Local subsidence exacerbates sea-level rise projections, for example, rates of observed sea-level rise in Vietnam are already 2.24 mm a^{-1} (1957–1989) (Thanh et al. 2004). For the Chinese coast, relative sea-level rise predictions allowing for subsidence are of the order of 40–100 cm by 2050, which is much higher than the IPCC global-mean projections (Li et al. 2004).

Although the past provides some insights into how the plains that characterise each of the megadeltas has evolved, it is not a perfect analogue for what will happen if the sea rises in the future even allowing for increased human modifications. During the long period of more-or-less stable sea level over the past 6000 years, low-lying delta plains have built out, and the surface elevation of these deposits has been lowered in relation to current sea level through a series of processes, such as subsidence, flexure of the underlying plate in response to sediment loading, compaction of sediments, and dewatering. This is shown schematically in Figure 10.2, where it is clear that a rise of sea level across the broad prograded delta plains that now exist will differ from the aggradation of coastal environments during the postglacial transgression (e.g. stages 1 and 5 in Figure 10.2 are not directly analogous). Geomorphological reconstruction gives some guide to anticipated changes, but modern delta morphology includes extensive areas that are extremely low-lying. Some parts of delta plains may be below the elevation reached by the sea at high tide on the coast, but may not be subject to inundation at present because of subtle topographic variations of the plains surface not distinguishable from SRTM DTMs or attenuation of high-tide level through complex tidally-influenced channels. Hence, a complex pattern of land loss and inundation seems likely in response to sea-level rise.

In considering possible impacts and adjustments to sea-level rise, Broadus (1993) promoted the concept of a shore-parallel retreat as the sea rose to facilitate analysis of possible impacts. In the case of Bangladesh, he estimated that a 1 m rise of sea level would cover 7% of the habitable land, involving 5% of the population, with a 5% decrease of GDP. This approach has been widely adopted, despite the provisos and caveats that Broadus advocated, namely the uncertainty about sea-level rise, the fact that it would be gradual in comparison with human capacity to adapt, that costs would lie well into the future and could therefore be discounted, perhaps amounting to only 1–2% of GDP, and that technological change might facilitate future economic savings. The concept of simple translation of the shoreline landwards as the sea rises, does not, however, capture the more complex changes that occur where variable topography is subject to diverse hydrodynamic processes.

By contrast, a companion paper by Brammer (1993) in the same volume expressed a contrary view, highlighting the geographical complexities of any detailed impact assessment. Based on considerable field experience, Brammer emphasised that the GBM delta plains are neither homogeneous nor static. Year to year variability and occasional catastrophic changes are on a much greater scale than the gradual changes expected to result from a slowly rising sea level. In contrast to the view of shore-parallel retreat propounded by Broadus, Brammer implied that sedimentation within the mangrove-lined tidal channels of the Sundarbans might enable them to keep pace with sea-level rise, but that there might be greater implications for the population of the GBM system as a result of increased flood levels in mid delta. Altered flooding would have significant impacts on the resident population affecting the number of rice crops per year and rice yields, and necessitating heightening of house mounds and defences for roads and other communications.

The sensitivity of rice cultivation to flood conditions under scenarios of sea-level rise, and the consequences for the intensity and patterns of rice cultivation, were outcomes resulting from recent modeling by Wassman et al. (2004) in relation to the Mekong delta. They found that rice production will vary according to regional and local differences in depth and duration of seasonal flooding, soil moisture properties, salinity, and irrigation practices. In contrast to an earlier study by Zeidler (1997) which considered that up to 20% of the rice producing area of Vietnam would be lost under sea-level rise, Wassman et al. (2004) used the Vietnam River System and Plains (VRSAP) model, a 2D finite difference hydraulic model, to simulate the changes in depth of wet season flooding under scenarios of sea-level rise. The modeling used observed hydrological parameters and the configurations and dimensions of channels as input, under scenarios of 20 and 45 cm higher sea level in the South China Sea. GIS-based interpolations of daily average water level were mapped in wet season months, to show spatial variations in the more extensive flooding and its impact on rice cultivation.

10.5.2 Coastal Flooding and SRES Climate and Socio-Economic Scenarios

The extent to which coastal areas experience flooding in the future will be a function of changes in the natural systems as a result of climate change, but will also depend on a series of socio-economic factors, such as the ability and willingness of communities to invest in adaptation measures. This issue has been examined by Nicholls (2004), who considered the sensitivity of coasts around the world to climate and global mean sea-level rise using the HadCM3 model, driven by the Special Report on Emission Scenarios (SRES) emission scenarios (Nakićenović et al. 2000).

Based on 1990 population estimates, there appear to be about 200 million people around the world (4% of the total population) living beneath the 1 in 1000 year storm surge elevation, and 450 million people living beneath the 10 m contour (Small and Nicholls 2003). Of these, on average about 10 million people per year experience coastal flooding due to storm surges (Nicholls 2004). The average annual people flooded (also termed the people at risk) will increase in the future as population increases, except where additional flood management and protection is undertaken. In the modeling, estimates of the likely number of people flooded were calculated initially without sea-level rise, and then compared with conditions of higher sea level (Nicholls 2004).

The SRES report supersedes the IS92 emission scenarios widely adopted by the IPCC to give an indication of ways in which the world's climate will be influenced by human adjustments. Four different pathways (A1, A2, B1, B2), termed storylines, have been outlined, being mutually consistent characterisations of how the world might develop, depending on environmental awareness and globalisation (Arnell et al. 2004). The A1 world is a global and economically-driven materialist/consumerist world, with increasing globalisation, rapid economic growth, and technological innovation. The A2 world is economically-driven but much more localised and heterogenous, with regional economic growth, but more diverse technological solutions and a gap between richer and poorer nations. The B1 world involves global co-operation, but has environmental policies and clean technologies rather than being entirely economically driven. The B2 world is environmentally conscious but more localised.

Population projections increase beyond the horizon of 2015 described above, and although global population continues to increase up to 2050 in all SRES cases, there is substantial divergence thereafter depending on the scenario. The number of people flooded will increase as populations expand, but will depend on the exposed population, the surge regime including the magnitude of relative sea-level rise, and the adaptive capacity of the people, especially their ability to manage flooding. Under scenarios of high population growth, significant subsidence due to poor delta management, and limited economic growth, the number of people who might experience flooding could be very high, implying a regional catastrophe. Under scenarios of lower population growth, lower subsidence, and higher economic growth (or access to other resources for adaptation) the flooding might be effectively managed.

Modeling was at a coarse national spatial scale, and did not consider the scale of individual deltas, let alone the considerable spatial variability that either the topographic (SRTM) or population (GPW-3) datasets demonstrate within Asian megadeltas. Data were examined using 192 polygons that represent coastal countries in the early 1990s, subsequently aggregated to regional and global levels (Hoozemans et al. 1993). This highly aggregated scale imposes limitations on the interpretation of specific results. In the absence of global databases on flood protection levels, the standard of protection was estimated using national GDP per capita as a direct measure of adaptive capacity. The issue of protection standards is particularly problematic and assumes a range of protection scenarios from optimistic to pessimistic. Economic development serves to increase vulnerability through destruction of natural resources, such as the mangrove forests that might otherwise have acted as a first line of coastal defence against storm surges. Richer nations generally enjoy a higher level of protection through engineered structures such as embankments and other flood defences. This can be demonstrated with reference to China. China has 12,000 km of dykes constructed along its coast, which are reinforced and heightened periodically. Coastal provinces are generally prosperous, contributing a major component of Chinese GDP, and it has been calculated that the costs associated with these protection and adaptation measures to counter a rise in sea level of 65 and 100 cm correspond to only 0.008 and 0.017% of GDP respectively (Li et al. 2004).

In the context of the 1990s, it is apparent that a large proportion of the population at risk of flooding occurs in South Asia (4.3 million) and East Asia (2.9 million). Additional people will be flooded in the future under each of the SRES scenarios as the populations described above expand through growth and in-migration. The most vulnerable socio-economic world is the A2, which has the largest population growth and the smallest increase in GDP per capita (Table 10.5). Without any rise in sea level, a further 18.4 million people worldwide,

Table 10.5. Regional incidence of flooding (millions of people per year) as modelled for the 2080s in terms of the HadCM3 SRES scenarios under lagged evolving protection (low subsidence and low population growth scenario), with the impact of a rise in sea level (in italics) in terms of additional people flooded. Sea-level rise varies from 22–34 cm, in response to SRES drivers

Region	1990	2080							
		A1	+SLR	A2	+SLR	B1	+SLR	B2	+SLR
South Asia	4.3	0.1	*0.6*	12.4	*13.6*	0.1	*0.2*	1.0	*1.3*
Southeast Asia	1.9	0.1	*0.5*	3.6	*0.4*	1.1	*0.1*	0.2	*0.1*
East Asia	2.9	0.0	*0.0*	0.8	*0.1*	0.0	*0.0*	0.1	*0.0*
Global total	10.3	0.4	*6.6*	18.4	*28.3*	1.4	*1.4*	3.0	*15.6*

Source: Nicholls 2004, Table 14.

12.4 million of them in South Asia, would on average be anticipated to experience flooding in any year (clearly in many individual years this would be exceeded). When sea-level rise is added to the modeling, an additional 13.4 million people in South Asia would be exposed to flooding (Table 10.5). It is important to note, however, that under alternative SRES scenarios the number of additional people at risk is considerably smaller. There are three implications of this generalised modeling: first, the development pathway will have a more profound effect on susceptibility to flooding through the 21st century and beyond than the anticipated rate of sea-level rise; second, human-induced subsidence could greatly exacerbate the impacts of global sea-level rise due to climate change; and, third, these megadeltas make a disproportionate contribution globally to coastal vulnerability, with or without sea-level rise.

10.5.3 Impact of Damming on Sediment Replenishment

Deltas are dynamic landforms maintained by a large supply of sediment. There are several potential storages that can be recognised on the sediment transport pathway from catchment source to receiving basin sink. Only recently have data become available to attempt to quantify the fractionation of the sediment between these storages across selected catchments. The availability of sediment in the headwaters for transmission downstream depends on rates of weathering, detachment, and entrainment. Its transport depends on the competence of flow, with the potential for sequestration on the floodplains of the alluvial channel (see Chapter 4). Sediment which does make it to the apex of the delta can then be fractionated in various ways. Within the delta, distributaries may be active or abandoned and the proportions of the load that they carry can vary. The extent to which flow goes overbank is also important in that it constrains deposition on the delta plain surface. Monsoon flooding of the plains with sediment-laden waters may occur, but increasingly flood mitigation works are aimed to contain flow within the channel by levées, or artificial embankments. The sediment that reaches the delta front may be carried further seaward, and be lost to deep water, or where currents or wave action are strong, it can be reworked onto the abandoned part of the delta plain helping to maintain delta form. In the case of the Indus and GBM there is a fan in deep water, on which sediment has accumulated. On other deltas, such as the Huanghe, heavily-sediment laden, hyperpycnal flows carry turbid waters down the delta front (Wright et al. 1986). In the case of the Mekong, mud is carried along the delta front and the muddy Camau Peninsula has built up by this longshore transport of mud in the wet season, with flow reversal and the upstream penetration of turbid waters in the dry season (Wolanski et al. 1998).

The pattern of sediment deposition also has considerable spatial variability, with sedimentation being more rapid in low-lying areas where water ponds as a result of flooding. Spatial variation in sedimentation has been demonstrated for the GBM system. Sandy river sediments, 3–20 m thick, dominate the upper delta plain. Tectonically-induced subsidence of around 1–4 mm a^{-1} in the Sylhet Basin (on the course of the former Brahmaputra River) has served to trap 70–80 Mt a^{-1} of sediment there. ^{210}Pb and ^{137}Cs dating of short cores indicates floodplain accretion of >10 mm a^{-1} in the modern braidbelt of the Brahmaputra River, sequestering >50 Mt a^{-1} of coarse sediments and 30 Mt a^{-1} of mud on the floodplains where accretion is 1–3 mm a^{-1} (Goodbred and Kuehl 1998). Determining sedimentation rates over the subaerial lower delta plains is problematic, and it is spatially variable, with preferential fine-grained sediment deposition of up to 7 m thick in localised depressions called *bils*. Sedimentation rates of 1–2 mm a^{-1} in the southern delta plains have generally been balanced by high sediment supply from the Himalayan source. Riverine sediment supplied primarily by the Meghna leads to accretion in the river-dominated Meghna Delta plain of 5–16 km^2 a^{-1} (a process which has been successfully enhanced by engineering works and planting of mangrove forests). Some fraction of this sediment appears to be advected alongshore by westward flowing currents to the tide-dominated Ganges Tidal Plain, and within mangrove forests of the Sundarbans sedimentation rates of up to 11 mm a^{-1} have been recorded (Allison and Kepple 2001). Nevertheless rapid subsidence, whether tectonic, flexural, or through compaction and dewatering of sediments, leads to overall recession of this tide-dominated coast, with relict distributary shoals on the innermost shelf and erosion of the shoreline by 3–4 km since 1792 (Allison 1998b). Around 30% of the sediment load of the rivers appears to be deposited on the subaqueous delta that has been prograding at 15–20 m a^{-1}.

Radiocarbon dating control on boreholes along the coast of several of the megadeltas has clarified the average rate at which deltas have been prograding over the past 2000 years. Over that period, the shoreline has migrated around 80 km at the mouth of the Huanghe, 100–150 km at the mouth of the Changjiang, 20–30 km at the mouth of the Red River, 30–40 km at the mouth of the Mekong, and 10–25 km at the mouth of the Chao Phraya. These are average rates, and even over this period human impact may have been occurring but is difficult to discriminate in terms of its effect on sediment dynamics. Rates may also vary as a result of natural factors.

Many of the rivers of China continue to bring down such large volumes of sediment that, despite human impact and extensive land claim, they continue to prograde, and the coast is characterised by net wetland renewal (rates of 21 km^2 a^{-1} on the Huanghe, 16 km^2 a^{-1} on the Changjiang and 11 km^2 a^{-1} on the Pearl) rather than net wetland loss (Li et al. 2004). Nevertheless, it is important to recognise that there is a threshold value of sediment input below which the system would cease to prograde and would start to retreat (see Chapter 4.2.4). For example, it has been suggested that the Huanghe requires 245–300 Mt of sediment annually to maintain delta building, if it receives less than this amount, a net loss of wetland would occur (Li et al. 2004). However, the current sediment discharge of the Huanghe is below this level due to the operation of the Xiaolangdi Dam since 1999, resulting in coastal erosion for all the modern delta including its river mouth (Jiongxin 2003).

In other cases the sediment load has clearly decreased (Syvitski et al. 2005); for example, the Kotri Barrage on the Indus River has substantially decreased flow and reduced sediment load (Milliman et al. 1984). The effect of dams is exacerbated by water extraction for irrigation and other use, and saltwater penetration up tidal channels and into agricultural land appears to be a consequence. Similarly, the Farakka Barrage on the Ganges has decreased dry-season flow on the channel downstream, in an effort to maintain a navigable channel in the Hooghly ensuring access to Calcutta. Access to Haiphong is no longer available for big ships because of the siltation of access channels. The Changjiang River has a load of around 5.3×10^8 t a^{-1} at the recently completed Three-Gorges Dam in its middle reaches (see Chapter 4.1). This dam will significantly reduce the supply of land-building sediment to the East China Sea, and there will be a need to recognise the effects of this and lags involved with movement of this sediment downstream in managing the coast. Prior to the operation of the Three-Gorges Dam, interference from numerous other dams has caused an approximately 73% decrease in the load that reaches the mouth, which has resulted in a critical local loss of saltmarsh. Similar problems are encountered in the Red and Mekong river deltas; the sediment discharge of the Red River was reduced by 60–70% due to the construction of the Hoa Binh Dam, and the Mekong already shows a 10% reduction south of Laos due to dams in China, with plans for further dam construction.

Although accumulation of sediment in basins beyond the mouths of these large river systems does not record variations in sediment flux over the Quaternary (Métivier and Gaudemer 1999), it is certainly clear that pathways and storages have changed significantly with sea-level fluctuations. Figure 10.7 illustrates this schematically. When sea level was lower during the last glaciation sediment appears to have been transported seawards and deposited on deep-water fans in the case of the Indus and GBM (Goodbred and Kuehl 1999). During the late Holocene, under conditions of relatively stable sea level close to its present level, there has been progradation of delta plains with fractionation of sediment across the surface of those plains, along the delta front and offshore. Those parts of the abandoned delta which no longer receive sufficient additional sediment experience erosion, accelerated where the delta has been experiencing subsidence (see Chapter 4.2.4). In cases where the majority of sediment is sequestered behind dams, however, the supply of

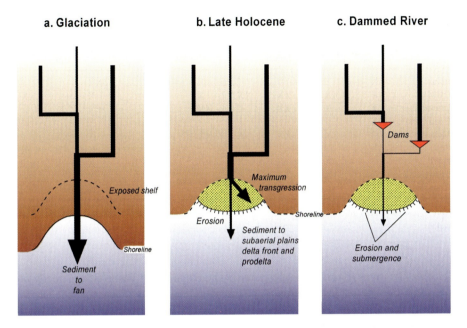

Figure 10.7. Fractionation of sediment in its transport from catchment to coast. At different stages in the development of deltas sediment has been fractionated and sequestered in different ways. At low sea level much sediment appears to have been carried through the system with substantial deposition on deep-sea fans, as in the Indus and Bay of Bengal. During mid and late Holocene complex deltas have developed with fractionation of sediment between the subaerial delta, the delta front and the prodelta. In the 20th/21st centuries human retention of sediment in upstream dams has been, or is likely to be, a major driver of delta change

sediment to the coast is diminished and erosion and submergence become widespread. The Nile is an example of a case where chronic erosion has resulted from the construction of a major dam (Milliman et al. 1989), but the retreat of sections of coastline along other deltas around the world can also be attributed to dam construction and the decrease of sediment supply.

10.5.4 The Role of Extreme Events and Related Hazards

Climate change, and associated sea-level rise, will be gradual, and the incremental effects will occur over many decades or longer. The Asian megadeltas are already subject to a series of extreme events and other hazards that have far-reaching impacts on delta populations and their livelihoods. Table 10.6 indicates several environmental hazards that are already of concern, and these impacts are likely to be exacerbated by environmental change.

Coastal erosion is already a prominent feature of many delta shorelines; in places it may be a natural component of the abandoned phases of delta development, but

Table 10.6. Summary of major hazards presently experienced on Asian megadeltas

Megdelta	Erosion	Cyclone	River floods	Storm surge	Saline intrusion	Subsidence	Dam	Acid soil	As
Indus	X	X	?	?	X	?	XX		
GBM	X	XX	XX	XX	X	X	X		XX
Irrawaddy	X	X	XX	?	XX	?	X		
Chao Phraya	XX					XX	X		
Mekong	?	X	XX	X	XX	?	X	XX	
Red	X	X		X		XX	X	X	X
Pearl	?	X	X	X		X	X	X	
Changjiang	X	X	XX	X		XX	X		
Huanghe	X		X			X	X		

Note: Acid soil refers to identified acid sulphate conditions; As refers to arsenic contamination of shallow wells.

in other situations it has been accelerated by human agency. Those cities directly on the coast may be subject to erosion of the shoreline. Many of the megadeltas are prone to damage from cyclones, both through wind damage and through the associated surge and other weather-related hazards. Flooding remains a problem in many of the low-lying areas, and in other cases reductions of freshwater flows have led to saltwater penetration and acid sulphate soil development on the plains. Each of these hazards may already threaten population centres on the Asian megadeltas, almost all are likely to be exacerbated by global climate change and the continued human development of the catchment and delta plains.

Seawater intrusion has already been observed along almost all of the deltas, with particularly pronounced intrusion where freshwater flows have been reduced. Dry season saltwater penetrates further up distributaries on many megadeltas as river flows decrease (through natural diversion or human interference) and this will be enhanced if flows are reduced by further damming (Wolanski et al. 1998). Typhoons play a major role and there have been large death tolls from the associated storm surges. The coast of China has experienced 390 storm surges between the period 1949–1986, more than 300,000 were killed in 1931, and flooding continues to result in human tragedies each year (Chen and Stanley 1998). Whereas storms in east Asia are accompanied by surges of typically 1–2 m, a maximum of 3 m has been experienced along the Vietnam coast (Thanh et al. 2004). The worst in southern China was a surge that reached 5.94 m in Guangdong province (Li et al. 2004). Embankments provide protection up until their design height (generally built for 1 in 50 to 1 in 100 years storm), but they also reduce the capacity of the waterbody to hold large floods, promoting sediment deposition in the channel at slack water, further reducing channel capacity. Flooding levels can also be accentuated by heavy rain. Surges associated with such storms in the Bay of Bengal over the past 200 years have resulted in a death toll in the GBM system that probably exceeds 1.3 million people, including losses of more than 500,000 in

1970 and 100,000 in 1991 (Nicholls et al. 1995, Ali 1996). Extreme events impact many other coasts; for example, Typhoon Linda hit Vietnam and Thailand in 1997, and a super cyclone caused widespread flooding and an estimated 10,000 deaths in Orissa in eastern India in 1999.

Land claim has been important and is still practiced in China, where placing rocks to retain muddy water at high tide encourages sedimentation and claims further land. Such coastal engineering practices have a long history and can be traced back to protection works in the Tang Dynasty around 1000 years ago. These reclamations are assisted because the large sediment supply means that build-out is of the order of 20–40 m a^{-1}. Channels have sluices on them to prevent inundation under surges. Ground subsidence has been known in Shanghai since 1921, but has been reduced since restrictions on extraction of groundwater were strictly enforced from 1966 onwards. The city centre subsided 2.63 m over the period 1921–1965; whereas Tianjin subsided 2.7 m over the period 1959–1993 (Li et al. 2004). Erosion, occurring on the coast of China is mainly due to reduced river supply of sediment, although some is due to sand mining, and only 10% can be attributed by these authors to sea-level rise.

10.6 KNOWLEDGE GAPS AND RESEARCH DIRECTIONS

This chapter has concentrated on the natural processes that shape megadeltas at a range of scales, and advocates integration of the study of physical processes with data on socio-economic factors, in an effort to develop a more integrated approach to planning and management of the very populous megadeltas in the Asia-Pacific region. Figure 10.8 indicates that there is a range of natural and socio-economic processes operating at different spatial scales; our understanding of most of these is incomplete.

10.6.1 Natural Processes

Geological investigations of Holocene (and longer-term) development of megadeltas provide insights into the directions and rates of processes and set the context within which to better manage deltaic processes so that resident populations can maximise the opportunities and minimise the hazards of their location. The delta plains that exist today represent particularly extensive near-horizontal coastal plains, most of which have been deposited during the mid and late Holocene. The former shoreline at the maximum transgression provides a useful boundary by which to define the delta extent. Whereas the long-term evolution of deltas can be reconstructed based on stratigraphy and radiocarbon dating, and the physics of fluid dynamics and sedimentary processes are understood at the microscale, the behaviour of individual deltas at decade-to-century scale remains poorly known, and the response to intrinsic thresholds with significance at planning and management timescales, such as distributary switching, is less reliably understood (Figure 10.6). There needs to be a greater focus on the extent to which

Natural Processes

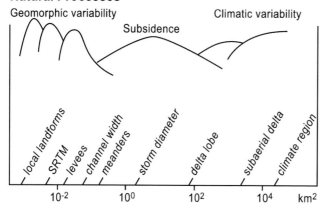

Socio-economic processes

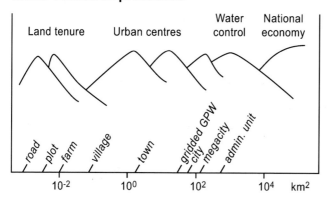

Figure 10.8. Schematic representation of spatial variability in the natural processes and human usage of megadeltas. Different physical processes are manifest at different spatial scales, and socio-economic processes also operate at different scales

sediment budgets of different components of a delta are influenced by different physical factors, particularly whether river, wave, or tide processes dominate. For example, the abandoned delta plain of most Asian megadeltas is tide-dominated, and behaves in a different way to the river-dominated active delta distributaries that may receive substantial sediment loads.

A geological perspective, as developed in this chapter, need not imply that engineering projects on major deltas should not proceed, but, as stressed by Milliman et al. (1989), it could guide how projects are undertaken. All actions in these deltas are likely to have implications for the distribution and pathways of sediment. Ongoing research will be needed to establish not only what these impacts are, but also to examine the natural archive of sedimentological data within cores, and to determine comparative rates of sedimentation under past, present, and future environmental conditions. In many cases it will be necessary for geoscientists to work with other specialists and to extend their consideration from sediment to other factors such as nutrients (see Chapter 4).

There has been increasing recognition that, in contrast to traditional engineering philosophies of 'controlling nature', it may be more effective and efficient to develop management that works more with nature (sometimes called 'natural systems engineering'), and to incorporate natural, or slightly modified, patterns of sediment movement to reduce the impacts of human activities (e.g. Nicholls et al. 1995). There will be a need for increased awareness of the sedimentological implications of actions in the future. In Asian megadeltas nutrient depletion may follow loss of sediment sources, and eutrophication associated with large populations remains a risk. Geochemical consequences of delta evolution have not received the attention they deserve; for example, proliferation of acid sulphate soils occurs once surface waters are excluded, by irrigation or drainage, and high levels of arsenic can occur in sediments as found in parts of the delta plains of the GBM and Red River systems.

Knowledge of the elevation of the land surface of these low-lying plains is a prerequisite for their better management. Traditional surveying has proved difficult across the intermittently flooded delta plains and survey control, against which sedimentation or subsidence rates might be assessed, has been problematic. New technologies offer sophisticated options to establish the subtleties of terrain variation. For example, the gross topography of Asian megadeltas can be compared, as in Figure 10.3, using digital terrain models based on shuttle imagery (SRTM). Remote sensing, such as the JERS and MODIS imagery, offers synoptic overviews of these poorly accessible areas, and airborne remote sensing techniques, such as laser (LiDAR) surveying can form the basis for terrain models at unprecedented resolution.

However, integrated geoscientific assessments of future delta evolution need to consider a range of natural processes, including subsidence, compaction, sediment supply, sea-level change, and changes in tidal amplitude, which show both temporal and spatial variability within each delta (Figure 10.8). Delta environments are generally treated as passive, but they are dynamic, as emphasised by Brammer (1993). On many deltas, channels can be expected to change their position, and banks to erode. Changes in flow through the distributaries of each delta will respond to both environmental changes and to man-made interventions. For example, saline intrusion in the Ganges tidal plain has been aggravated by the damming of former Ganges distributaries and the abstraction of water. There is a clear need for better morphodynamic models of distributary processes, such as hydrodynamics, and sedimentation (see Chapter 4.4.2).

The trajectory of regional climate change through the 21st century and beyond is uncertain (Figure 10.8), but modest increases in temperature, precipitation, and sea level seem likely in much of the Asia-Pacific region. Change in the future is assured, but it will be a considerable challenge to determine which components, or how much, of the observed future changes can be attributed to natural climate variability and how much to human influence. The nature of the impact of future sea-level rise remains uncertain, but appears to involve increased potential for loss of property, flood risk, and loss of life, as well as other stresses such as the realisation of acid sulphate soil development following flood control. Too often potential sea-level rise impacts are viewed as linear; linear assessments are based primarily on simplistic assumptions ignoring the complexity and variability of landscape dynamics. However, significant non-linear effects can be expected in response to threshold levels, such as embankment and dyke heights, as well as the elevations of natural levées. This has been tragically demonstrated in the wake of Hurricane Katrina and its impact on New Orleans in August and September 2005. Although it is possible to continue to build up flood defences to counter gradual sea-level rise or regional subsidence, the impact when an event does breach the defences, may be substantially more catastrophic. Cities such as Shanghai and Bangkok, already known to have undergone subsidence and protected by flood defences, will become still more vulnerable in the future.

10.6.2 Societal Response

Spatial variability in natural processes, such as geomorphology, hydrology, and salinity patterns, provides the physical framework over which is draped considerable spatial variability in human use of the landscape, including settlement, communications, land use, and tenure (Figure 10.8). The GBM system is by far the largest and contains a population of well over 100 million people, and the most rapidly expanding megacity (Dhaka). Other deltas have millions of people, and are also experiencing rapid urbanisation with several more cities either already megacities or likely to become so during the first half of the 21st century. Each of these factors varies at spatial and temporal scales that differ within and between deltas. Combining natural resource and socio-economic datasets, such as DTM and population datasets (as in Figure 10.4) is a powerful method to examine exposure to risk, but it must be remembered that natural and anthropogenic processes operate at differing scales, and rarely is information available at the optimal scale for detailed vulnerability assessment.

The topography of the Asian megadeltas is summarised in Figure 10.3, using SRTM DTMs. This shows the broad pattern of natural landscape variation but each comprises a more locally variable mosaic of landforms than is captured by the space-based global overview data. Similarly, the distribution of people, estimated from gridded population statistics, also involves spatial clustering at scales that are not fully amenable to analysis at the scale of global data. Thus, while the combination of the two datasets gives broad generalisations of the enormous numbers of people already at risk of flooding, more detailed vulnerability assessments will need to be based on data that can be appropriately synthesised across scales (Figure 10.8). The Asian megadeltas are already subject to considerable spatial variability in both physical and human landscapes, and this complexity will be compounded, and the impacts generally exacerbated, as a result of changing climate.

Low-lying areas appear most vulnerable to inundation, but their susceptibility is a function of whatever protective measures are instigated to reduce risk. Any consideration of sensitivity to these impacts is highly dependent on socio-economic scenarios and the different possible development pathways. There is no question that much of coastal Bangladesh is very vulnerable; much already experiences a range of stresses, and sea-level rise is likely to be an additional stress on an already heavily impacted coast. This is also true of the other megadeltas; the natural processes within these deltas are now inextricably linked with human use of resources. Rice production has gone from subsistence, to self-sufficiency and in many areas rice is now exported, based on judicious cropping of rice in relation to flood levels. Choice in the adoption of several possible productive activities across the delta plains in addition to agriculture (e.g. forestry, fish production, prawn farming, or salt production) increases the potential adaptive capacity of communities. Research on multiple land use across megadeltas is in its infancy. For example, changes to flooding conditions as a result of change in precipitation or sea level can be modeled, as shown by Wassmann et al. (2004) for the Mekong Delta, but implications for rice production or adoption of alternative land-use options have yet to be explored in detail. Detailed modeling of this kind provides insights into impacts of a higher sea level, but does not include a series of additional impacts, such as the significant observed effects of increase of dry season tide levels and saline intrusion, which may also affect rice productivity. Nor does the modeling include the effects of extreme events such as storm surges, or the geomorphological changes in channel morphology that may result from changed processes.

More research is needed to investigate the roles of extreme events in these systems. Nevertheless, whereas weather-related hazards, such as those listed in Table 10.6, exert occasional acute impact on deltaic coasts, it seems inescapable that human activities are often a chronic stress on the system. Synergistic effects between several stresses will often increase the vulnerability of the low-lying delta plains generating impacts that are sometimes quite unexpected. Finding the appropriate balance between flood control and sediment delivery remains a challenge for Asian megadeltas, in common with large deltas in developed countries such as the Mississippi. As Asian societies become increasingly prosperous, adaptation becomes more affordable; for the poorest societies, such as those in Bangladesh, there remains low adaptive capacity, and it is highly uncertain how development will modify this situation (e.g. Nicholls 2004).

Human activities also involve many negative effects, including changes to supply of sediment, water and nutrients, as well as loss of habitats and spread of pollutants. There is a need for more research into the ecological footprint of megacities and urban agglomerations on the environment, as well as an understanding of their role in globalisation, as discussed in Chapter 7. With improving living standards, there is likely to be increased demand for flood protection, and less tolerance of flood damage. It is clear that susceptibility to flooding is very sensitive to the extent of flood management and in this respect socio-economic factors often over-ride climatic factors. Water management has a long and proud history in these deltas but damming of the upstream rivers poses a major threat, with reduced sediment and nutrient supply, as well as altered water conditions. This is also usually associated with increased saline intrusion, which can be a dramatic impact in itself.

Sedimentation in one sector of the delta may offset subsidence, but both processes occur at rates that equal or exceed present rates of sea-level change. Human use of these deltaic environments, for whatever purposes, whether urbanisation, agriculture, or subsistence, will have economically significant impacts if the balance between sediment demand and supply is upset, even without further accentuation through global climate changes. By encouraging long-term perspectives on delta utilisation it may be possible to undertake sustainable development that will minimise future vulnerability. Given that all the Asian megadeltas are so sensitive to a range of natural and human-induced drivers of change, an integrated assessment approach is fundamental to understanding delta vulnerability to climate and other change.

The complexity of delta systems with the multitude of interacting natural and socio-economic factors suggests that we need to build on the earlier study of Milliman et al. (1989) and begin to consider all the factors that will shape future delta evolution. This indicates the need to collect much more data on these megadeltas, and to undertake multidisciplinary studies. Vulnerability analyses need to be focused on detailed local factors such as topography and integration with flood levels, land use, and other relevant factors. Existing population data rarely adequately represent the extent of urbanisation and the rapid growth of megacities and urban agglomerations in the megadeltas, and their ecological footprint needs to be examined (see Chapter 7). Such assessments are challenging from a number of perspectives, but are essential if scientific understanding is to be translated into useful guidance for coastal management.

ACKNOWLEDGEMENTS

The material for this chapter was assembled and discussed during meetings of the International Geological Correlation Project IGCP-475 "Deltas in the Monsoon Asia-Pacific region", Asia-Pacific Network (APN) "Mega-Deltas of Asia", and CCOP (co-ordinating committee for Geoscience programmes in East and Southeast Asia) "DelSEA" project which jointly organised the international conference on "Deltas: a conceptual model and its application to future delta vulnerability" in Thailand in January 2004, and the international conference on "Deltas: geological modeling and management" in Vietnam in January 2005.

REFERENCES

Ali A (1996) Vulnerability of Bangladesh to climate change and sea level rise through tropical cyclones and storm surges. Water, Air and Soil Pollution 92: 171–179

Allison MA (1998a) Geologic framework and environmental status of the Ganges-Brahmaputra delta. Journal of Coastal Research 14: 826–836

Allison MA (1998b) Historical changes in the Ganges-Brahmaputra delta front. Journal of Coastal Research 14: 1269–1275

Allison MA, Kepple EB (2001) Modern sediment supply to the lower delta plain of the Ganges-Brahmaputra River in Bangladesh. Geo-Marine Letters 21: 66–74

Allison MA, Khan SR, Goodbred SL, Kuehl SA (2003) Stratigraphic evolution of the late Holocene Ganges-Brahmaputra lower delta plain. Sedimentary Geology 155: 317–342

Arnell NW, Livermore MJL, Kovats S, Levy PE, Nicholls R, Parry M, Gaffin SR (2004) Climate and socio-economic scenarios for global-scale climatic change impact assessments: characterising the SRES storylines. Global Environmental Change 14: 3–20

Aung N (1993) Myanmar coastal zone management. In: McLean R, Mimura, N. (eds) Vulnerability assessment to sea level rise and coastal zone management: Proceedings of the IPCC Eastern Hemisphere Workshop, Tsukuba, Japan, 3–6 August 1993, pp 333–340

Bannerjee M, Sen PK (1988) Paleobiology and environment of deposition of Holocene sediments of the Bengal Basin, India. In: Paleoenvironment of East Asia from the mid-Tertiary: Proceedings of the Second Conference. Hong Kong Centre of Asian Studies, University of Hong Kong, pp 703–731

Brammer H (1993) Geographical complexities of detailed impact assessment for the Ganges-Brahmaputra-Meghna delta of Bangladesh. In: Warrick, RA, Barrow EM, Wigley TML (eds) Climate and Sea Level Change: observations, projections and implications. Cambridge University Press, pp 246–262

Brammer HB (1996) The Geography of the soils of Bangladesh. University Press Ltd, Dhaka, Bangladesh

Broadus JM (1993) Possible impacts of, and adjustments to, sea level rise: the cases of Bangladesh and Egypt. In: Warrick RA, Barrow EM, Wigley TML (eds) Climate and Sea Level Change: observations, projections and implications. Cambridge University Press, pp 263–275

Büdel J (1966) Deltas: a basis of culture and civilization. In: Scientific Problems of the Humid Zone Deltas and their Implications: Proceedings of the Dacca Symposium. UNESCO, pp 295–300

Chen Z, Stanley DJ (1993) Yangtze delta, eastern China: 2: Late Quaternary subsidence and deformation. Marine Geology 112: 13–21

Chen Z, Stanley DJ (1995) Quaternary Subsidence and River Channel Migration in the Yangtze Delta Plain, Eastern China. Journal of Coastal Research 11: 927–945

Chen Z, Stanley DJ (1998) Rising Sea Level on Eastern China's Yangtze Delta. Journal of Coastal Research 14: 360–366

Chen Z, Wang ZH (1999) Yangtze Delta. China: Taihu lake-level variation since 1950's, response to sea level and human impact. Environmental Geology 37: 333–339

Chen Z, Song BP, Wang ZH, Cai YL (2000) Late Quaternary Evolution of the Subaqueous Yangtze Delta: stratigraphy, sedimentation, palynology and deformation. Marine Geology 162: 423–441

Chen Z, Saito Y, Kanai Y, Wei T, Li L, Yao H (2004) Low concentration of heavy metals in the Yangtze estuarine sediments, China: a diluting setting. Estuarine, Coastal and Shelf Science 60: 91–100

Chen Z, Wang Z, Schneiderman J, Tao J, Cai YL (2005) Holocene climate fluctuations in the Yangtze delta of eastern China: the Neolithic response. The Holocene 15: 915–924

de Groot TAM (1999) Climate shifts and coastal changes in a geological perspective. A contribution to integrated coastal zone management. Geologie en Mijnbouw 77: 351–361

Douglas I (2005) The Mekong River Basin. In: Gupta A (ed.) The Physical Geography of Southeast Asia. Oxford University Press, pp 193–218

Emmel FJ, Curray JR (1982) A submerged late Pleistocene delta and other features related to sea level changes in the Malacca Strait. Marine Geology 47: 197–216

Galloway WE (1975) Process framework for describing the morphologic and stratigraphic evolution of deltaic depositional systems. In: Broussard ML (ed.) Deltas: models for exploration. Houston Geological Society, pp 87–98

Geyh MA, Kudrass H-R, Streif H (1979) Sea-level changes during the late Pleistocene and Holocene in the Strait of Malacca. Nature 278: 441–443

Goodbred SL (2003) Response of the Ganges dispersal system to climate change: a source-to-sink view since the last interstade. Sedimentary Geology 162: 83–104

Goodbred SL, Kuehl SA (1998) Floodplain processes in the Bengal Basin and the storage of Ganges-Brahmaputra river sediment: an accretion study using ^{137}Cs and ^{210}Pb geochronology. Sedimentary Geology 121: 239–258

Goodbred SL, Kuehl SA (1999) Holocene and modern sediment budgets for the Ganges-Brahmaputra river system: evidence for highstand dispersal to flood-plain, shelf, and deep-sea depocenters. Geology 27: 559–562

Goodbred SL, Kuehl SA (2000a) The significance of large sediment supply, active tectonism, and eustasy on margin sequence development: Late Quaternary stratigraphy and evolution of the Ganges-Brahmaputra delta. Sedimentary Geology 133: 227–248

Goodbred SL, Kuehl SA (2000b) Enormous Ganges-Brahmaputra sediment discharge during strengthened early Holocene monsoon. Geology 28: 1083–1086

Han M, Hou J, Wu L (1995) Potential impacts of sea level rise on China's coastal environment and cities: a national assessment. Journal of Coastal Research 11: 79–90

Hanebuth T, Stattegger K, Grootes PM (2000) Rapid flooding of the Sunda Shelf: a late-glacial sea-level record. Science 288: 1033–1035

Higham CFW, Lu TLD (1998) The origin and dispersal of rice cultivation. Antiquity 72: 867–877

Holmes DA (1968) The recent history of the Indus. Geographical Journal 134: 367–381

Hoozemans FMJ, Marchand M, Pennekamp HA (1993) A global vulnerability analysis: vulnerability assessment for population, coastal wetlands and rice production on a global scale. 2nd edition. Delft Hydraulics, The Netherlands

Hori K, Saito Y, Zhao QH, Cheng XR, Wang PX, Sato Y, Li CX (2001) Sedimentary facies of the tide-dominated paleo-Changjiang (Yangtze) estuary during the last transgression. Marine Geology 177: 331–351

Hori K, Saito Y, Zhao QH, Wang PX (2002) Architecture and evolution of the tide-dominated Changjiang (Yangtze) River Delta, China. Sedimentary Geology 146: 249–264

Hori K, Tanabe S, Saito Y, Haruyama S, Nguyen V, Kitamura A (2004) Delta initiation and Holocene sea-level change: example from the Song Hong (Red River) Delta, Vietnam. Sedimentary Geology 164: 237–249

Jiongxin X (2003) Sediment flux to the sea as influenced by changing human activites and precipitation: example of the Yellow River, China. Environmental Management 31: 328–341

Kazmi AH (1984) Geology of the Indus delta. In: Haq BU, Milliman JD (eds) Marine Geology and Oceanography of Arabian Sea and coastal Pakistan. van Nostrand Reinhold, New York, pp 71–84

Kremer HH, Le Tissier MDA, Burbridge PR, Rabalais NN, Parslow J. Crossland CJ (2005) LOICZ II: Science Plan and Implementation Strategy. IGBP

Kuehl SA, Hariu TM, Moore WS (1989) Shelf sedimentation off the Ganges-Brahmaputra River system: evidence for sediment bypassing to the Bengal Fan. Geology 17: 1132–1135

Li C, Fan D, Deng B, Korotaev V (2004) The coasts of China and issues of sea level rise. Journal of Coastal Research, Special Issue 43: 36–49

Li P, Qiao P, Zheng H, Fang G, Huang G (1991) The environmental evolution of the Zhujiang (Pearl River) Delta in the last 10000 years. China Ocean Press, Beijing

Mathers S, Zalasiewicz J (1999) Holocene sedimentary architecture of the Red River Delta, Vietnam. Journal of Coastal Research 15: 314–325

Métivier F, Gaudemer Y (1999) Stability of output fluxes of large rivers in South and East Asia during the last 2 million years: implications on floodplain processes. Basin Research 12: 293–303

McLean RF, Tsyban A (2001) Coastal zones and marine ecosystems. In: McCarthy JJ, Canziani OF, Leary NA, Dokken DJ, White KS (eds) Climate change 2001: impacts, adaptation, and vulnerability. Cambridge University Press, pp 343–379

Milliman JD, Meade RH (1983) World-wide delivery of river sediment to the oceans. Journal of Geology 91: 1–21

Milliman JD, Quraishee GS, Beg MAA (1984) Sediment discharge from the Indus River to the ocean: past, present and future. In: Haq BU, Milliman JD (eds) Marine Geology and Oceanography of Arabian Sea and coastal Pakistan. van Nostrand Reinhold, New York, pp 65–70

Milliman JD, Syvitski JPM (1992) Geomorphic/tectonic control of sediment discharge to the ocean: the importance of small mountainous rivers. Journal of Geology 100: 525–544

Milliman JD, Broadus JM, Gable F (1989) Environmental and economic implications of rising sea level and subsiding deltas: the Nile and Bengal examples. Ambio 18: 340–345

Milliman JD, Qin YS, Ren M-E, Saito Y (1987) Man's influence on the erosion and transport of sediment by Asian rivers: the Yellow River (Huanghe) example. Journal of Geology 95: 751–762

Morgan JP, McIntire WG (1959) Quaternary geology of the Bengal Basin, East Pakistan and India. Geological Society of America Bulletin 70: 319–342

Nakićenović N, Alcamo J, Davis G, de Vries B, Fenhann J, Gaffin S, Gregory K, Grübler A, Jung TY, Kram T, La Rovere EL, Michaelis L, Mori S, Morita T, Pepper W, Pitcher H, Price L, Raihi K, Roehrl A, Rogner H-H, Sankovski A, Schlesinger M, Shukla P, Smith S, Swart R, van Rooijen S, Victor N, Dadi Z (2000) Emissions Scenarios. A special report of working group III of the Intergovernmental Panel on Climate Change. Cambridge University Press, 599 pp

Nguyen VL, Ta TKO, Tateishi M (2000) Late Holocene depositional environments and coastal evolution of the Mekong River Delta, southern Vietnam. Journal of Asian Earth Sciences 18: 427–439

Nicholls RJ (1995) Coastal megacities and climate change. Geojournal 37: 369–379

Nicholls RJ (2004) Coastal flooding and wetland loss in the 21st century: changes under the SRES climate and socio-economic scenarios. Global Environmental Change 14: 69–86

Nicholls RJ, Mimura N, Topping JC (1995) Climate change in South and South-east Asia: some implications for coastal areas. Journal of Global Environmental Engineering 1: 137–154

Nilsson C, Reidy CA, Dynesius M, Revenga C (2005) Fragmentation and flow regulation of the world's large river systems. Science 308: 405–408

Paw JN, Thia-Eng C (1991) Climate changes and sea level rise: implications on coastal area utilisation and management in SE Asia. Ocean and Coastal Management 15: 205–232

Phienwej N, Nutalaya P (2005) Subsidence and flooding in Bangkok. In: Gupta A (ed.) The Physical Geography of Southeast Asia. Oxford University Press, pp 358–378

Ren ME, Shi YL (1986) Sediment discharge of the Yellow River (China) and its effect on the sedimentation of the Bohai and the Yellow Sea. Continental Shelf-Research 6: 785–810

Ruosteenoja K, Carter TR, Jylha K, Tuomenvirta H (2003) Future climate in world regions: an intercomparison of model-based projections for the new IPCC emissions scenarios. Finnish Environment Institute, Helsinki, 83 pp

Saito Y, Wei H, Zhou Y, Nishimura A, Sato Y, Yokota S (2000) Delta progradation and chenier formation in the Huanghe (Yellow River) Delta, China. Journal of Asian Earth Sciences 18: 489–497

Saito Y, Yang ZS, Hori K (2001) The Huanghe (Yellow River) and Changjiang (Yangtze River) Deltas: a review on their characteristics, evolution and sediment discharge during the Holocene. Geomorphology 41: 219–231

Small C, Nicholls RJ (2003) A global analysis of human settlement in coastal zones. Journal of Coastal Research 19: 584–599

Somboon JRP (1988) Paleontological study of the recent marine sediments in the lower central plain, Thailand. Journal of Southeast Asian Earth Sciences 2: 201–210

Stanley DJ, Chen Z (1996) Neolithic settlement distributions as a function of sea level-controlled topography in the Yangtze Delta, China. Geology 24: 1083–1086

Stanley DJ, Hait AK (2000) Holocene depositional patterns, neotectonics and Sundarban mangroves in the western Ganges-Brahmaputra Delta. Journal of Coastal Research 16: 26–39

Stanley DJ, Warne AG (1993) Nile Delta: recent geological evolution and human impact. Science 260: 628–634

Stanley DJ, Warne AG (1994) Worldwide initiation of Holocene marine deltas by deceleration of sea-level rise. Science 265: 228–231

Suter JR (1994) Deltaic coasts. In: Carter RWG, Woodroffe CD (eds) Coastal evolution: Late Quaternary shoreline morphodynamics. Cambridge University Press, Cambridge, pp 87–120

Syvitski JPM, Vörösmarty CJ, Kettner AJ, Green P (2005) Impact of humans on the flux of terrestrial sediment to the global coastal ocean. Science 308, 376–380

Ta TKO, Nguyen VL, Tateishi M, Kobayashi I, Saito Y, Nakamura T (2002a) Sediment facies and late Holocene progradation of the Mekong River Delta in Bentre Province, southern Vietnam: an example of evolution from a tide-dominated to a tide- and wave-dominated delta. Sedimentary Geology 152: 313–325

Ta TKO, Nguyen VL, Tateishi M, Kobayashi I, Tanabe S, Saito, Y (2002b) Holocene delta evolution and sediment discharge of the Mekong River, southern Vietnam. Quaternary Science Reviews 21: 1807–1819

Tanabe S, Saito Y, Sato Y, Suzuki Y, Sinsakul S, Tiyapairach S, Chaimanee N (2003a) Stratigraphy and Holocene evolution of the mud-dominated Chao Phraya delta, Thailand. Quaternary Science Reviews 22: 789–807

Tanabe S, Ta TKO, Nguyen VL, Tateishi M, Kobayashi I, Saito, Y (2003b) Delta evolution model inferred from the Holocene Mekong Delta, southern Vietnam. In: Sidi FH, Nummedal D, Imbert P, Darman H, Posamentier HW (eds) Tropical deltas of southeast Asia: sedimentology, stratigraphy, and petroleum geology. SEPM Special Publication, Tulsa, OK, pp 175–188

Tanabe S, Hori K, Saito Y, Haruyama S, Vu VP, Kitamura A (2003c) Song Hong (Red River) Delta evolution related to millennial-scale Holocene sea-level changes. Quaternary Science Reviews 22: 2345–2361

Thanh TD, Saito Y, Huy DV, Nguyen VL, Ta TK, Tateishi M, (2004) Regimes of human and climate impacts on coastal changes in Vietnam. Regional Environmental Change 4: 49–62

Tjia HD (1996) Sea-level changes in the tectonically stable Malay-Thai peninsula. Quaternary International 31: 95–101

Tobler W, Deichmann U, Gottsegen J, Maloy K (1997) World population in a grid of spherical quadrilaterals. International Journal of Population Geography 3: 203–225

Umitsu M (1985) Natural levées and landform evolutions in the Bengal lowlands. Geographical Review of Japan 58 (Ser. B): 149–164

Umitsu M (1993) Late Quaternary sedimentary environments and landforms in the Ganges Delta. Sedimentary Geology 83: 177–186

Wassmann R, Hien NX, Hoanh CT, Tuong TP (2004) Sea level rise affecting the Vietnamese Mekong Delta: water elevation in the flood season and implications for rice production. Climatic Change 66: 89–107

Wells JT, Coleman JM (1984) Deltaic morphology and sedimentology with special reference to the Indus River Delta. In: Haq BU, Milliman JD (eds) Marine Geology and Oceanography of Arabian Sea and coastal Pakistan. van Nostrand Reinhold, New York, pp 85–100

White NJ, Church JA, Gregory JM (2005) Coastal and global averaged sea level rise for 1950 to 2000. Geophysical Research Letters 32, L01601, doi 10.1029/2004GL021391

Wolanski E, Nhan NH, Spagnol S (1998) Sediment dynamics during low flow conditions in the Mekong River Estuary, Vietnam. Journal of Coastal Research 14: 472–482

Woodroffe CD (1993) Late Quaternary evolution of coastal and lowland riverine plains of Southeast Asia and northern Australia: an overview. Sedimentary Geology 83: 163–173

Woodroffe SA, Horton BP (2005) Holocene sea-level changes in the Indo-Pacific. Journal of Asian Earth Sciences 25, 29–43

Wright LD (1985) River Deltas. In: Davis RA (ed.) Coastal Sedimentary Environments. Springer-Verlag, New York, pp 1–76

Wright LD, Coleman JM (1973) Variations in morphology of major river deltas as functions of ocean wave and river discharge regimes. Bulletin of the American Association of Petroleum Geologists 57: 370–398

Wright LD, Coleman JM, Erickson MW (1974) Analysis of major river systems and their deltas: morphologic and process comparisons. Technical Report No. 156. Louisiana State University, Louisiana, pp 1–114.

Wright LD, Yang Z-S, Bornhold BD, Keller GH, Prior DB, Wiseman WJ (1986) Hyperpycnal plumes and plume fronts over the Huanghe (Yellow River) Delta front. Geo-Marine Letters 9: 97–106

Yokoyama Y, De Deckker P, Lambeck K, Johnston P, Fifield LK (2001) Sea-level at the Last Glacial Maximum: evidence from northwestern Australia to constrain ice volumes for oxygen isotope stage 2. Palaeogeography, Palaeoclimatology, Palaeoecology 165, 281–297

Zeidler RB (1997) Continental shorelines: climate change and integrated coastal management. Ocean and Coastal Management 37: 41–62

CHAPTER 11

NEW DIRECTIONS FOR GLOBAL CHANGE RESEARCH RELATED TO INTEGRATED COASTAL MANAGEMENT IN THE ASIA-PACIFIC REGION

NICK HARVEY[1] AND NOBUO MIMURA[2]
[1]*Geographical and Environmental Studies, Adelaide University, Australia*
[2]*Center for Water Environment Studies, Ibaraki University, Japan*

11.1 INTRODUCTION

Integrated coastal management (ICM) has some examples of success in the Asia-Pacific region but, as noted in Chapter Three, global change has not yet emerged as a major coastal management issue for the region. There have been attempts to deal with some biophysical outcomes of global change such as research into new methods of assessing regional coastal vulnerability to sea-level rise for the Eastern Hemisphere (McLean and Mimura 1993, Mimura and Yokoki 2001). However, there is not much evidence that such approaches have so far been incorporated into coastal management practices except where targeted donor agency programs are used for specific individual countries.

There is less evidence of research into the regional socio-economic impacts of the globalisation of economic development and resource use. Scientific debate on the global value of coastal goods and services (Costanza et al. 1997, Wilson et al. 2003) has triggered research in a number of mostly developed countries but with some exceptions (e.g. Adger et al. 2001) there has been little research on the relevance of such valuations for the Asia-Pacific region. For example, the extensive shrimp culture in the region, developed in part at the expense of mangrove forests, has been fuelled by a demand largely from outside the region where there are higher labour costs and greater environmental controls on coastal resource use. On a broader scale the high proportion of the global fish catch from the Asia-Pacific region is disproportionate to the quality of fish consumed within, rather than exported from, the region. Another issue, noted in Chapter Eight, is the widespread poverty in the rural sector of the Asia-Pacific region, notwithstanding the intensive use of coastal resources in that sector.

The intensity of coastal resource use by people in the Asia-Pacific region and the rapidly growing coastal population reinforce the need for accurate predictions of future climate changes as a basis for forecasting impacts and developing appropriate response strategies. For the coastal zone the most important climate change variables have been listed in Table 11.1 and include: increases in air temperatures and sea-surface temperatures; sea-level rise; and changes in the magnitude, frequency, and distribution of interdecadal (e.g. Pacific Decadal Oscillation – PDO) interannual (e.g. El Nino-Southern Oscillation – ENSO) and seasonal extreme events (e.g. typhoons/cyclones, droughts). These climate change variables are additional to other global change issues relating to economic and trade impacts on the coastal resources of the region. Another important issue for the whole region is to identify the sectors and areas most vulnerable to these climatic and oceanic conditions using a multisectoral and integrated approach to managing and living on the coast.

Table 11.1. Global climate change variables and regional impacts

Global climate change variables	*Regional coastal impacts*
1) Increase in air temperature	1) Ecosystem stress
2) Increase in sea-surface temperature	2) Coral bleaching
3) Sea-level rise	3) Coastal inundation, coastal erosion, salinization
4) Changes in magnitude and frequency of PDO and ENSO, typhoons and droughts	4) Increased river flooding, increased storm erosion, changes in fluvial sediment delivery

Future global warming and accelerated sea-level rise in the Asia-Pacific coastal zone will restrict the activities of the people who occupy it and the ability of the environment to sustain them. Some of the most important of these impacts (see Table 11.1) in the Asia-Pacific region will arise from coastal inundation, exacerbation of river flooding, beach erosion, groundwater salinization, and coral bleaching resulting from increasing sea surface temperatures. These effects will trigger a series of other impacts on managed and non-managed environments, human society and the economy.

Each of the chapters in this book has focused on key coastal issues where further research is needed on the potential impacts of global change. This final chapter attempts to draw on the conclusions and recommendations from these chapters and together with material presented in the APN coastal Synthesis Report (Harvey et al. 2005) to identify possible future research directions. The intention is that the findings from the proposed global change research will be of relevance for coastal policy makers in order to improve ICM practices in the region. It is recognised that this objective for new research directions presents a number of challenges if it is to be successful such as:

- provision of appropriate research funding;
- time to produce meaningful research outcomes;

- presentation of results in a format of direct practical relevance to policy makers; and
- political will to incorporate global change issues within ICM frameworks.

This chapter first comments on some of the findings regarding ICM in the region, in particular methods for assessing best practice in ICM and governance and funding issues which are hampering its progress. The chapter then uses recommendations from each of chapters Four to Ten to develop a list of priority global change research directions.

11.2   LESSONS FROM ICM PRACTICE IN THE REGION

It is clear that the Asia-Pacific region has a diversity of coastal management practices but there has been a move to adopting integrated coastal management (ICM) in a number of countries. Even where ICM programs have been developed in selected countries using donor funding such as the PEMSEA examples, they have met with limited success (Chua 1998). Case studies from Indonesia and the Philippines have focused on the sustainable nature of ICM programs and have concluded that the long-term (5–7 years) cyclical development of successful ICM programs (see Olsen 2003) is not possible where funding dries up before there are 'in-country' self sufficient mechanisms for continued support of ICM programs (White et al. 2005, Christie et al. 2005). The definition of ICM used in this book avoids the term coastal 'zone' for reasons outlined in Chapter One and places the emphasis on managing 'human' use of resources.

Integrated Coastal Management is a continuous and dynamic process incorporating feedback loops which aims to manage human use of coastal resources in a sustainable manner by adopting a holistic and integrative approach between terrestrial and marine environments; levels and sectors of government; government and community; science and management; and sectors of the economy (Harvey 2004 p 568).

Sustainability of coastal resource use is the central objective of ICM and requires patterns and intensities of resource use today that do not compromise the ability of future generations to meet their needs. As explained in Chapter Three, there is a difference in emphasis between the sustainable management of human coastal resource use and what has been termed 'sustainable coastal development' and the 'improvement of the quality of life'. Although sustainable coastal development and improved quality of life for coastal dwellers are valuable outcomes arising from proper management of coastal resource use, they tend to reflect short-term anthropocentric goals rather than long-term environmental goals.

The recent evolution of coastal management in the Asia-Pacific Region has, in large part, been driven by external forces. The prevailing paradigms of management, which evolved in the west over the last two decades, have been promoted as a solution to issues of resource depletion in the region. These paradigms have subsequently been promoted through international agreements, foreign aid, NGO activity, and research. External forces have also contributed to resource depletion. Globalisation of trade, combined with the rapid growth of regional economies, has fuelled changes in patterns of resource access, use, and consumption. These changes are exerting chronic stress on coastal resources, particularly coral reefs, mangroves, and fisheries. Demand for coastal space is increasing, leading to the privatisation of coastal lands and the decline of common property resource management by local communities.

Approaches to coastal management in the Asia-Pacific Region are as diverse as the states are numerous. In most states, at a local level, there still exists a range of traditional customs and taboos that govern access to coastal resources. These practices underlie and in many cases co-exist with recent forms of coastal management that are statute and regulation based, derived from western models of centralized environmental management. These traditional approaches have shown remarkable resilience over time. They are flexible and responsive to local circumstances. The goal of these approaches is to sustain resources, such as fisheries, by modifying rates and patterns of harvest depending on local resource availability. This approach is entirely consistent with modern concepts of sustainability, but seldom recognized and protected. In contrast, many states in the region have adopted western models of coastal management. These have tended to be formalized and prescriptive, highly centralized, and hierarchical.

It appears that no countries in the region have a single agency responsible for ICM and only three have interagency or inter-ministerial coordination so that it is not surprising, that the implementation of ICM has been slow. However, there has been a lot of effort to develop special area programs or management plans, usually with external assistance such as the ASEAN-US CRM program and subsequent PEMSEA program. Recognition of traditional customs and approaches to coastal resource management is not always provided for by governments but it is clear that local communities are often actively involved in management. Colonial systems of regulation have in some cases subsumed traditional approaches, to the detriment of local communities whose importance in coastal management is recognised throughout the literature. Some nations have failed to provide for significant community involvement or are unable to do so because of recent conflict or extreme poverty. In other cases low or negligible levels of community involvement are the product of the prevailing political system, where political influence is highly centralized and democratic institutions are weak or incompletely developed.

Experience has demonstrated that there are significant barriers to implementing environmental policy towards ICM, even in the developed nations of the region. At the national level, there are major disincentives to achieving sustainable patterns of resource use, including the lack of incentives for business and government, the time horizons of elected officials, and the centralisation of management. ICM can be criticised for legitimising the reorganisation of society and space, for the primary purpose of stimulating national and international economic development. Pre-existing resource management regimes are displaced by ICM regimes, which facilitate the further subdivision of community resource management systems by encouraging national and global capital penetration. ICM regulation creates space for the new global economy, by organising the coast into new arenas for investment and by politically and spatially marginalising pre-existing resource users. This process has been reinforced by some major donors and lending agencies.

ICM is clearly not a panacea to problems of resource depletion in the region. Without safeguards that protect the relationship between local communities and associated coasts ICM may even be counter-productive. Community management responsibility and authority should be retained as the basis for effective ICM. The direct transfer and application of methods of management developed in the west, for example, marine protected areas, is unlikely to work without significant community support. Management should occur at an appropriate scale, recognizing local, regional, and national communities of interest. National and local policies and programs should be complementary and benefit both constituencies. In many states this will require a major reconsideration of the importance and role of local communities.

11.3 LESSONS FROM PAST GLOBAL CHANGES

As noted in Chapter Six, the region's coasts are affected by natural cycles of global change, the most recent of which followed the peak of the last glacial cycle at around 18,000 years ago. Since then the global rise of sea level has affected the coasts of the Asia-Pacific region but not in a uniform manner because the sea-level signature is superimposed on other geotectonic and geomorphological processes. These processes produce significant regional coastal differences as illustrated in Chapters Four, Five, and Ten. Coasts along convergent plate boundaries in Southeast Asia, Papua New Guinea and island arcs of the southwest Pacific have evolved under dynamic tectonic activity and many of these are still experiencing active tectonic processes although, as noted in Chapter Five, the rates of these processes tend to be an order of magnitude less than the projected rate of sea-level rise for the next century. Central Pacific islands away from the plate boundaries mostly experience subsidence making them vulnerable to future sea-level rise but in some places islands have been uplifted.

Elsewhere the tectonic origin of major coastal features is evident such as the Gulf of Thailand which originated from plate collision starting in the early Tertiary period followed by the development of the broad low-lying basin at the head of the Gulf. Development of the large river systems of East Asia relates, in part, to long-term geological processes and has resulted in focal points for major fluvial discharge. Chapter Four discusses the important linkages between the catchment and coast using Syvitski's (2005) sediment load calculations for the region to show that past (prior to human influence) amounts of sediment being brought to the coast were significantly higher than today. This is directly related to Chapter Ten which examines the coastal deposition of this sediment from some of the larger rivers into what has been termed megadeltas. The delta plains have mostly been deposited during the mid and late Holocene extending from the former shoreline at the maximum sea-level transgression. These have developed at the mouths of the Indus, Ganges-Brahmaputra-Meghna, Irrawaddy, Chao Phraya,

Mekong, Red, Pearl, Changjiang, and Huanghe rivers. Chapter Ten provides insights into the directions and rates of past processes which can help better manage modern deltaic processes.

The sea-level transgression began some time after 18,000 years ago and reached its maximum extent in the mid-Holocene, resulting in the drowning of the Asia-Pacific coast causing major coastal changes such as on the Sunda Shelf where the rising sea cut land bridges to isolate the current islands. Elsewhere, the impact of this sea-level transgression and warmer temperatures allowed the establishment and expansion of coral reefs in the region. Subsequently, these reefs have been affected by a stabilisation of sea level and in many places a relative sea-level fall during the latter part of the Holocene.

It is important to recognize the differences in coastal evolution and understand past processes in order to better inform current coastal management practices. In particular, it is necessary to be aware of the different rates associated with different processes and use these as a context for rates of predicted future changes. For example, it has already been mentioned that many of the tectonic rates are an order of magnitude or less compared with the projected rate of sea-level rise. However, it should be noted that past rates of sea-level rise associated with the postglacial sea-level transgression were higher than the projected rates for future sea-level rise. Similarly there is a difference between the rates for geological and geomorphological processes compared with rates of human impact on the coast.

11.4 HUMAN IMPACT ON THE COASTS OF THE ASIA-PACIFIC REGION

The importance of understanding past and present coastal changes has to be placed in the context of modern coastal processes and rates of change along with human impact on the coasts of the Asia-Pacific region. Population increase, urbanisation, pressure on rural coasts, catchment modification, pollutant outputs, and degradation of coastal ecosystems are all issues which will exacerbate any future global change impacts on the coast. Most of these issues have been dealt with in detail in previous chapters so that here they are discussed briefly to highlight some of the major issues.

11.4.1 Urbanization and Population Pressure on the Coast

The problems of future global change faced by cities in the Asia-Pacific coastal zone will be exacerbated with increasing population and urbanisation. On the one hand cities are major contributors to greenhouse gas emissions while on the other hand coastal cities become more vulnerable to the consequences of global warming, accelerated sea-level rise, and changes in the patterns and dimensions of extreme events.

About 55% of the world's coastal cities with populations between 1 and 10 million people are in Asia and about 83% of the megacities are also in Asia with the fast-growing ones in the tropics. By 2015, half of the world's coastal megacities will be in Asia.

Cities attract people as they offer economic, social and cultural opportunities, and are strongly associated with an improved quality of life. However, in many developing countries of the Asia-Pacific region, urbanization is still saddled with problems of insufficient infrastructure, water shortage, poor sanitation, air pollution, and traffic congestion. Much of the migrant poor can only find shelter in squatter colonies that grow around the cities.

The development of coastal cities in the Asia-Pacific region has negative impacts on the marine and coastal environments. These cities contribute domestic and industrial wastewater and solid waste, hazardous waste, and waste gases. The more significant impacts on the coastal environment include:

- degradation of seawater and sediment quality;
- toxicity to fish, shrimp, and algae;
- eutrophication, which results in 'red tides'; and
- wetland destruction, habitat modification, and biodiversity loss.

With increasing population and urbanisation in the coastal zone, it would be necessary to identify the potential urban 'hotspots' that require attention to various types of problems:

- 'grey' agenda issues, e.g., water supply, sanitation, and infectious diseases;
- combination of 'grey' agenda and 'brown' agenda issues, e.g. air and water pollution, and other negative aspects of industrial processes; and
- 'green' agenda issues, e.g., consumption-related problems, ecosystem health, ozone depletion, and greenhouse gas (GHG) emissions.

The identification and prioritisation of issues are important initial steps to ease the problems as many will worsen with global warming and accelerated sea-level rise. 'Grey' and 'brown' agenda issues are faced by rapidly growing cities, such as Bangkok, Jakarta, and Shanghai, and other cities including Hanoi, Ho Chi Min City, Vientiane, Yangon, and Phnom Penh. A few large cities, such as Tokyo, Seoul, and Taipei, are more concerned with 'green' agenda issues.

The problems of increasing population and urbanisation pressure cannot be treated in isolation from the rest of the coastal environment. Many are cross-sectoral and some are trans-boundary in nature. Policy-makers need to know the seriousness of the problems, and their relationships to global warming and accelerated sea-level rise in the coastal context, to mobilise resources to achieve maximum effect. Some, if not many, of these problems have to be included within the context of appropriate coastal management. In addition, poor monitoring protocols and facilities, lack of adequate policy and regulatory measures, and institutional weakness contribute to the complexity of the problems.

Human impact on the region's rural coasts is high although the drivers for the high resource use do not necessarily come from the rural communities themselves where poverty is a major issue. It is clear that over-fishing occurs in the major fishing grounds as shown by figures presented for biomass reduction in Chapter Eight. Another rural impact comes from the very heavy use of fertiliser by world standards in farming practices in the region. This has resulted in extensive areas of coastal eutrophication. Perhaps the most visible impact on coastal ecosystems in rural areas has been the dramatic loss of mangrove forests in the region which have been converted to shrimp culture and to a smaller extent for habitation and infrastructure development. Many of the human impacts on the rural coastal areas of the region, also directly affect the residents of the region because they have limited access to high quality produce, the wages are low, there is restricted access to quality water, and the degradation of rural coastal ecosystems such as mangroves and coral reefs makes the rural coastal communities even more vulnerable as shown by the 2004 Indian Ocean tsunami.

11.4.2 Impact on Rivers, Catchments, and Deltas

In the Asian-Pacific region, the interactions between the catchment and the adjacent coastal areas are characterized by rapid changes in freshwater, sediment, nutrient and pollutant fluxes from land, and by extremely high pressure exerted on the catchment and estuarine-coastal environments by intense human activities. As noted in Chapter Four, modeling of sediment flow from the major rivers of the region has shown the dramatic effect of human impact (Syvitski 2005). This is highlighted by the situation on the Yellow River where extensive dam construction has reduced water flow and prevented sediment from reaching the sea.

Woodroffe et al. (2006) note in Chapter Ten that the Asian megadeltas have a history of human exploitation, and support large populations with strong urbanisation trends. The megadeltas have been affected by water control, from flood mitigation and groundwater extraction to irrigation and channelisation, exacerbating compaction, subsidence, and acid sulphate soil development. They note that upstream dams, water abstractions and diversions have impacted on sediment supply to the deltas causing increased coastal erosion.

Woodroffe et al. (2006) stress the need for a better understanding of natural processes and global change impacts in order to design long-term management strategies for the megadeltas. For example, the GBM delta system contains well over 100 million people and includes the most rapidly expanding megacity (Dhaka). There are also millions of people in the other deltas which are experiencing rapid urbanisation with either current or future megacities. Most of coastal Bangladesh is already vulnerable with projected sea-level rise likely to put further pressure on an already heavily impacted coast (Chapter Ten). Similarly with other megadeltas the natural deltaic processes are linked with human use of resources. Human impact has affected the supply of sediment, water, and nutrients, loss of habitats, and spread of pollutants. Many areas are sensitive to flooding and its management so that human impact factors often over-ride climatic factors. Damming of the upstream rivers poses a major threat, with reduced sediment and nutrient supply, as well as altered water conditions. This is also usually associated with increased saline intrusion.

Rivers in Southeast Asia deliver a large percentage of the global freshwater and sediment discharges into the coastal zone so that management of their catchments is important in controlling the environmental characteristics on the adjacent coasts. Modifications to the coastal environments, in turn, affects the catchment areas. For example, the Three Gorge Dam in China will considerably change the patterns of water and sediment transport, with different monthly discharge distributions. As a result, the growth rate of the Changjiang River delta will be reduced or even subject to coastal erosion. Furthermore, salt intrusion into the lower reaches of the river will be intensified, the catchment geomorphology (river channel geometry) will be in a non-equilibrium state, and the ecosystem of the catchment-coast system will be affected.

11.4.3 Pollutant Impacts in the Asia-Pacific Region

Assessment of nutrient and pollutant discharges from the region's rivers is complicated. The catchment areas of the region are generally densely populated, with an extremely large scale of rice-based agriculture. Artificial fertilizers and pesticides are used extensively, resulting in heavy nutrient (P and N) loads and Persistent Organic Pollutants (POPs). In addition, the majority of the region is under pressure from development. Chemical industries and other pollutant-generating factories have been established in many developing countries. An unknown but large amount of pollutants is, therefore, discharged into the rivers and coastal waters every day.

There are two major threats from land-based sources of pollution to the marine environment; 1) pollution from land-based activities; and 2) physical, ecological, and hydrological modification of critical habitats. Pollution from land-based activities threatens water quality, critical habitats, and sustainable use of resources. Sewage, sedimentation, and agricultural activities are identified as primary threats to critical species and habitats as well as non-living resources.

The causes of these threats includes a number of factors such as: policy, regulation and enforcement; technical capacity; data gaps; inadequate infrastructure; economic valuation of the resources; land tenure; resource pricing; loss of traditional management systems; development pressures; lack of community awareness and education; lack of community involvement; lack of planning; and other management issues.

11.4.4 Impacts on Coastal Ecosystems

Coral reefs, mangrove forests, and sea grass meadows are the three most productive coastal ecosystems in the Asia-Pacific region. Millions of people depend upon them for their livelihood. They are important because they supply a wide range of goods and services including:
- food products, building materials and fuel wood;
- natural protection to shorelines;
- nurseries for other species; and
- resources for tourism and recreation.

Demands for such goods and services is growing, but the capacity of many of these ecosystems to meet this demand has been substantially reduced. While there are pristine coral reefs, mangrove forests, and seagrass meadows in the region, many are in serious decline and others have been completely destroyed. All three ecosystems are intertidal or immediately sub-tidal, that is they are intimately connected to the land, and are therefore subject to land-sourced sediment and nutrients and land-based pollutants. Other human associated activities that degrade these ecosystems include mining of reefs, deforestation of mangroves for firewood and timber, and clearance of sea grass beds. Superimposed on these local factors are those associated with global warming and sea-level rise, as well as those associated with the globalisation of economic activity such as tourism and trade, starkly exemplified by the conversion of mangrove forest to shrimp ponds.

The Asia-Pacific region has the highest concentration of coral reefs, mangrove forests, and sea grass meadows in the world, and also the greatest number and diversity of reef, mangrove and seagrass species and communities. On the other hand, the human pressures on, and the scale of degradation and destruction of these ecosystems is also greater than elsewhere.

Coral reefs, mangrove forests, and seagrass meadows contain a great biodiversity and are subject to continual natural change. However, their sustainability is threatened where the ecosystems have been substantially degraded or destroyed, and much of their biodiversity lost. In such cases the replacing habitat may provide some alternative products and/or services for the local community, though frequently the range of products is substantially reduced and/or production from the area is not for local consumption but is exported out of the region.

The conversion of a multi-use community resource, like a mangrove forest, to a single-use resource such as aquaculture, does have impacts on local food security and human health, as well as access to a previously 'common' resource. Research in the region clearly shows that degradation of reefs, mangroves, and sea grass beds tends to harm rural populations more directly than urban populations, and has its most direct and severe impact on poorer people.

While it may be possible to reduce the depletion and degradation of reefs and mangroves through rehabilitation and reseeding projects, the scale of such reduction is likely to be minor in relation to destructive effects. Moreover, the loss of resilience and buffer capacity of these coastal ecosystems impairs the capacity of reefs, mangroves, and sea grass meadows to recover, and as a consequence the quality and quantity of their provision of goods and services is reduced.

11.5   KNOWLEDGE GAPS AND RESEARCH DIRECTIONS

11.5.1   Urbanisation, Population Pressures, and Socio-economic Trends: Urban and Rural Coasts

A key issue for research in the urbanised coasts of the region is developing methodologies for assessing the sustainability of the rapidly growing urban areas and specifically the megacities. A major research question relates to the sustainability of megacities in the Asia-Pacific and actions needed to assure their sustainability. This requires a focus on strategic planning and management techniques which are able to integrate the complexity of demographic and socio-economic trends in the region along with other issues such as livelihood, environmental quality and conservation of heritage buildings.

Such research into development planning should account for the costs of preventing the destruction of natural resources and pollution and the costs of protecting the underprivileged in the urban areas as noted in Chapter Seven (Wong et al. 2006). There is a need for research into multi-layered governance and integrated policy responses at all scales of governance. It is also necessary to integrate environmental issues into urban research, to include catchments, coast and marine environments. This research needs to recognize various scales and rates of urban processes such as the physical environment within cities, the environmental systems of cities and their regional hinterlands, and the environmental implications of the global network of megacities (see Chapter 7). As noted by Wong et al. new techniques are available for this type of research. There is also a need for integration of research inputs from various groups of scientists to contribute to regional programmes of urban research.

Further research is needed on the rural coasts of the region to investigate human-environment dynamics at the appropriate spatial and temporal scales. This is particularly important because of the complexity of local, national and global drivers and the interaction with the adjacent catchments upstream and the marine environment further offshore.

There is also a need for research to understand what scale of economies can improve the quality of life of rural coastal dwellers while sustaining the ecological functioning of ecosystems. There is also a need to conduct research on coastal change, combining the predicted impacts of globalized trade and climate change. LOICZ (www.loicz.org) provides an excellent template for pursuing integrated coastal research through its science plan and implementation strategy which may be adapted at multiple scales by research and policy communities with local to international research interests.

11.5.2   Megadeltas

There is a need for research into the vulnerability of megadeltas to inundation due to rising sea level. This research requires investigation into both physical protective measures and socio-economic scenarios from different development pathways, such as forestry, fish production, prawn farming, or salt production, which will increase the potential adaptive capacity of communities. Detailed modeling of multiple land use across megadeltas is needed to provide insights into impacts of a higher sea level. Such models should address additional impacts, such as the increase of dry season tide levels, saline water intrusion, increased storm surges, and geomorphological changes in channel morphology. They should also include potential impacts of extreme events and threshold levels of physical protective measures such as levees. The failure of such structures in the wake of Hurricane Katrina in the United States in 2005 illustrates the need for proper planning and the danger of increased risk associated with potential sea-level rise. Research into long-term planning of delta utilisation may result in more sustainable development that will minimise future vulnerability.

It is necessary to conduct research into the behaviour of individual deltas at decade-to-century scales and the response to thresholds at the planning and management timescales, such as distributary switching and engineering works. As noted by Woodroffe et al. (2006) the complex delta systems need research which will consider all the factors that shape future delta evolution (Chapter 10). This requires more data collection on the megadeltas, and multidisciplinary studies. Woodroffe et al. (2006) suggest that vulnerability analyses need to focus on detailed local factors such as topography and integration with flood levels, land use, and other relevant factors. The combination of natural and socio-economic datasets, such as DEM and population datasets provide powerful research tools to examine risk, and new technologies, such as digital terrain modeling based on shuttle imagery (SRTM) coupled with remote sensing, can offer synoptic overviews of poorly accessible areas. Such assessments are challenging from a number of perspectives, but are essential if scientific understanding is to be translated into useful guidance for coastal management.

11.5.3 Catchment-Coast Interactions

In order to enhance our ability to predict future changes in the behaviour of the catchment-coast system, more research is needed into processes and mechanisms of how changing climatic conditions will affect river discharge and sediment transport. Development of models to predict the changes in these processes is a major challenge for future research, which will create a useful base for integrated management of catchment-coast systems. Such models would need to include the catchment hydrological cycle, sediment yield and transport to coastal waters, river channel morphological evolution, and nutrients/pollutants emissions, and water quality changes in the river catchments. They would also need to include estuarine and adjacent coast sediment accretion and erosion patterns; saline water intrusion patterns in the estuary; influences of sediment, nutrient and pollutant inputs on estuarine and adjoining continental shelf ecosystems; and influences of any engineering schemes on the natural system.

More research is also needed on process modeling approaches for material discharges; river channel and delta evolution; geochemical tracing methods to define material retention; modeling sedimentary records to identify the evolutionary history of the catchment-coast system; simulation modeling of catchment-coast system behaviour; generation of scenarios for catchment-coast changes and artificial intelligence tools for integrated catchment-coast management. Output from this research can then be applied to development planning and management of the catchment-coast system. Research should also be conducted on the best management systems for protection of coastal wetland and coral reef ecosystems; estuarine and coastal environmental changes; potential for coastal land reclamation; use of aggregate resources in the river channel and coastal areas; flood defence; water quality improvement and ecosystem health; sustainable utilization of biological resources; and regional management of future catchment-coast development.

11.5.4 Coastal Pollution

Although the effect of the nutrient and pollutant discharges may be evaluated by a scenario-based modeling approach, there is a need to obtain quantitative information on the nutrient and pollutant inputs.

Important issues in association with the water, sediment, nutrient and pollutant discharges from the Asia-Pacific region include:
- Patterns of catchment hydrological cycling and freshwater discharges, under the influences of climate change, water use, and dam construction;
- Sediment yield and input to the coast in response to dam construction;
- Modifications to the estuarine-deltaic areas due to catchment development, in terms of morphodynamics and seawater intrusion;
- River catchment morphological evolution in response to changes in the hydrological cycling and river mouth morphology;
- Quantity and spatial distribution of nutrient and pollutant emissions; and
- Catchment and coast water quality, in relation to nutrient and pollutant emissions.

The timing of material input and budgeting to the coast will be an important component of this research. The application of such research outcomes will be of most relevance where there is a recognition of the need for research into appropriate integrated catchment management. This is important because of the limited resources of the individual countries and is particularly relevant for trans-boundary river catchments involving a number of countries such as the Mekong or the Ganges-Brahmaputra-Meghna.

11.5.5 Coastal Ecosystems

Key global change impacts include increased sea surface temperature and sea-level rise as well as a possible increase in the frequency and intensity of both storms and climate-regime oscillations such as ENSO. While some of the impacts may be difficult to determine, and while some benefits may accrue from plant growth and biomass increases from elevated carbon dioxide in the atmosphere, there is little doubt that the majority of global change impacts on coastal ecosystems will be negative. In some situations ecosystem adjustment and adaptation to global change forces is geographically constrained, landward migration of mangroves being impeded by the presence of coastal infrastructure is one example.

Throughout the region subsistence agricultural production systems and artisanal fisheries are stressed as a consequence of high coastal population growth, pressure from commercial activities as well as extreme natural events such as floods and droughts. In addition to these stresses coastal agriculture and food security can be impacted by sea-level rise in at least two ways:
- coastal land may be permanently inundated, making it unsuitable for agricultural production; and
- land may be subject to periodic inundation from extreme events such as king tides, or intrusion of salt water into freshwater lenses contaminating the freshwater lens.

Theoretical adaptations to these stresses include the conversion of coastal lands to mariculture, and the growth of salt tolerant species. But such adaptations are not likely to be undertaken rapidly, nor without the provision of agricultural and fisheries extension services.

Many of the medium-term impacts of climate change and sea-level rise on coastal ecosystems and resources are already familiar to coastal dwellers within the Asia-Pacific region, and some have experience in coping with them. However, the additional global change stress associated with external economic pressures on those ecosystems and resources does bring into sharp relief the question of adaptation, food security, and sustainability.

11.5.6 Coastal Geomorphic Change

Among the continental countries of the Asian continent within the region of study, coastal evolution is probably least well known in Burma and India, with still much more research needed in Cambodia, Korea, Thailand, and Vietnam before the understanding can reach the level it has achieved elsewhere in the region. There is a general need to improve understanding of Quaternary coastal evolution in the region in order to provide a background of coastal processes and rates of change in the past as a context to the predicted impacts of climate change on the coast, in particular the projected sea-level rise.

Along the island coasts of the Asia-Pacific region, more research is needed on the large islands of Southeast Asia, such as Borneo, Java, Sulawesi, and Sumatra. While research has focused on the more geologically conspicuous parts of these islands, comparatively little systematic study of coastal evolution has been made, a comment that also applies to most of the smaller islands in Southeast Asia and to the Philippines. There has been a marked imbalance in research targeting the pre-Holocene evolution of the coasts of active island arcs in the western Pacific. For example, hardly any research has been undertaken in Solomon Islands where, along with many other Pacific Island groups in such locations (such as Tonga and Vanuatu), much of the key work was reported more than 30 years ago.

Another important area for research is the need for monitoring and documenting recent coastal changes. In particular, there is potential to link coastal changes over the last few thousand years with climate change data from the geological record. The importance of current coastal monitoring is to document geomorphic changes in order to assess whether they are linked to any identifiable climate change events.

11.6 FUTURE DIRECTIONS FOR GLOBAL CHANGE COASTAL RESEARCH IN ASIA-PACIFIC REGION

The various chapters in this book have attempted to present the current scientific understanding of some of the key states, drivers, and responses of the coastal environment in the Asia-Pacific region. The region contains a wide diversity of coastal types, a range of past and current coastal processes, and is being subject to large scale and often rapid changes. These are brought about by global, regional and even local changes by both natural and socioeconomic drivers. Climate change and accelerated sea-level rise will create impacts on the region's coasts, which are superimposed on existing human impacts created by population growth and urbanization, particularly the increase in megacities. In the face of these complex issues, integrated coastal management (ICM) has been promoted as a methodology primarily to protect coastal resources, but also to achieve sustainable coastal development and improve the quality of life for coastal inhabitants. The successful development of ICM needs a firm scientific basis for all components of states, drivers, and responses. Such information should be provided in an integrated manner, and to this end, we need a closer relationship between scientists and other stakeholders such as policy-makers, politicians, local dwellers, and fishermen.

The role of scientific research is increasing in many ways and its importance for management of the Asia-Pacific coasts has been recognised in the APN Coastal Synthesis and other fora. In this final section of the book, we attempt to present recommendations for the future directions of coastal research which is particularly relevant for addressing global change in the region. The recommendations for research are discussed under three headings; 1) Understanding coastal processes and change in the Asia-Pacific region, 2) Coastal research related to global change, and 3) Research to develop relevant ICM including adaptation to global change.

11.6.1 Understanding Coastal Processes and Change in the Asia-Pacific Region

Understanding the underlying processes and mechanisms for coastal change is of fundamental importance for both scientific research and management practices, including the geomorphological and geological context, interaction of river catchment and coasts, material flows through rivers and coastal zones including freshwater and sediments, coastal ecosystems, and impacts of human activities. However, we lack sufficient knowledge on many of these aspects for a number of countries within the region, particularly the developing countries. Without such basic knowledge and data, we cannot estimate how seriously the region's coasts will be affected by the global change, and where and when these changes will occur. It appears that many countries and people in the region will be ill-equipped to address coastal change issues related to future climate change without the background of such scientific research. We emphasize the importance of promoting fundamental coastal research immediately, including observation and monitoring, in order to provide a scientific basis for management purposes. Research is needed in at least three key areas:

1. *coastal ecosystems*
   - to understand the current state and distribution of those coastal ecosystems that (help) sustain coastal populations in the Asia-Pacific region;
   - to understand their (potential) role in sustaining future coastal peoples;

- to identify key areas where rapid change is occurring or is threatened; and
- to develop management strategies (such as ecosystem restoration) for the future productivity of these ecosystems.

2. *coastal pollution*
   - to assess the degree and sources of pollution affecting the Asia-Pacific coasts;
   - to investigate appropriate strategies for improving the condition of these polluted coasts with a view to their role in sustaining humans in the future; and
   - to contribute to global strategies for (atmospheric) pollution reduction.

3. *problem solving*
   - to generate and share appropriate data which will assist problem solving for future global change issues;
   - to develop solutions which are appropriate to the cultures and environments of the Asia-Pacific region;
   - to share the results of new approaches to coastal management; and
   - to trial new technologies for coastal management.

11.6.2 Coastal Research Related to Global Change

There are seven key issues where there is a critical need for better understanding of the effects of future warming and sea-level rise. First, there is a need to develop scenarios for climate changes relevant to the region and then to examine impacts on the natural environment and on human society. However, because of the complex interactions it is also important to conduct research into multiple stresses on the environment and society. Another area for research is how to achieve sustainable development under conditions of global warming and accelerated sea-level rise. Research is also needed into the identification of process thresholds and their implications, as well as adaptation to global change. These seven issues are outlined in further detail below:

1. *future scenarios of global change*
   - to use as a realistic basis for forward planning for the Asia-Pacific coast;
   - to extend the acquisition of data by monitoring, data referring to both environmental and human (societal) change; and
   - to continue to develop sub-regional models of future change for key areas such as megacities.

2. *Impacts on the natural environment*
   - to identify where increased SST and sea-level rise may alter coastal ecosystems; and
   - to identify where future accelerated sea-level rise will render coastal areas in the Asia-Pacific region unusable by humans for the purposes they are currently used for.

3. *Impacts on human society*
   - to assess adaptation strategies including adjusting human lifestyles and reorganizing the geography of coasts through informed management of vulnerable areas; and
   - to assess impacts on human health, and need for lifestyle adaptations.

4. *Multiple stresses on environment and society*
    - to develop models of multiple stresses projected to affect people occupying Asia-Pacific coasts;
    - develop projections of change in both time and space; and
    - use the above to generate data as an essential component of integrated coastal planning and management.
5. *Achieving sustainable development under conditions of global warming and accelerated sea-level rise*
    - develop pathways for mainstreaming climate change into national development strategies; and
    - incorporate sea-level and climate change into sustainable development policies and targets
6. *The identification of process thresholds and their implications*
    - conduct research into process thresholds for the prediction and anticipation of future climate changes, especially sea-level rise; and
    - Investigate strategies for dealing with rapid change, and consequent disruption to Asia-Pacific coasts.
7. *Adaptation*
    - Conduct research into improved impact and coastal vulnerability assessment;
    - improve discussion of common and specific adaptation options; and
    - enhance human resilience and adaptive capacity.

11.6.3 Scientific Research on Appropriate ICM Methodologies for Global Change

There is much scientific research relating to global change on Asia-Pacific coasts that is not being used to inform coastal managers. Understanding of the pathways of decision-making about coastal management will aid the effective use and application of scientific understanding and data. As noted from the APN Synthesis Report, one of the APN's goals for its global change research related to developing science-policy linkages. However, this goal was not achieved for most of the APN funded projects. It is also clear from ICM effectiveness studies for the region that there is scope for more research into the development and application of appropriate criteria for assessing ICM effectiveness, governance issues related to ICM, and funding and sustainability issues. Future research is needed in a least five key areas:

1. *Linkages and mechanisms for ICM in the Asia-Pacific region*
    - to improve understanding of the linkages between research, policy, and coastal management in the region;
    - to investigate the effectiveness of policy and top-down approaches to management in countries of the region;
    - to identify the most influential people for management of coastal resources at community (or sub-regional) level in different countries; and
    - to build the capacity of those people to make informed decisions about the future of the coasts for which they are responsible.

2. *Enhancement of ICM in the Asia-Pacific region*
   - to develop more appropriate, more targeted and more effective strategies for ICM;
   - to encourage the development and adoption of country-specific (or sub-region specific) systems of ICM;
   - to encourage the recognition, protection, and dissemination of traditional coastal-management practices.
3. *Dialogue between decision-makers in the Asia-Pacific region*
   - mechanisms for discussion of commonalities (in environments, challenges, solutions) within and between countries;
   - pathways for cooperation to address transboundary issues; and
   - development of region-level initiatives to fulfil specific goals.
4. *Science communication about future global change on Asia-Pacific coasts*
   - to develop communication strategies that are culturally appropriate;
   - to develop a positive message about practical strategies for the future;
   - to make the message accessible through vernacular languages; and
   - use concepts and examples that are familiar to target audiences.
5. *Research focus on most vulnerable Asia-Pacific coastal environments*
   - atoll islands and other low-lying inhabited islands where in-island adaptation options are severely restricted;
   - delta coasts where people occupy low-lying areas of land and are subject to environmental stresses from both the sea and rivers; and
   - areas of rapidly increasing population where there will be insufficient land area to accommodate future populations and insufficient resources to sustain them.

11.7 CONCLUSION

It is clear that there is considerable scope for further global change research focusing on coastal management issues on the Asia-Pacific coast. However, as shown from a number of coastal global change research projects in the region there have been past weaknesses in terms of science-policy linkages and science communication strategies. New research projects need to be structured so that outputs better inform coastal managers and policy makers in the region and global change becomes recognised as a major issue for coastal management. As part of any communication strategy for policy-makers it is also necessary to have a general understanding of global change issues through appropriate education.

Unfortunately, much of what has been communicated to the public, has focused on the negative aspects of global change, which may be meaningful to scientists but may cause anxiety for the general public, who are not always able to contextualise global change issues. The public needs to be informed and persuaded to develop, implement, and sustain appropriate adaptation strategies for that part of the coast they occupy and help manage. It is suggested that discussions and information material concerning global change should emphasize appropriate and feasible adaptations to global change; no-regrets adaptation options; understanding of previous human adaptations to the effects of global change in the region, and that traditional knowledge can be usefully employed in the future.

Education must also be appropriate. A large proportion of the general public in the Asia-Pacific region are not familiar with regional geography or global change issues in places other than their own local area and do not speak English as a first language. However, the language of global change is largely English and the concepts are largely derived from English systems of thinking so that much will be gained by producing appropriate educational material for various parts of the Asia-Pacific region in vernacular languages.

There needs to be a recognition that while ICM has been endorsed internationally as an appropriate mechanism for managing coastal resources, there is no simple model applicable to all Asia-Pacific countries. Although there are some sub-regional groupings in ICM practice, many programs have not yet had sufficient time to demonstrate their success. There is also a diversity of governments in the Asia-Pacific region with a continuum of top down – bottom up management approaches. In some countries it may require targeting 'people of influence' at the community level, be they traditional chiefs, elected leaders, or religious leaders, and giving them the information necessary to guide informed and appropriate decisions concerning the coasts in their particular 'spheres of influence'. Capacity building may need to target 'people of influence' rather than simply assume that people in government decision-making positions are the only ones able to influence coastal management.

As noted at the start of this chapter, new research directions for coastal global change research present a number of challenges such as: 1) research funding, 2) timely production of meaningful outcomes, 3) results of direct practical relevance to policy makers, and 4) the political will to include global change issues within ICM frameworks. It is not realistic to expect that the political will to incorporate global change into ICM programs will occur before the scientific community has demonstrated the significance of the issue, the imperative to act quickly and has effectively communicated this to both the general public and the policy-makers.

REFERENCES

Adger WN, Kelly PM, Tri NH (2001) Costs and Benefits of Mangrove Conversion and Restoration. In: Turner RK, Bateman IJ, Adger WN (eds) Economics of Coastal and Water Resources: Valuing Environmental Functions. Kluwer Academic Publishers, Dordrecht, The Netherlands

Christie P, Lowry K, White AT, Oracion EG, Sievanen L, Pomeroy RS, Pollnac RB, Patlis JM, Eisma R-LV (2005) Key Findings from a Multidisciplinary Examination of Integrated Coastal Management Process Sustainability. Ocean & Coastal Management 48: 468–483

Chua TE (1998) Lessons Learned from Practicing Integrated Coastal Management in Southeast Asia. Ambio 27(8): 599–610

Costanza R, d'Arge R, de Groot R, Farber S, Grasso M, Hannon B, Limburg K, Naeem S, O'Neill RV, Paruelo J, Raskin RG, Sutton P, van den Belt M (1997) The Value of the World's Ecosystems and Natural Capital. Nature 387: 253–260

Harvey N (2004) Integrated Coastal Management. In: Goudie AS (ed.) Encyclopedia of Geomorphology. Volume 1. Routledge, London; New York, pp 567–570

Harvey N, Rice M, Stevenson L (eds) (2005) Global Change Coastal Zone Management Synthesis Report. Asia-Pacific Network for Global Change Research, Kobe, Japan, 37 pp

McLean R, Mimura N (eds) (1993) Vulnerability Assessment to Sea-Level Rise and Coastal Zone Management. Proc. the IPCC Eastern Hemisphere Workshop, 468 pp

Mimura N, Yokoki H (eds) (2001) Global Change and Asia-Pacific Coasts. Proc. APN/SURVAS/LOICZ Joint Conference on Coastal Impacts of Climate Change and Adaptation in the Asia-Pacific Region, APN and Center for Water Environment Studies, Ibaraki University, 285 pp

Olsen, SB (2003) Frameworks and Indicators for Assessing Progress in Integrated Coastal Management Initiatives. Ocean & Coastal Management 46: 347–361

Syvitski JPM, Vörösmarty CJ, Kettner AJ, Green P (2005) Impact of Humans on the Flux of Terrestrial Sediment to the Global Coastal Ocean. Science 308: 376–380

White AT, Christie P, D'Agnes H, Lowry K, Milne N (2005) Designing ICM Projects for Sustainability: Lessons from the Philippines and Indonesia. Ocean & Coastal Management 48: 271–296

Wilson MA, Costanza R, Boumans R, Liu S (2003) Integrating Assessment and Valuation of Ecosystem Goods and Services Provided by Coastal Systems. In: Wilson JG (ed.) The Intertidal Ecosystem. Royal Irish Academy, Dublin, pp 1–28

Wong PP, Boon-Thong L, Leung MWH (2006) Hotspots of Urbanization and Population Growth in the Asia-Pacific Region. In: Harvey N (ed.) APN Coastal Zone Management Synthesis. Chapter 7. Springer

Woodroffe C, Nicholls R, Saito Y, Chen Z, Goodbred S (2006) Landscape Variability and the Response of Asian Megadeltas to Environmental Change. In: Harvey N (ed.) APN Coastal Zone Management Synthesis. Chapter 10. Springer

INDEX

Absolute poverty, rural coastal research, 226
Accretion and erosion, saline water intrusions, 75
Adaptation, mangrove planting, 154
Agenda 21, 1, 2, 41, 50, 138
Aggregate sources, coastal aggregate mining, 262
Agricultural chemical pollutant loading, 245
Agricultural chemical use, 244
Agricultural inputs, agricultural productivity, 207–209
Agricultural pollution, DDT, commercial agriculture, 242
APN goals, *APN* coastal synthesis, 8, 316, 329
APN, importance of region, 7–8, 49, 331
Appropriate education, ICM practice, research challenges, 332–333
Aquaculture, Asia-Pacific farmers, rural coastal dwellers, 205
Aquaculture, mangrove clearance, Gulf of Thailand, 219
Aquaculture, population growth, 28–29, 203–205
Areas of research, tsunami, sea-level rise, 108–110
Arsenic, heavy metals, TBT, organotin compounds, 265
Artificial shoreline protection, 141
Asia-Pacific coastal population, Asia-Pacific fisheries, fishing capacity, 199–203

Bangladesh, coastal changes, 128
Bangladesh, GBM delta, population estimates, 292–293
Bangladesh, mangrove areas, human response, 155
Biological oxygen demand, 20
Book structure, 11–14
Bourewa Lapita site, 109

Capacity building, legislative framework, 269
Capture fisheries, 29
Case study-Japan, 129–132, 145–149
Case study-Manila Bay, 151–152
Case study-pacific islands, 140
Catch, consumption, trade, high-value finfish, 200

Catchment–coast interactions, LOICZ, 67–90
Cereal production, meat production, agricultural production, 207–208
Changjiang River, pollution, 70, 71, 72, 73, 83, 86, 302
Changjiang, urbanisation, 294
China, economic growth, rural–urban migration, pace of urbanisation, 173
China's urbanisation, megacities, urban regions, globalisation, 174
Chinese coast, 241
Chlorinated pesticides, PCBs, 259
Climate change, coastal hazards, coastal erosion, 48
Climate change response, sedimentation, sea-level rise, inundation, 296
Climate change, sea-level rise, 31–34, 100–102, 296–298
Climate change variables, regional impacts, coastal synthesis, 316
Coastal Bangladesh, extreme events, 309
Coastal characteristics, 95
Coastal cities, grey, green, brown issues, human impact, 321
Coastal degradation, chemical pollution, pollution prevention, 236
Coastal diversity, 28, 50
Coastal evolution, scales, regional changes, 93–112
Coastal governance, 137
Coastal governance outcomes, ICM assessment, ICM effectiveness, 54, 59
Coastal management definitions, sustainable development, 1, 331
Coastal management diversity, ICM implementation, ICM barriers, 318
Coastal management issues, diversity of coastal management, 40
Coastal management research directions, global change and ICM practice, 62–63
Coastal management, CZAP conferences, 3, 39
Coastal pollution, nutrient discharges, coastal ecosystems, 327–328
Coastal pollution, problem solving, global change scenario, impacts, 330

335

Coastal rural population, developing countries, 198–199
Coastal urban agglomeration, economic growth, 177
Community involvement, coastal management practices, 50
Community management, sea-level rise, tectonics, plate collision, 319
Community-based management, 35, 58
Convergent plate boundary islands, 105, 106, 130
Coral bleaching, marine fisheries, 27, 28–29
Coral reefs, mangroves, seagrass meadows, research directions, 323–328
Coral, mangrove, seagrass, 5, 25–28
Country features of population, 166
Cycles of ICM development, 43

Decade-to-century timescales, catchment–coast, process modelling, 326
Delta evolution, geochemical tracing, sediment record modeling, 83–84
Delta front, sediment deposition, radiocarbon dating, 301
Digital terrain models (DTMs), plains topography, SRTM data, 283–288
Dioxin, PTS chemicals, pollution hot spots, 257
Distributary switching, delta morphodynamics, dewatering, compaction, 289
Distribution of rural poor, 225
DOC study, dynamite fishing, 250
Drivers of coastal management, 45–50
DTMs, Asian megadeltas, 285

East Asia, major pollutants, chemical pollution, coastal projects, 238
Economic growth, coral reef, mangrove, 126
Ecosystem health, sustainable resource use, regional development, 88–89
Ecotoxicological data, inorganic compounds, mine discharges, 260–263
Effective management units, Southeast Asia coastal management, 153
Effectiveness of coastal management, ICM success, 59–62
Engineering mitigation measures, 136
Engineering schemes, processes and mechanisms, 77–81
Environmental issues, overfishing, Gulf of Thailand, South China Sea, 211
ESCAP, population research, megacities, 187–191

Eutrophication, fertiliser, nutrient loading, 215
Eutrophication, sewage disposal, CCA leaching, wood treatment, 263–265

Farming capital, developing countries GDP, 206
Fertilizers, nuclear waste pollution, 267–268
Fish kills, coastal development, habitat modification, mangrove clearing, 215
Fish-based nutrition, aquaculture, environmental costs, 203
Flood exposure, damming, sediment replenishment, 300
Flooding patterns, delta fronts, delta populations, 292–295
Food production, developing country tourism, 209–211
Future human–coast interaction, 127
Future research, 156

Ganges-Brahmaputra-Meghna (GBM) delta, distributary switching, 277, 288–289
GEF, PEMSEA, research, megacity conferences, 189
Global change, 1–14, 17–35, 315–333
Globalisation definition, Asia-Pacific globalisation, 175
Green, brown and gray environmental issues, Bangkok, 184–186
Gulf of Thailand, small-island nations, 212–214

Hazardous chemicals, nutrient pollution, chemical pollution, 235
Holocene sea level, coral reefs, 26–28, 96, 105, 106
Human agents of change, 94
Human influence, geological evolution, 279
Human occupation, prehistory, human impact, sediment load changes, 291
Human response to coastal change, human environment interaction, 117–128
Huon Peninsula, research gaps, 111
Hydrological changes, sediment retention, 78, 89

ICM barriers, tropical ICM, ICM indicators and frameworks, 60
ICM, coastal process and change, coastal ecosystems, 329
ICM comparative assessment, 54–59
ICM enhancement, science communication, vulnerable communities, 332

Index 337

ICM practice, ICM approaches, 54–56, 62, 317–319
ICM practice, ICM sustainability, coastal management evolution, 317
ICM underlying assumptions, community involvement, 44
Industrial pollutant loadings, 254
Insularity index, continental island southeast Asia, 103–105
Integrated coastal management (ICM), 34–35, 41–44, 54–59, 315–333
Integrated coastal management (ICM), coastal management terms, 34–35, 41–44, 54–59, 315–333
Integrated management, further research, 62, 89, 190, 273
International conventions and agreements, globalisaton, 46
International tourists, developing country destinations, tourism sector, 210
Island arcs, geomorphological influences, 99, 108, 319, 328

Jakarta, Mumbai, Shanghai, 184

Land claims, research directions, natural processes, 305–308
Land loss, inundation, shore-parallel retreat, sedimentation, 297
Land-based pollution, habitat modification, coastal ecosytems, 323
Land-based pollution, management deficiencies, 273
Land-ocean areas, 271
Landscape variability, Asian megadeltas, environmental change, 277–310
Large rivers, human activities, dams, 70, 309
Large rivers, modern river sediment, pre-human sediment discharge, 70, 279
Large rivers, suspended sediment load, 278
Late Holocene human–coast interaction, falling sea level, 123–124
Liveable cities, environmentally sound cities, urban poor, 181
Low-value finfish, crustacean products, 200

Major hazards, cyclones, flooding, storm surges, seawater intrusion, 304
Management of environmental change, coastal land reclamation, 86–87
Managing channel aggregate resources, flood defences, 87–88
Mangrove clearance, coastal reclamation, 142

Mangrove clearing, regional comparison of mangrove clearance, 217
Mangrove forests, terrigenous and marine sediment, 25, 102
Mangroves, coral reefs, 20, 26–28, 85–86, 216
Marine pollution, Pacific Island coastal impacts, 232–236
Material fluxes, catchment–coast characteristics, 69
Mega urban regions, urban area rapid expansion, 178
Megacities, Asian megadeltas, 295
Megacities, megacity diversity, urban conglomeration, 176
Megacity research, remote sensing, GIS-based analysis, 190
Megadelta plain elevation, delta chronology, fluvial landscape, 284
Migration, internal migration, rural–urban migration, 167
Mining impacts, mine discharges, 261
Modern human–coast interaction, 124–126
Modern humans and Pacific coast, lowland agriculture, 121, 125
Mollusc production, food fish consumption, 202
Monitoring and observation, EEZ and pollution, 271, 272
Monsoon, drivers of coastal change, geotectonic influences, 97–99
Multiple stress, sustainability, adaptation, ICM methodologies, 331

Natural systems, geoscientific assessments, regional climate change, 307
Nature of Asia-Pacific coasts, postglacial sea-level rise, climate, 96, 101, 102, 145, 280, 281
New Guinea tsunami, 143
New research directions, ICM, coastal values, 315–333
Non-government orgaisations, 47–48
Noxious wastes, animal-related wastes, sewage, 247
Nutrients and pollutants, coastal ecosystem, large and small catchments, 76, 79
Nutrients and pollutants, hydrological cycling, 69, 73, 79

Oil spillages, dissolved organic carbon (DOC), 250
Oil spills, 24–25, 250
Organochlorine compounds, TBT, fuel products, 265

Pacific coastal management, 4–7, 8–9, 17–35, 117–28
Pacific intraplate islands, atoll islands, 107–108
Pacific island colonisation, 121
Pacific islands, sea-level rise, artificial structures, 153
PEMSEA, 35, 49, 54, 56, 61, 189, 317
PEMSEA, comparative assessment of ICM practice, 54–59
PEMSEA ICM sites, 55
Per capita water consumption, sanitation access, 223
Persistent organic pollutants (POPs), oil spills, 24, 255–260
Persistent organic pollutants (POPs), organic compounds, agrochemicals, 240, 241,242
Pollutants, coastal pollution, 231
Pollution, 20–25, 236–237
Poorest inhabitants, land fertility, coastal water productivity, 227
POPs, PCBs, waste quantities of POPs, 256
Population aging, urbanisation, city boundary definition, 169–170
Population and urbanisation, research programmes, 186–187
Population concentration, 6
Population density, GDP per capita, early human–coast interactions, 119–122
Population growth decline, population strategies, 164–165
Population growth, fertility, mortality, fertility decline, 165
Population growth, urbanisation, hot spots, 163–192
Population migration, inter-regional migration, aging population, 169
Population mobility, rural exodus, 168
Population projections, flood risk, 299
Postglacial sea-level rise, oceanic migration, 102, 145
Potassium nitrate, industrial discharges, heavy metal, 252
Prehistory, sea-level changes, 130
Process modeling, 82
PTS pesticides, non-point source pollution, ship-based discharges, 244, 249
PTS, DDT, toxicological studies, 258

Radionuclide sources industrial effluents, fish processing, 267–268
Rapid urbanisation, ecological footprint, environmental crises, 181
Red River Delta, Mekong Delta, 286, 287
Red tide, hazardous chemicals, 20, 22, 241, 321
Reef-fringed uplifted islands, 106
Regional chemical use, 239
Regional planning, wetland and coral reef protection, 85–86
Regional research, rapidly-rising coasts, 110–111
Research gaps, areas of research, 139
Research methods and techniques, 81–85
Response constraints and opportunities, insularity index, 118–119
Rice cultivation, sea-level rise, coastal flooding, socio-economics, 298
Rising land, sea-level variation, 104
River-dominated, wave-dominated, tide-dominated, deltaic plain, 286
Rivers, catchments, deltas, sediment discharges, 322
Rural area reclassification, urban primacy, China's urbanisation, 172–174
Rural coasts, tsunami, poverty, 197
Rural population, urbanisation, urban transition, 172
Rural poverty, natural resource use, 224

Science-policy linkages, 63
Scientific collaboration, coastal degradation, community participation, 49
Scientific research institutes, 138
Scope of book, definition of region, 8–9
Scope of region, 10
Sea level, areas of inundation, 296–298
Sea level, small islands, 19, 133, 142, 212–214, 296–298
Seabed mining, EEZ, coastal pollution, 272
Seafood nutrition, mangrove removal, 204
Seagrass, 28
Sea-level changes, landbridges, 133
Sea-level changes, population densities, inundation, 133
Sea-level changes, terrigenous and marine sediment, 133
Sea-level rise, case study-Indonesia, 125–126
Sea-level rise, geomorphic change, coastal evolution, coastal research, 328–332
Sea-level rise, institutional response, coastal hazards, 135
Sea-level transgression, coastal evolution, human impacts, urbanisation, 319
Sea-level transgression, Holocene sedimentation, 281
Seawater DDT, chemical loading, agrochemicals, 242

Index

Sediment fractionation, extreme events, hazards, coastal erosion, 302–303
Sediment load, sediment flux, Quaternary, 302
Sediment mobility, morphodynamic processes, 79
Sediment yield, channel morphology, 74
Sedimentation, subsidence, human impacts, 307
Settlement patterns, shellmounds, 147, 148
Sewage discharges, BOD generation, 263
Shellmounds, human adaptation, 146
Shellmounds, postglacial transgression, 146
Ship garbage, ship discharge, 246
Ship-generated waste, waste reception demand, 247
Shoreline location, intertidal sediments, delta morphology, 282–283
Short timescales, shoreline erosion, climate and sea-level change, 99, 100, 107, 124
Simulation techniques, future management scenarios, artificial intelligence, 84–85
Small Pacific Island nations, biomass, 213
Societal response, megadelta topography, population distribution, 308
Solid waste, plastics, by-product wastes, 233
Southeast Asian lowlands, case study-Manila Bay, 151–152
Spatial variability, spatial scales, geological perspective, 305
State of environment, deltas, 18, 83, 277–310
Subsidence, groundwater extraction, 150
Sugar milling impacts, POPs, persistent toxic substances (PTS), 255
Sundarbans, Bangladesh, tsunami, 218
Suspended sediment loads, sea-level change, 280
Sustainable cities, integrated policy, 191
Suva Harbour, waste management, oil pollution, 249

Temporal variability, human landscape, 290
The region, definition of coastal zone, book structure, 9–11
Three Gorges Dam, 70, 71, 143, 302
Traditional knowledge, Pacific stakeholder groups, 52
Tropical cyclones, 117
Tsunamis, case study-New Guinea tsunami, 143–144
Tuna, artisanal catch, Bay of Bengal, 214

Underprivileged, governance, integrated policy, rural coasts, megadeltas, 325
UNEP, APFED, APMA, IGES, research, 188
Uplift and subsidence, 98
Urban dichotomy, mega urban regions, 178–180
Urban poor, poor women, environmental issues, 183

Waste disposal, household rubbish, 233
Waste outlets, coastal disposal, coastal impacts, 237
Waste streams, POPs, toxic chemicals, 253
Water consumption, human consumption, water stress, 221
Water issues, water withdrawal, 220
Water quality, material input, 76
Water sources, consumption levels, water source access, 222
World cities, Tokyo, Seoul, 180
WSSD, coastal values, 1, 2, 41

Coastal Systems and Continental Margins

1. B.U. Haq (ed.): *Sequence Stratigraphy and Depositional Response to Eustatic, Tectonic and Climatic Forcing*. 1995 ISBN 0-7923-3780-8
2. J.D. Milliman and B.U. Haq (eds.): *Sea-Level Rise and Coastal Subsidence*. Causes, Consequences, and Strategies. 1996. ISBN 0-7923-3933-9
3. B.U. Haq, S.M. Haq, G. Kullenberg and J.H. Stel (eds.): *Coastal Zone Management Imperative for Maritime Developing Nations*. 1997 ISBN 0-7923-4765-X
4. D.R. Green and S.D. King (eds.): *Coastal and Marine GeoInformation Systems: Applying the Technology to the Environment*. 2001 ISBN 0-7923-5686-1
5. M.D. Max (ed.): *Natural Gas Hydrate in Oceanic and Permafrost Environments*. 2000 ISBN 0-7923-6606-9; Pb 1-4020-1362-0
6. J. Chen, D. Eisma, K. Hotta and H.J. Walker (eds.): *Engineered Coasts*. 2002 ISBN 1-4020-0521-0
7. C. Goudas, G. Katsiaris, V. May and T. Karambas (eds.): *Soft Shore Protection*. An Environmental Innovation in Coastal Engineering. 2003 ISBN 1-4020-1153-9
8. D.M. FitzGerald and J. Knight (eds.): *High Resolution Morphodynamics and Sedimentary Evolution of Estuaries*. 2005 ISBN 1-4020-3295-1
9. M.D. Max, A.H. Johnson and W.P. Dillon: *Economic Geology of Natural Gas Hydrate*. 2006 ISBN 1-4020-3971-9
10. Nick Harvey (ed.): *Global Change and Integrated Coastal Management: The Asia-Pacific Region*. 2006 ISBN 1-4020-3627-2